Advanced Control and Estimation Concepts, and New Hardware Topologies for Future Mobility

Advanced Control and Estimation Concepts, and New Hardware Topologies for Future Mobility

Editor

Francis F. Assadian

MDPI • Basel • Beijing • Wuhan • Barcelona • Belgrade • Manchester • Tokyo • Cluj • Tianjin

Editor
Francis F. Assadian
University of California Davis
USA

Editorial Office
MDPI
St. Alban-Anlage 66
4052 Basel, Switzerland

This is a reprint of articles from the Special Issue published online in the open access journal *Energies* (ISSN 1996-1073) (available at: https://www.mdpi.com/journal/energies/special_issues/control_estimation_hardware_topologies_mobility).

For citation purposes, cite each article independently as indicated on the article page online and as indicated below:

LastName, A.A.; LastName, B.B.; LastName, C.C. Article Title. *Journal Name* **Year**, *Volume Number*, Page Range.

ISBN 978-3-0365-3473-2 (Hbk)
ISBN 978-3-0365-3474-9 (PDF)

Contents

energies

MDPI

Editorial

Advanced Control and Estimation Concepts and New Hardware Topologies for Future Mobility

Francis F. Assadian

Department of Mechanical and Aerospace Engineering, University of California Davis, Davis, CA 95616, USA; fassadian@ucdavis.edu

Current transportation and, in general, mobility have a negative impact on our environment. It is well known that our transportation makes up a large portion of the GHG emissions around the world. In California alone, transportation is the largest energy-consuming and greenhouse-gas-emitting sector, making up slightly over 40% of the state's GHG emissions.

The three technology pillars for tackling this issue with cost-effective solutions are (a) Electrification, (b) Advanced Control and Optimization, and (c) Virtual sensing. This Special Issue presents the latest research and work of various world-renowned academic institutions, through 10 papers, on the impact of these aforementioned pillars on Future Mobility.

Wang et al. [1] investigated the impact of differential drive assist steering (DDAS) on the steering assistance and the driver steering effort. In this paper, the authors proposed a unique control strategy to account for the tire nonlinear characteristics and hence, results in improved tracking accuracy and subjective steering feel.

Topić et al. [2] developed a simulation tool to support decision-making processes for planning city bus transport electrification. The simulation tool is designed to use real driving cycles and techno-economic data to compute the optimal powertrain type, number of e-buses and required charging stations, and predict the total cost of ownership, including investment and return on investment cost/benefit analysis.

Nazari et al. [3] focused their research work on an innovative low-voltage hybrid device that enables engine boosting and downsizing in addition to mild hybrid functionalities. The proposed configuration uses a planetary gear set and a brake to permit the power split supercharger (PSS) to share a 9 kW motor between supercharging the engine and direct torque supply to the crankshaft. Utilizing an adaptive equivalent consumption minimization energy management strategy, the proposed combined system shows a fuel consumption reduction of 18.4% over the standard FTP75 cycle.

Jang et al. [4] designed an algorithm for a shuttle bus with four in-wheel electric motors to optimize the front and rear axle tractions. In addition to optimizing traction, they have considered the input terminal voltage changes during the driving cycle. They showed that the proposed algorithm could improve the energy consumption by 6%, using 'Manhattan Bus Driving Cycle', when compared to the conventional vehicle.

Zhang et al. [5] investigated optimal power management of a fuel cell hybrid small UAV for the maximization of flight time in an uncertain (stochastic) environment. The power management accounts for the limits on the rate of change of fuel cell power output while maximizing the flight time duration. The simulation results indicate the capability of up to 2.7 h of flight time.

Filipozzi et al. [6] formulated the estimation of tire normal force as an input estimation. Then, two observers are proposed to solve this problem by using a standard quarter-car suspension model. The first observer is based on a newly developed concept called YCOO (Youla Controller Output Observer). The second observer is the Kalman filter approach

Citation: Assadian, F.F. Advanced Control and Estimation Concepts and New Hardware Topologies for Future Mobility. *Energies* **2022**, *15*, 1568. https://doi.org/10.3390/en15041568

Received: 2 February 2022
Accepted: 2 February 2022
Published: 21 February 2022

utilizing an Unbiased Minimum Variance filter. A detailed comparative analysis between these two approaches is presented in this work.

Anselma [7] utilized a multi-objective offline optimal control approach as a speed target generator for the following vehicle adaptive cruise control (ACC). This offline approach is based on dynamic programming. The intent of this offline optimal control approach is to minimize energy consumption while enhancing passenger comfort. The approach is flexible enough to be implemented in a wide range of powertrain configurations from conventional to power-split HEV.

Li et al. [8] studied road traffic flow models, and speed prediction methods for New Energy Vehicle are provided. Furthermore, the influence of vehicle lateral dynamics, including correlation control methods, for vehicle speed prediction is reviewed. The authors provide a potential application of the aforementioned concepts for the next generation of intelligent transportation systems.

Arasteh et al. [9] investigated a comparative analysis of three different smart brake-by-wire actuators. These three actuators consisted of Electro-Hydraulic Brakes, Electro-Mechanical Brakes, and Electronic Wedge Brakes. The authors proposed an objective metric based on energy usage, maximum power requirement, and dynamic responsiveness. They modeled the plants using an energetic modeling method and designed robust controllers utilizing Youla parameterization technique. The authors then performed both linear and nonlinear optimization on the controlled plants and compared the actuator performances based on the specified objective function.

Mallon et al. [10] explored the modeling and control of a lithium-ion battery and ultra-capacitor hybrid energy storage system for an electric vehicle for improved battery lifespan and energy consumption. They developed an optimal aging-aware energy management strategy that controls both battery and ultra-capacitor aging and compared these results to strategies that control only battery aging, strategies with no aging, and non-optimal strategies. A case study on an electric bus showed a 28.2% improvement in battery lifespan while requiring only a 7.0% decrease in fuel economy.

The contributions from the research works included in this Special Issue offer new data, information, and findings to continue the R&D effort in the field of Future Mobility, with the aim of stimulating the research community to further contribute to the development of the field.

I wish to thank the authors for their contributions to this Special Issue.

Funding: This research received no external funding.

Conflicts of Interest: The author declares no conflict of interest.

References

1. Wang, J.; Wang, X.; Luo, Z.; Schuler, J.; Assadian, F. Active Disturbance Rejection Control of Differential Drive Assist Steering for Electric. *Energies* **2020**, *13*, 2647. [CrossRef]
2. Topić, J.; Soldo, J.; Maletić, F.; Skugor, B.; Deur, J. Virtual Simulation of Electric Bus Fleets for City Bus Transport Electrification Planning. *Energies* **2020**, *13*, 3410. [CrossRef]
3. Nazari, S.; Siegel, J.; Middleton, R.; Stefanopoulou, A. Power Split Supercharging: A Mild Hybrid Approach to Boost Fuel Economy. *Energies* **2020**, *13*, 6580. [CrossRef]
4. Jang, I.-G.; Li, C.-S.; Hwang, S.-H. Energy Optimization of Electric Vehicles by Distributing Driving Power Considering System State Changes. *Energies* **2021**, *14*, 594. [CrossRef]
5. Zhang, J.; Kolmanovsky, I.; Amini, M.R. Stochastic Drift Counteraction Optimal Control of a Fuel Cell-Powered Small Unmanned Aerial Vehicle. *Energies* **2021**, *14*, 1304. [CrossRef]
6. Filipozzi, L.; Assadian, F.; Kuang, M.; Johri, R.; Velazquez, J. Estimation of Tire Normal Forces including Suspension Dynamics. *Energies* **2021**, *14*, 2378. [CrossRef]
7. Anselma, P.G. Optimization-Driven Powertrain-Oriented Adaptive Cruise Control to Improve Energy Saving and Passenger Comfort. *Energies* **2021**, *14*, 2897. [CrossRef]
8. Li, L.; Coskun, S.; Wang, J.; Fan, Y.; Zhang, F.; Langari, R. Velocity Prediction Based on Vehicle Lateral Risk Assessment and Traffic Flow: A Brief Review and Application Examples. *Energies* **2021**, *14*, 3431. [CrossRef]

9. Arasteh, E.; Assadian, F. A Comparative Analysis of Brake-by-Wire Actuators Using Optimization Strategies. *Energies* **2022**, *15*, 634. [CrossRef]
10. Mallon, K.; Assadian, F. A Study of Control Methodologies for the Trade-Off between Battery Aging and Energy Consumption on Electric Vehicles with Hybrid Energy Storage Systems. *Energies* **2022**, *15*, 600. [CrossRef]

energies

MDPI

Review

Velocity Prediction Based on Vehicle Lateral Risk Assessment and Traffic Flow: A Brief Review and Application Examples

Lin Li [1], Serdar Coskun [2,*], Jiaze Wang [1], Youming Fan [1], Fengqi Zhang [3] and Reza Langari [4]

1 School of Traffic and Transportation, Northeast Forestry University, Harbin 150040, China; lilin@nefu.edu.cn (L.L.); 2651223890@nefu.edu.cn (J.W.); fanyouming@nefu.edu.cn (Y.F.)
2 Department of Mechanical Engineering, Tarsus University, Tarsus, Mersin 33400, Turkey
3 School of Mechanical and Precision Instrument Engineering, Xi'an University of Technology, Xi'an 710048, China; zfqdy@126.com
4 Department of Mechanical Engineering, Texas A&M University, College Station, TX 77840, USA; rlangari@tamu.edu
* Correspondence: serdarcoskun@tarsus.edu.tr

Abstract: Forecasting future driving conditions such as acceleration, velocity, and driver behaviors can greatly contribute to safety, mobility, and sustainability issues in the development of new energy vehicles (NEVs). In this brief, a review of existing velocity prediction techniques is studied from the perspective of traffic flow and vehicle lateral dynamics for the first time. A classification framework for velocity prediction in NEVs is presented where various state-of-the-art approaches are put forward. Firstly, we investigate road traffic flow models, under which a driving-scenario-based assessment is introduced. Secondly, vehicle speed prediction methods for NEVs are given where an extensive discussion on traffic flow model classification based on traffic big data and artificial intelligence is carried out. Thirdly, the influence of vehicle lateral dynamics and correlation control methods for vehicle speed prediction are reviewed. Suitable applications of each approach are presented according to their characteristics. Future trends and questions in the development of NEVs from different angles are discussed. Finally, different from existing review papers, we introduce application examples, demonstrating the potential applications of the highlighted concepts in next-generation intelligent transportation systems. To sum up, this review not only gives the first comprehensive analysis and review of road traffic network, vehicle handling stability, and velocity prediction strategies, but also indicates possible applications of each method to prospective designers, where researchers and scholars can better choose the right method on velocity prediction in the development of NEVs.

Keywords: new energy vehicles; speed prediction; macroscopic traffic model; traffic big-data; deep learning; vehicle lateral dynamic and control; unresolved issues; application of speed prediction

Citation: Li, L.; Coskun, S.; Wang, J.; Fan, Y.; Zhang, F.; Langari, R. Velocity Prediction Based on Vehicle Lateral Risk Assessment and Traffic Flow: A Brief Review and Application Examples. *Energies* **2021**, *14*, 3431. https://doi.org/10.3390/en14123431

Academic Editor: Francis F. Assadian

Received: 16 May 2021
Accepted: 7 June 2021
Published: 10 June 2021

Publisher's Note: MDPI stays neutral with regard to jurisdictional claims in published maps and institutional affiliations.

1. Introduction

Today's world is undergoing unprecedented changes in the 21th century. A new round of technological revolution and industrial transformation is in the ascendant, and intelligent networked new energy vehicles (NEVs) have become the strategic direction of global industrial development [1]. In February 2020, China's 11 national ministries and commissions, including the National Development and Reform Commission, the Ministry of Industry and Information Technology, the Ministry of Public Security, and the Ministry of Transport, jointly issued the Smart Vehicle Innovation and Development Strategy in its "Strategic Vision", which points out that: "Looking forward from 2035 to 2050, China's standard intelligent vehicle system will be fully completed and more complete. The vision of a safe, efficient, green, and civilized intelligent vehicle system has been gradually realized, and the intelligent vehicle can fully meet the people's growing needs for a better life."

The application of traffic information in the design of hybrid electric vehicles (HEVs) energy management is one of the important objects, wherein the adjustment of power distribution ratio based on traffic information is achieved by predicting the change of vehicle speed [2,3]. In order to achieve the goals of "safe", "efficient", and "green" for future intelligent networked NEVs, it is of great significance to predict and plan their future speed [4,5], because the above three goals are closely related to the speed, as shown in Figure 1. "Efficient" means that the vehicles in the traffic flow move at high speeds, which makes the roads more efficient. A vehicle driving at a higher speed increases the probability of road traffic accidents, so in order to drive "safely", the vehicle often slows down in advance in the condition of high traffic risk. This process leads to the change of vehicle speed and the problem of optimal vehicle speed planning. Accurate speed prediction is the key to energy management, reduce emissions, and improved energy-saving control of NEVs [6,7], that is, "green" driving. Therefore, it is of great significance to accurately predict and reasonably plan vehicle speed for balancing the relationship among "safe", "efficient", and "green" of vehicles in the future traffic system.

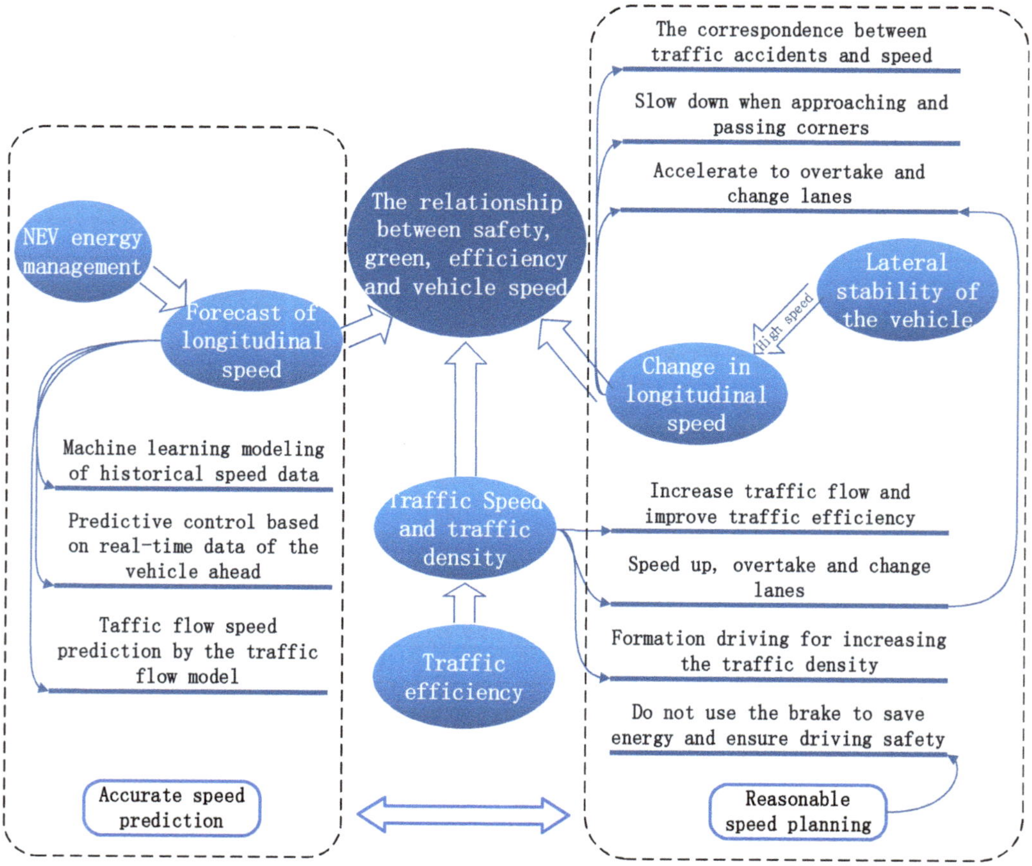

Figure 1. The relationship between "safety", "efficiency", "green", and speed of intelligent networked vehicles.

The automobile industry is now undergoing an electrification revolution. Take China as an example, in 2018, the production and sales of NEVs reached 1.27 million and 1.256 million units, respectively. In 2019, the production and sales of NEVs reached 1.242 million units and 1.206 million units, respectively. From January to December in 2020, the production of NEVs reached 1.366 million units, and the sales volume reached 1.367 million units [8].

The above data is briefly described in Figure 2. On the whole, China's new energy vehicle production and sales are greatly affected by government policy. In 2019, due to the retreat of subsidy policies, the production and sales of NEVs decreased compared with 2018. Due to the impact of the epidemic, China restored the subsidy policy of NEVs in 2020, and the production and sales increased accordingly.

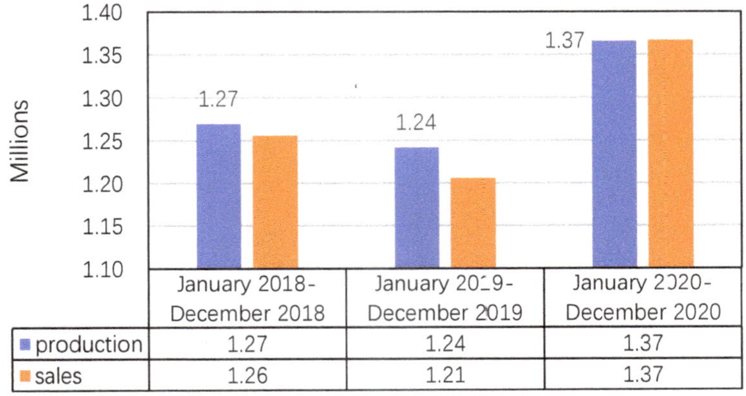

Figure 2. Comparison of production and sales volume of new energy vehicles in China in recent three years.

1.1. The Relationship among Velocity Prediction, Traffic Environment, and Vehicle Handling Dynamics

In order to achieve the goals of "fast", "safe", and "energy savings" for future intelligent networked NEVs, a basic challenge is to model the road traffic environment and accurately predict the evolution of traffic flow [9]. Accurate speed prediction is the key to energy management and energy-saving control of NEVs, and the driving condition determines the energy consumption and driving safety simultaneously. Therefore, it is very important to predict and select working conditions. Drivers control vehicles' speed, and in fact interfere with the prediction of the speed. The level of human driving skills plays a key role in the prediction error of longitudinal and lateral speed. In the context of intelligent connected vehicles, a vehicle can automatically perceive the surrounding environment and drive themselves. The vehicle will have a clear understanding of its future traffic environment and its handling stability, which will increase the accuracy of speed prediction. Therefore, the velocity prediction method is based on vehicle dynamics concerned for intelligent and connected NEVs in this paper.

The road traffic network has temporal and spatial, self-organization, and random characteristics. Analyzing the dynamic evolution law of the road traffic network and the traffic operation situation is helpful to fully understand the complex characteristics of the road network, and one can grasp the mechanism of traffic bottlenecks and traffic accidents [10]. The traffic dynamic evolution law model under the mixed traffic network is constructed to realize the rapid identification of traffic behavior characteristics, which provides technical support for further prediction and evaluation of vehicle speed and road traffic network security risks.

Speed prediction is not only related to the inherent characteristics of traffic flow but also related to the judgment of drivers or intelligent vehicles for traffic risks. Combined with intelligent networked vehicles, the road network abstract model and traffic risk prediction and evaluation model with low computational resource occupancy and more flexibility will lay a foundation for vehicle velocity/routing planning technology under multiple constraints and more effective traffic management methods. Therefore, the establishment of a traffic risk prediction and assessment model based on real-time information is of great importance for improving the safety of vehicles driving in bad weather conditions such as rain and snow. How to effectively integrate the existing traffic information collection methods to analyze the evolution law and operating situation of the road traffic network in real-time, and to accurately assess and predict traffic network security risks is a key research topic in the field of road traffic safety [10]. The framework of road traffic network security risk identification is shown in Figure 3.

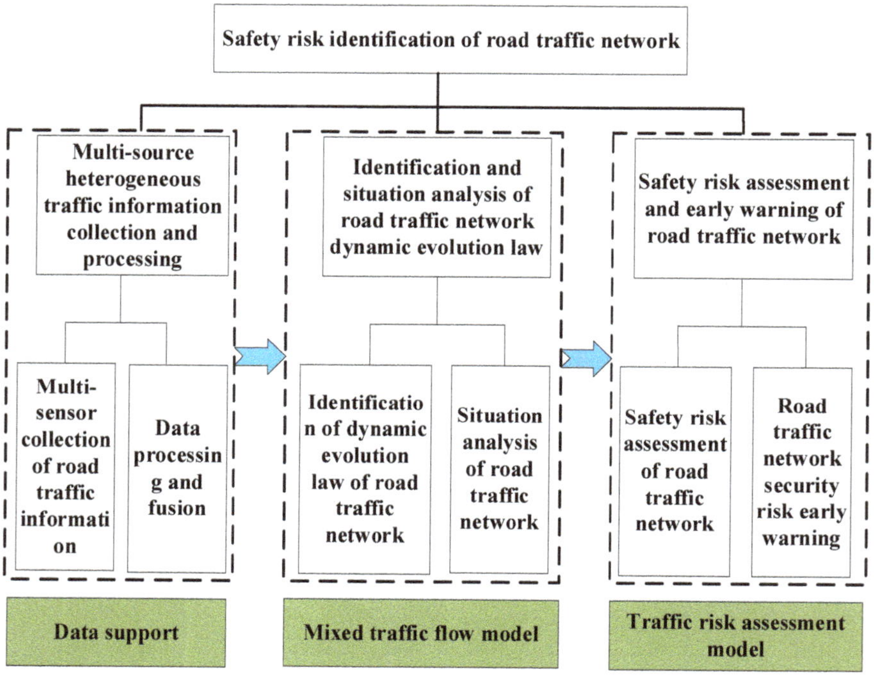

Figure 3. Research framework of road traffic network security risk identification.

The safety and reliability of traffic have an important influence on users' speed choices [11]. Road traffic network safety risk prediction and assessment is a process of accurate identification, effective assessment, and early warning of risks in road network, which is the core link of road traffic network risk identification [10]. Previous studies of road network safety risk assessment are mostly based on traffic accident simulation or later on the basis of traffic accident data analysis and the qualitative assessment, such as road network risk impact factor assessment, road network vulnerability assessment, and vehicle collision risk factor prediction in case of emergency or extreme weather, but it is difficult to find the safety risk impact factors before the accident on time [10]. Traffic hazard identification and risk quantitative assessment before accidents are the key technologies to break through the risk identification of road transportation network [10].

Under bad weather conditions such as rainy and snowy, research on road traffic risk prediction and assessment has a practical significance for vehicle speed forecast and planning. In order to predict traffic risk, it is necessary to explore the evolution law of road traffic system, which is quantified by the traffic flow model. The evaluation of traffic risk needs to be established with the lateral dynamic model and stability controller model of the vehicle. Vehicles with different handling characteristics need a personalized vehicle traffic risk prediction and assessment. Therefore, we will review the influence of various stability control methods on speed prediction. Next, we will introduce the current research progress of macroscopic traffic flow model and artificial intelligence big data model. At the same time, we will explore the tendency of these two-model combined with lateral traffic risk assessment model on the prediction and assessment of traffic risk. It will give a direction of the solution for basic theoretical problems of real-time accurate prediction and assessment of traffic risk before the accident, with the goal of improving vehicle safety and saving energy for the application of safe path and energy-saving speed planning of intelligent networked vehicles.

1.2. Contributions and Paper Structure

Although there are many studies on energy management strategies (EMSs) for HEVs [12–14], there is almost no work that has focused on velocity prediction methods combined with traffic risk prediction and assessment of HEVs. In the literature, studies on driving prediction of predictive energy management of plug-in hybrid electric vehicles are put forward [15] without introducing traffic risk prediction and assessment of HEVs.

Thus, the inspiration behind this article is to conduct a brief review on vehicle speed prediction based on traffic environment and vehicle lateral risk assessment. Prospective designers of NEVs will benefit from a number of approaches in the field where they can better establish their solutions. In response to the above analysis, the contributions of this paper are as follows:

(1) Firstly, we reviewed the macroscopic-traffic-flow-model-based prediction method, which could help improve vehicle speed prediction. This mathematical model has a quick solving speed, and it is easy to integrate the advantages of the information from various traffic sensors and communication systems. This makes the amount of interference quickly back the forecast results in real-time traffic.
(2) Secondly, we reviewed forecasting method based on traffic data, helping to facilitate the possible integration of multiple prediction algorithms. On combining model-based forecasting methods and data-based forecasting methods, the hybrid prediction method has high computing efficiency, covering more data, and updating online at the same time. This paper can help improve the prediction method's instantaneity, accuracy, and robustness.
(3) Thirdly, different from available studies, since the vehicle lateral dynamics and correlation control methods are emerging techniques for velocity prediction, we provide a list of studies for potential applications in velocity prediction in NEVs.
(4) Fourthly, a questionnaire section about the influence of various traffic flow models and vehicle lateral dynamics is given, and the application field of speed prediction algorithms along with missing points provides deep insight for prospective designers.
(5) Lastly, a set of application examples are given, wherein three applications are introduced considering various traffic flow models, vehicle lateral dynamics, and speed prediction methods.

This survey is structured as follows. A review of vehicle speed prediction methods for NEVs, with an emphasis on macroscopic traffic flow models, data-based traffic flow models, and influence of vehicle lateral dynamic on speed prediction is introduced in Section 2. In Section 3, research status and analysis of the development of vehicle speed prediction methods for NEVs are established. The application field of speed prediction is discussed in Section 4. Lastly, conclusions and future trends are summarized in Section 5.

2. Review of Vehicle Speed Prediction Methods for NEVs

The velocity of a vehicle is closely related to its traffic environment. Therefore, it is important to accurately predict the change of traffic flow parameters for improving the accuracy of vehicle speed prediction. We now introduce macroscopic traffic flow models, data-based traffic flow models, and the influence of vehicle lateral dynamics on speed prediction in this sequel.

2.1. Macroscopic Traffic Flow Model and Vehicle Velocity Prediction

Macroscopic traffic flow models use a mathematical model to describe traffic dynamics. The mathematical model evaluates the unmeasured area based on real-time input data. This model is usually based on empirical relationship, in which the parameters to be evaluated are either obtained from external calculation of historical data or generated from internal evaluation of an algorithm. The macroscopic traffic flow model method has been widely used in traffic state assessment.

Literature [16,17] introduces in detail the current developments and trends of the above macro traffic flow model, as well as traffic modeling, evaluation, and control methods based on intelligent networked vehicles. The macroscopic traffic flow model method has the following advantages: First, this method explains the mechanism of traffic, expands the observation data, and provides additional information. Therefore, this method can use less data to predict accurate traffic conditions. Second, it has higher interpretability. This means that even if the prediction is not accurate, the reason may be found in a certain confidence interval. Third, it can be directly integrated with traffic control practices, such as using model predictive control. The macroscopic traffic flow model method also has the following shortcomings: First, inaccurate or uncalibrated models will lead to poor performance of traffic state assessment results. Therefore, in actual application, the macroscopic traffic flow model traffic state assessment method must be carefully selected and calibrated. In this case, checking the validity of a model or calibrating a model requires a big set of data [18]. Second, since the macro-traffic flow model cannot be adjusted adaptively, it cannot reflect the changes in traffic flow brought by random traffic interference alone. Therefore, in the absence of real-time traffic information, it is more suitable for the prediction of traffic flow that does not change rapidly, such as the prediction of traffic parameters in expressways and urban loops.

Traffic flow (simulation) models are used to characterize complex traffic flow systems in order to understand, describe, and predict traffic flow [19]. It is a basic tool for analysis and experimental research of transportation systems. The traffic flow model is not only used in the traditional fields of traffic system design, testing, management, and personnel training. With hot research direction on intelligent vehicles and intelligent transportation systems, it has also been used to evaluate and predict the state of the transportation system [18]. Macroscopic traffic flow models can be classified according to the degree of detail they represent in traffic systems. This classification can be implemented by considering different levels of traffic entities in their respective flow models. Literature [20] classifies the traffic flow model as follows:

1. Submicroscopic models (describing in detail the equations of vehicle subunits and their interactions with surrounding vehicles).
2. Micro model (describing the distinction and tracking of individual entities in detail).
3. Mesoscopic model (medium detail description).
4. Macro model (less description of individuals).

The microscopic traffic flow model not only describes the time and space behavior of system entities (vehicles, drivers, etc.), but also describes the interaction between them in detail. For example, a lane change behavior of each vehicle in the flow is described as a series of driver decisions.

Similar to the microscopic model, the submicroscopic model describes the characteristics of a single vehicle in the traffic flow. However, in addition to the detailed description of the driving behavior, the control behaviors (transmission shifting, ESP, etc.) of the vehicle in response to the surrounding conditions are also modeled. Moreover, the sub-modules of the vehicle are also modeled by mathematical equations.

The mesoscopic traffic flow model neither identifies nor tracks individual vehicles, but it lists the behavior of individual vehicles (for example, in the form of probability). In view of this, traffic is represented by small groups of traffic entities. The model does not describe the behavior and interaction of these groups in detail. For example, the lane change of a single-vehicle is described as a transient event. The decision to change lanes is based on parameters such as the relative density of the lane and the speed difference. Some mesoscopic models analogous to the theory of gas motion have been pushed out. These gas motion models describe the dynamic distribution of model velocity.

The macroscopic traffic flow model describes the traffic flow at an overall level without identifying its constituent details. For example, traffic flow is aggregated (represented by flow, density, and speed). The behavior of individual vehicles (lane changes, etc.) is usually not displayed. A macroscopic traffic flow model assumes that it is appropriately allocated to the road lanes and uses approximate methods to achieve it. Macroscopic traffic flow models are usually classified based on the number of their partial differential equations. Usually, one side of the equation is the model representation, and the other side is the order of the equation. Here are some typical traffic flow models and their extended models [18].

2.1.1. One-Dimensional Flow Model

The Lighthill-Whitham-Richards (LWR) model [21] is a famous innovation in the mesoscopic traffic flow model. It uses a conservation law of vehicles in the traffic flow and assumes that the traffic parameters (traffic speed and density) follow a fundamental diagram (FD) in equilibrium. This traffic model has the ability to identify traffic congestion and distinguish between traffic congestion and free flow. Moreover, because of its relatively simple equations, it can be quickly calculated and solved. However, due to the limitations of the model, it cannot reproduce some complicated phenomena, for example: unstable flow and the stop-and-go phenomenon.

After the LWR model, higher-order models were developed in order to show phenomena different from equilibrium traffic state (usually referred to as micro disturbance and non-equilibrium state), higher-order models usually use another momentum equation to describe the evolutionary relationship of traffic speed instead of the FD. The Payne-Whitham (PW) model [22] is the first well-known high-order modeling attempt. The PW model and its extension [23] successfully reproduce some well-known traffic phenomena, such as hysteresis, reduced capacity. However, it still has some shortcomings, such as negative speed [17].

The Aw-Rascle-Zhang (ARZ) model [24] is another high-order model. It allows the equilibrium state of traffic to be transformed into other states. It also overcomes the limitations of the PW model. It can be derived from the LWR model. The ARZ model is further developed and extended to the following models: general second-order model [25], phase change model [26], and generalized ARZ (GARZ) model [27].

Recently, the theory of explaining the traffic flow model has made new progress. Some models can be represented using Hamilton-Jacobi partial differential equations (HJ-PDE). HJ-PDE has been studied in detail in the field of partial differential equations and physics [28], and its theory can effectively solve this kind of model [29]. For example, the traffic density state variable $\rho(t, x)$ of the traditional LWR model can be transformed into the state variable $N(t, x)$ of the accumulation flow. In addition, due to the nature of the accumulated flow, this model can implicitly express vehicle trajectory and travel time. The relationship between macro and micro can be shown. Then, a Lagrangian coordinate system was proposed and used accumulated flow and vehicle trajectories in the mesoscopic traffic flow model [30]. Similar topics have also been discussed in other traffic flow models [31].

2.1.2. Multi-Lane Models, Multi-Class Models, and Random Models

Actual traffic often has multiple lanes and multiple categories, different from the model introduced in Section 2.1.1. The studies [32,33] proposed some models that consider multi-lane and multi-category traffic behavior. For these extended models, modeling lane change behavior is one of their main challenges. Regarding lane changes, human behavior is quite important.

For the random properties of the traffic and the uncertainty of input/output data, a stochastic model [34] based on the LWR model is proposed. In the references [35,36], the source of randomness in dynamics explains the heterogeneity of vehicles, which means the relationship between multi-class models and random models. Moreover, the work [37] developed an LWR stochastic model based on HJ PDE.

2.1.3. Development Trend

The traffic flow model is not only used in the traditional fields of traffic system design, testing, management, and personnel training [20]. As the research on intelligent vehicles and intelligent transportation systems is a hotspot, it is also used to evaluate and predict the state of the transportation system [18]. It extends from ensuring the stability and energy efficiency optimization of a single vehicle to the safety and efficiency of the entire system.

The main principle of the prediction algorithm is to combine numerical simulation, traffic model, real-time data, and historical data to predict the evolution of future traffic conditions. Designing a fast, scalable, and accurate road traffic forecasting tool is the key to overcome the lack of forecasting ability of the existing traffic management information system [18], and it can be applied to the prediction and planning of vehicle path and speed in the future. At present, there is no hybrid traffic flow model that combines the advantages of macro traffic flow model and data-based traffic flow model. The hybrid traffic flow model is of great significance to improve the accuracy, robustness, and real-time performance of prediction. In the future, the traffic flow model will not be used individually, but a multi-layered hybrid model, as well as possible complementary and combined use of model-based traffic flow models and data-based learning models.

2.2. Data-Based Traffic Flow Model and Vehicle Velocity Prediction

Traffic parameter prediction method based on big data and machine learning has attracted the research interest of many scholars in recent years [38,39]. The traffic management department realizes the prediction of traffic flow by collecting and analysising the current and historical traffic data. Big data analysis can effectively predict the occurrence of traffic accidents. Big data analysis mainly solves the following three problems: data storage, data analysis, and data management [40]. The collected traffic big data is trained into a prediction model by machine learning method to analyze the evolution trend of traffic. Machine learning models can be divided into: supervised learning, unsupervised learning, reinforcement learning, deep learning and entity-based algorithms [40]. The labeled training data is used for supervised learning algorithms. Linear regression, decision tree, neural network and support vector machine are typical supervised learning methods.

The data-based traffic flow model can be divided into historical data-based traffic flow model and real-time big data-based traffic flow model. Among them, the traffic condition assessment method that widely relies on historical data uses a statistical method or machine learning method to find the relationship between historical data. Traffic conditions are evaluated based on this correlation and real-time data, meaning that it does not require prior knowledge of explicit modeling in the macroscopic traffic flow model. This method usually requires a big amount of historical data.

The traffic flow model based on historical data has the following advantage: less time for model selection and calibration. The disadvantages are: first, based on historical data means that the model may fail when unexpected events occur or when a relatively long trend is predicted. Second, the computational consumption required for training and learning will be very high. Third, the method can be regarded as a "black box", which means it is unable to properly explain the model decisions [18]. Fourth, if the information from real traffic flow is greatly different from that stored in the data used for the training model, the prediction accuracy will not be guaranteed.

Compared with the macro traffic flow model and the traffic flow model method based on historical data, the method based on real-time big data is defined as a method that does not rely on the empirical relationship that appears in the macro traffic flow model but depends on real-time data flow. This means that this method relies less on the prior knowledge of transportation. Its advantage is that it is robust to uncertain phenomena or unpredictable accidents. In an era of ubiquitous sensors (smart phones) and the emergence of a large number of intelligent networked vehicles, methods based on real-time big data streams may become popular in the future [41,42]. This method is more suitable for the prediction of urban road network traffic flow.

2.2.1. Research Status of Traffic Flow Evolution Using History Traffic Data and Artificial Intelligence Methods

It has become a hot research direction using deep learning method to predict the change of traffic parameters, for example in [43–47]. The unsupervised incremental machine learning, deep learning, and deep reinforcement learning was adopted by Dinithi Nallaperuma et al. to structure an expansive smart traffic management platform [43]. It can successfully model traffic flow with fluctuation; however, the method proposed in the literature is not effective for predicting traffic flow with high frequency fluctuation. An improvement to this problem is to increase the amount of data used to train the deep learning model. The restricted Boltzmann Machine method was used to predict traffic [44]. This method has a better nonlinear fitting ability and high prediction accuracy for typical chaotic time series. Di Zang et al. [45] solved the task of long-term traffic speed prediction for elevated highways by coupling convolutional long-short-term memory and convolutional neural network (CNN) into a single framework.

For the problem of uncertain data used in the training model, a common solution is to combine fuzzy rules with deep learning [48] and neural network [49]. By introducing the fuzzy representation into the deep learning model to lessen the impact of data uncertainty, a deep convolutional network model was established to explore the spatiotemporal connection of traffic flow to promote traffic flow prediction in [48]. The experimental results show that the combination of deep learning and fuzzy theory can improve the prediction accuracy, compared with other methods, such as autoregressive integrated moving average (ARIMA), deep learning-based prediction model for Spatial-Temporal data, CNN, fully convolutional neutral network, and fatigue detection convolutional network. The Takagi-Sugeno system was used for fuzzy reasoning, and two learning processes were proposed to update the membership function of the fuzzy system [49]. The proposed method has advantages over the six traditional models, such as artificial neural network, support vector machine, ARIMA model, and vector autoregressive model.

In addition to representing uncertainty with fuzzy rules, another approach is to point out exactly what the uncertainties are and then label those uncertainties with contextual factors. The relationship between traffic flow values in a time interval is investigated based on a combination of contextual factors from historical data [50]. From the analysis results, forecasting accuracy can be better improved by the proposed new method. On the other hand, the design is slightly inferior to the conventional method due to inconsistent points. This can be interpreted to the high volatility degree associated with low-traffic-flow periods.

There are several ways to improve the prediction accuracy of deep learning methods: training data are screened [51,52] or training parameters are optimized [53]. A deep belief network (DBN) model and a kernel extreme learning machine classifier is combined as a prediction model, wherein the important features of the traffic flow data are extracted through DBN at the bottom of the network. To predict the traffic flow, the extracted results are inputted into the kernel extreme learning machine classifier [52]. Automatically use those highly correlated spatiotemporal points to train the deep learning network, and reduce the use of less correlated data [51]. This explains the interaction between past and future data to some extent. Traffic flow theory and its application on urban transportation networks with more efficient deep learning architectures is a promising study field [51]. This is the valuable research direction recommended by this paper.

2.2.2. Research Status of Velocity Prediction Using Real-Time Traffic Data

Real-time prediction model of vehicle travel speed is helpful to improve vehicle safety, maneuverability, and fuel economy. In order to achieve these effects, an accurate velocity prediction model needs to be established and can be successfully implemented in the real system. There are two main types of vehicle speed prediction models: the prediction model based on Markov chain and the prediction model based on recursive neural network. Markov chains are stochastic data-driven models that predict future states from state transition matrices and current states. Transient probabilities are aggregated into state transition matrices. The structure of measurement based on transfer conditions is intuitive and easy to implement. There are three kinds of Markov chain models: interval coding, fuzzy coding, and velocity constraint model [54]. The speed prediction models are based on recurrent neutral network (RNN) including standard RNN, long short-term memory, and gated recurrent unit (GRU) models [54]. Among the three models based on Markov chain, the model combining the fuzzy coding method and the constraint model has the highest prediction accuracy. Of the three RNN-based models, GRU has the highest prediction accuracy due to the appropriate structure of long-term dependent learning by combining the amount of previously determined state data [54].

2.2.3. Development Trend

More research study needs to be performed in the following aspects for future data-based vehicle speed prediction methods [15]:

- Multiple prediction algorithms integration.
- Online correction and update technique.
- Balance between running online and computing burdens.
- Driver model and driving style recognition.
- Multi-source information integration.

2.3. Influence of Vehicle Lateral Dynamic on Speed Prediction

The influence of vehicle lateral dynamics on vehicle speed prediction is based on the traffic risk assessment from drivers or intelligent vehicles. Traffic risk assessment models can be divided into two types: longitudinal and lateral models. The longitudinal traffic risk assessment model mainly evaluates vehicle collision accidents caused by untimely braking or insufficient braking force. It includes a risk assessment model based on traffic accident data or simulation data [55,56] and a model-based risk assessment model [57–59]. Model-based work often requires parameters and empirical assumptions, while data-based methods only focus on extracting the relationship between images and road safety without considering other influencing factors such as drivers; thus, it has certain limitations.

Lateral traffic risk assessment models often build lane change risk models based on road traffic environment and lane change path factors, such as building a highway exit lane change risk model based on a proportional advantage model [60]. The existing literature rarely considers the impact of the traffic environment and vehicle lateral handling characteristics [61]. How to dynamically and accurately analyze the impact of traffic conditions on the prediction of vehicle velocity change and intermittent road safety (vehicle lateral stability) is a huge challenge for road safety analysis in practical application [60].

In order to study the stability characteristics of different vehicles and generate their stability criteria (lateral traffic risk assessment model), it is necessary to study the structural characteristics of vehicles and their handling stability control methods. For example, distributed drive control (also known as torque vector control) enhances the vehicle's dynamic performance [62], so its traffic risk assessment should be different from that of ordinary front-wheel steering vehicles. In a similar work, Kun jiang et al. [63] presented a method to estimate and predict individual tire forces based on a vehicle dynamics model and observer with low-cost sensors and driver assistance map, which is in close relation to the speed of the vehicle. Yu-Chen Lin et al. [64] developed an ecological cruise control based on an adaptive prediction-based control strategy. The design guarantees the safety of driving, riding comfort, as well as fuel efficiency simultaneously when running on roads with curves and up-down slopes. However, the above two literature did not involve the personalized assessment of vehicle traffic risk, nor did they pay attention to the relationship between traffic risk and predicted speed.

Some researches combine vehicle lateral safety with longitudinal speed prediction. To this end, a three-degree-of-freedom vehicle lateral dynamics model with lateral load transfer ratio index is derived for a rollover speed prediction model in [65]; similarly [66,67]. Lin Li et al. [68] combined the energy management of HEVs with vehicle speed prediction and planning before entering the corner. Pan Song et al. [69] demonstrated an improved optimal speed adjusting method based on the vehicle handling stability and path following performance. Med Krid et al. [70] presented a model predictive control (MPC) strategy for an active anti-roll system, which aims to minimize the load transfer during cornering and the consumed energy by the actuators. Hongbin Ren et al. [71] formulated a quadratic optimization problem for an integrated control of longitudinal speed and lateral motion control based on longitudinal progression maximization and lateral path tracking error minimization.

2.3.1. Influence of Vehicle Stability Control Methods on Speed Prediction

In order to reach the destination faster, the driver or the intelligently connected vehicles will choose a higher speed, which means that the vehicle's speed is greater than the speed of the surrounding vehicles (traffic), so it needs to overtake and change lanes frequently. In bad weather conditions such as rain and snow, this can bring the risk of accidents such as vehicle skidding. Moreover, in a cold region's road traffic environment, traffic accidents caused by overtaking and lane change on snow and ice slippery road surface are common. In addition, vehicle steering characteristics, driving characteristics, control methods, mass, the height of center of mass, wheelbase, tire lateral stiffness, and other factors affect vehicle handling characteristics in a road traffic environment. Therefore, in order to assess the traffic risk caused by sideslipping and predict the speed variation of the vehicle, a method of lateral traffic risk assessment and speed prediction based on vehicle handling characteristics should be established. This lies in the connection between personalized lateral traffic risk assessment demand and vehicle speed prediction demand. Analyzing the mechanism of mutual coupling between the two demands is beneficial to improve the accuracy of risk assessment prediction and vehicle speed prediction at the same time, so as to achieve the goal of improving road traffic safety and energy saving.

Modern vehicles are equipped with advanced vehicle lateral control strategies, such as active front steering (AFS) and direct yaw moment control (DYC) to enhance vehicle lateral motion control. In recent years, there have been continuous publications on vehicle lateral dynamics control, such as [72–74]. At present, the research literature on distributed electric drive is limited, and it mainly focuses on the theme of four wheel independently actuated electric vehicle. The research content mainly focused on the innovation and application of control methods, such as [75–77]. The uncertainty of tire-road friction factor and vehicle load distribution affect the lateral stability and handling performance of the vehicle. The vehicle dynamics solution for this problem is through a combination of active rear wheel steering (ARS) (also known as "four-wheel steering") and DYC, or through AFS combined with DYC. The study [78] proposed a cooperative control method of AFS and DYC based on optimal guaranteed performance to achieve stability and better vehicle maneuverability. Literature [79] considered the lateral dynamics stability of the vehicle under the condition of time-varying vehicle longitudinal speed. In the work [80], an augmented linear variable parameter model based on combined proportional integral control rate and a robust gain-scheduling state-feedback controller is proposed. It minimizes the energy-to-peak control performance of the AFS/DYC system. Other researchers have proposed the combination of front-wheel active steering and direct yaw moment control to promote vehicle handling and stability in the literature [81]. The innovation of the above control methods focuses on the vehicle, closely following the reference path in the interference environment, but the speed prediction problem is not studied. Studies that combine speed prediction with vehicle lateral dynamics are usually found in driving decision problems, such as [82]. A receding horizon method based on mixed logic dynamics constraints with the objectives of the steering wheel of the driver, longitudinal speed control, and lateral lane tracking performance is established through a safety-guaranteed optimization model.

2.3.2. Development Trend

There is no control strategy to restore the stability of a vehicle after the tire forces on all four wheels are saturated, because there are no more controllable external forces acting on the vehicle. In this case, the way to avoid danger is to anticipate the danger in advance and take actions to avoid it. This is reflected in advance during the deceleration of vehicles, that is, the "speed prediction and planning" mentioned above. Therefore, it can be said that accurate prediction and planning of vehicle speed is very important for both vehicle energy efficiency and vehicle handling stability.

The existing distributed drive control theory research mainly focuses on the four-wheel independent control of electric vehicles. As the degree of vehicle electrification increases, there will be more and more research on the comprehensive consideration of distributed drive control and energy management of hybrid vehicles [65].

The advantages, disadvantages, and applications of the above traffic velocity and vehicle speed forecasting methods are shown in Table 1.

2.4. *The Relationship between Energy Management and Velocity Prediction of HEVs*

The United States Energy Information Administration forecasts that oil and other liquid fuels will continue to dominate the transportation industry from 2010 to 2040 although it is noted that their share will decrease significantly (for example, from 98% to 80%) [83]. In other words, fossil energy will still be the main energy resource under the current situation that diesel vehicles account for the majority. Carbon dioxide emissions produced by the transportation industry account for 22% of all emissions, which leads to climate change issues such as global warming [84]. To address air pollution, climate change, and energy shortages, vehicle engineering researchers and policymakers are looking for sustainable alternatives that are less dependent on oil and cause less pollution. HEVs is one of the most promising alternatives.

The main purpose of energy management is to allocate energy demand among different energy sources in order to maintain battery state-of-charge (SoC), optimize energy efficiency, and reduce fuel consumption and emissions, among other related purposes. Huang, Y. et al. [85] divides energy management strategies into two categories: offline and online. The benchmark for this classification is whether the algorithm can operate in real-time, because energy management is designed to be applied in real-time, or as a benchmark to prove the effectiveness of other approaches. Online energy management algorithms can be divided into rule-based energy management and optimization-based energy management [85], as shown in Figure 4.

Table 1. Advantages and disadvantages of different velocity prediction methods (VPM).

VPM	Advantages of VPM	Disadvantages of VPM	Suitable Application Scenarios	Ref.
Traffic Flow Model-based VPM.	• Self-explanatory about the mechanism of traffic. • Predicting accurate traffic conditions with less data. • It has a higher interpretability. • It can be directly integrated with traffic control practices. • It is easy to combine with various traffic sensors and communication system information.	• Inaccurate or uncalibrated models can result in poor performance of traffic condition assessments. • Since the traffic flow model cannot be adjusted adaptively, it cannot reflect the change of traffic flow caused by random traffic interference independently.	It is suitable for predicting traffic flow that does not change rapidly, such as traffic parameters within expressways and urban trunk road.	[21–27,32–37]
History data-based VPM.	Less time is devoted to model selection and calibration.	• Models can fail when unexpected events occur or when they are forecasting a relatively long trend. • The computational cost of training and learning can be very high. • Unable to properly explain the model decisions. • If the real traffic flow is significantly different from the information stored in the data used for the training model, the prediction accuracy will not be guaranteed.	Suburban roads where the traffic situation is relatively simple and the traffic flow is small.	[43–53]
Real-time data-based VPM.	It is robust to uncertain phenomena or to unpredictable accidents.	High requirements for traffic sensors and other infrastructure.	Forecast of traffic flow in urban road network.	[54]
Traffic safety-based VPM.	It is convenient to integrate with path planning of autonomous vehicles.	• Due to the variety of models and control methods, it is difficult to fully reflect the lateral dynamics characteristics of different vehicles. • Coupling relation between traffic risk and speed decision.	Automatic driving condition. Advanced driving assistance.	[62–71]

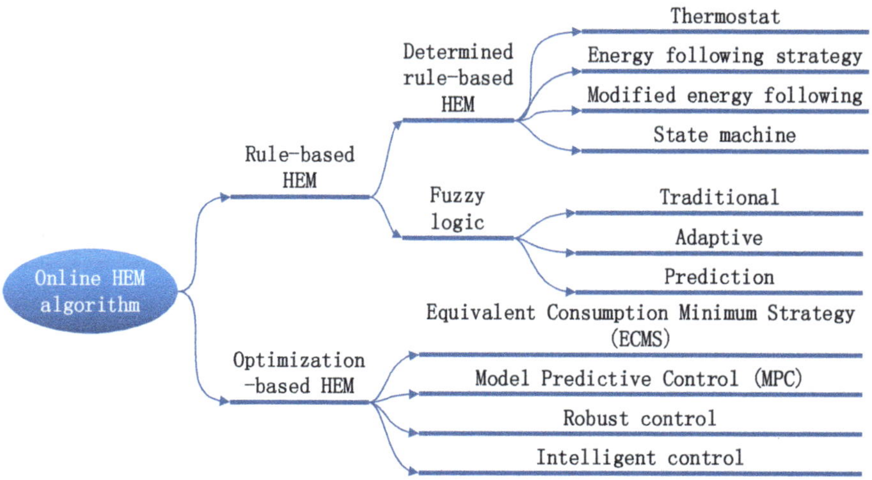

Figure 4. Online classification of energy management algorithms.

Velocity prediction and the quality of prediction results have great impacts on the performance of corresponding predictive energy management strategies (PEMSs) [15]. Velocity prediction should pay attention not only to the accuracy of the prediction, but also the length of the forecast time.

2.4.1. The Accuracy of Velocity Prediction in the Effectiveness of Energy Management

It is common to study an improved velocity prediction in the existing energy management methods of HEVs. Velocity prediction is usually associated with hybrid systems to achieve the objectives of minimum energy consumption, increasing battery life, improving driving safety, etc. Typical examples are given below:

A driver-oriented velocity prediction is created by a deep fuzzy predictor utilizing fuzzy granulation technology, and vehicle speed and acceleration are learned by transition probabilities of a finite-state Markov chain. A chaos-enhanced accelerated swarm optimization is presented with the dual-loop online intelligent algorithm to optimally determine power distribution between two power sources [86]. Markov chain and fuzzy C-means clustering are proposed for cooperative velocity forecasting that is composed of predictive sub-models to deal with various driving patterns. Forecasted velocity profiles are blended via the entire sub-models using quantified fuzzy membership degrees to obtain the final prediction results [87].

Under different short-term velocity horizons, i.e., 5 s, 10 s, and 15 s horizons, a deep neural network is utilized. At the same time, to calculate the optimal power-split at each MPC decision step, the dynamic programming method is applied [88]. Another strategy to speed forecast is introduced via a multi-stage neural network in [89]. ARIMA (autoregressive integrated moving average)-based data-driven strategy to predict short-term speed and road gradient in real-time is demonstrated [90].

Based on the driving power distribution under different driving cycles, a reinforcement learning controller (RLC) trained by the Q-learning algorithm is studied. A multi-step Markov speed prediction model-based RLC is embedded into a stochastic MPC to find optimal battery power in the predicted time [91]. To achieve the cross-type knowledge transfer between deep learning-based EMSs, a transfer-learning-based method is designed [92].

Another method is the pattern sequence-based speed predictor for accurate short-term speed prediction [93]. It is important to highlight that there are numerous speed prediction strategies based on the driver model. The main idea is that the speed prediction is obtained with optimization of the engine torque, the brake force, and the gearshift schedule, taking into account safe driving distance and traffic speed limits [94].

Through literature research, it can be concluded that accurate speed prediction can improve the above-mentioned objectives. The objective of speed planning and energy management is generally conflicted with driving safety cost, energy consumption cost, and battery life loss when the EMS is defined as a co-optimization problem over the moving horizon [95]. To better understand the influence of horizon length in speed prediction, the following section is introduced.

2.4.2. The Time Length of Velocity Prediction

For the prediction length of traffic parameters, it is relatively easy to accurately predict the change of traffic parameters within a short period of time, because the existing literature assumes that the traffic parameters remain unchanged within a short period of time [15]. However, short-term prediction of traffic parameters has limitations in practical application. For example, when short speed prediction is applied to the energy management of HEVs, it will lead to suboptimality of the solution [15]. In addition, short-term traffic parameter prediction cannot adequately provide the information required for vehicle path planning, and incorrect path planning will lead to the complete loss of fuel economy improved by predicted energy management [15].

These are the current challenges in accurately predicting traffic parameters: (1) The contradiction between prediction accuracy and computation burden (for example, the number of parameters to identify past traffic flow segments); (2) How to choose the duration of data collection and traffic parameter prediction so as to balance the contradiction between prediction accuracy and application requirements?

2.4.3. Development Trends

The goals of energy management for hybrid vehicles are as follows [65]:

1. The engine operates in an efficient range.
2. Making the vehicle friction process (between the tire and the road, and friction braking) as little as possible.
3. Increasing battery life. Battery SoC is not too high or too low (usually between 40% and 90%), and corresponds to saturated (90% to 100%) and insufficient (0% to 40%) states.

The second goal is the core issue of vehicle energy management, that is the "speed forecast and planning" that should be as accurate as possible in the future. The goal of "speed forecast and planning" is not limited to individual vehicles, but should be extended to the entire transportation system. In future, vehicular communication networks (vehicle-to-everything (V2X)), i.e., internet of vehicles and traffic infrastructure (mainly referring to traffic sensors, communication network, and big data analysis and prediction), should be established, and improvements should be made in vehicle intelligence vehicle electrification—braking should be mainly completed by regenerative braking of the motor. One goal of NEVs is to focus on braking without the frequent use of brakes, that is, to pursue accurate prediction and planning speed. Braking is only used in case of emergency.

3. Questions Raised

From the above research status and development analysis, the following key issues need to be resolved urgently: (1) Establishing a mixed traffic flow model to measure the overall operation of the comprehensive transportation network. The mixed traffic flow model is a complex transportation system model that breaks through the traditional single road network level and integrates multiple types of road networks. (2) Most safety risk assessments are based on accident data analysis, and the identification and quantitative assessment of risk points before accidents need to be improved. At the same time, the vehicle speed prediction method based on vehicle lateral dynamics needs to be studied carefully. Therefore, it is an interesting research direction to study the vehicle speed prediction method, which integrates vehicle handling stability and mixed traffic flow model with its mechanism and implementation method for improving vehicle safety and energy savings. It will have an important impact on improving traffic efficiency, reducing traffic risks, and improving energy utilization on a theoretical basis, revealing the dynamic evolution law and traffic operation situation of road traffic network, mastering the formation mechanism of traffic bottlenecks and traffic accidents. Further, the safety risk of road traffic network needs to be evaluated, so as to lay the foundation for early prediction and avoidance of accidents under adverse weather conditions such as rain and snow. In terms of practical engineering applications, the results provide personalized traffic risk assessment for vehicles with dynamic and handling characteristics, which can improve vehicle safety (safe path planning method) and energy utilization efficiency (energy-saving speed planning method), and promote the application and market-oriented development of intelligent vehicles and intelligent transportation systems.

3.1. Traffic Flow Model (Both Macro and Data-Based)

The macroscopic traffic flow model has experienced a relatively long period of development, and it is still continuing. In recent years, machine learning methods have been used to study and predict traffic parameters. Both model-based and data-based traffic flow models have their own advantages. Their comparison and possible combination and complementarity will be a research direction in the future.

For macroscopic traffic flow models, the following questions need to be answered:

1. What causes errors in the macro traffic flow model?
2. What determines the magnitude of the error?
3. How to improve the model to reduce the error?
4. What is the cause of the error in the prediction delay in the time axis?

For data-based traffic flow models, we have the following questions:

1. How does the macroscopic traffic flow model and the data-based model affect the error of prediction results?
2. How can combining the above two methods reduce prediction error?
3. How can neural networks correct the prediction delay of macroscopic traffic flow models?

3.2. Influence of Vehicle Lateral Dynamic on Speed Prediction

For the influence of vehicle lateral dynamic on speed prediction, we have the following questions:

1. What are the vehicle handling stability factors causing the speed prediction error?
2. How do traffic velocity and traffic density affect drivers' decision-making with different handling characteristics?
3. Energy, time, and safety are often conflicted. Their weights vary depending on the driver. What is the mechanism by which we get optimal path and speed?
4. Planning the longitudinal speed of intelligent vehicles to improve traffic efficiency, traffic safety, and energy utilization efficiency is a key scientific problem.

4. The Application Field of Speed Prediction

The development of modern intelligent vehicles and intelligent transportation systems requires that different research directions in the past be integrated and comprehensively considered to meet the requirements of simultaneous realization of multiple goals in vehicle design and transportation fields. The comprehensive consideration of vehicle "lateral handling stability" and "optimal energy efficiency" is becoming a trend. The power and steering structures of the existing NEVs are shown as Figures 5–7. The existing literature mainly focuses on the optimal multi-objective control problem of the handling stability and energy efficiency of pure electric vehicles [96,97], and its structure is shown in Figure 5. Energy management studies are common in the design of HEVs, for example: [98].

The stability control of HEV is usually obtained from a series hybrid vehicle similar to the four-wheel independent drive pure electric vehicle, for example: [99]. However, the optimal multi-objective control problem of handling stability and energy efficiency of parallel HEVs and Split HEVs is lacking. In Figure 7, the feature of this arrangement is that the front and rear wheels can be steered separately, and the two front wheels are driven by a wheel motor separately. "Lateral vehicle dynamics" is concerned with the handling stability of the vehicle, namely vehicle "safety".

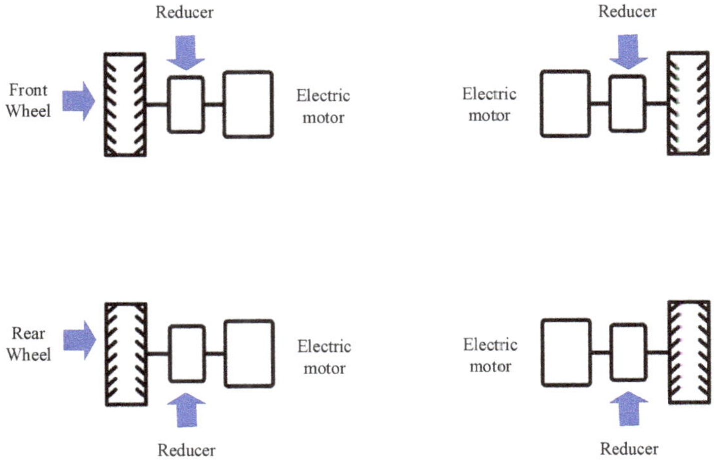

Figure 5. Power system layout structure of a four-wheel independent control electric vehicle.

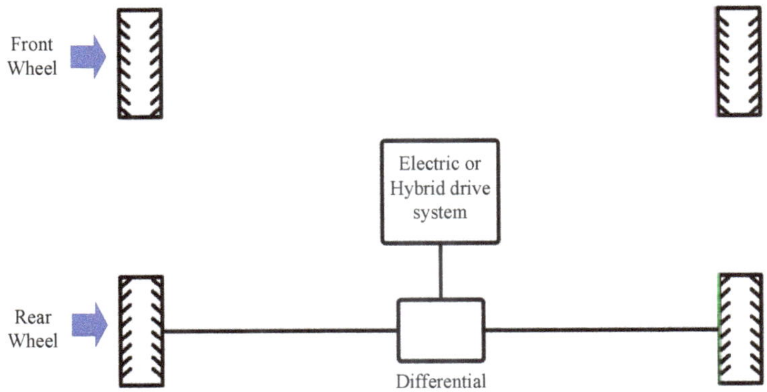

Figure 6. The layout structure of the power system of an ordinary front-wheel steering HEV bus.

Traffic flow parameters such as traffic velocity and traffic density have a direct impact on the speed prediction of vehicles. As shown in Figure 8, traffic parameters are collected by roadside sensors (such as millimeter-wave radar, cameras, etc.) and high-altitude unmanned aerial vehicles. Accurate traffic historical data and real-time data are very important for the accurate prediction of traffic velocity. The data collected by the sensors is applied to macroscopic traffic flow models and learning models to predict traffic velocity in short and relatively long horizons.

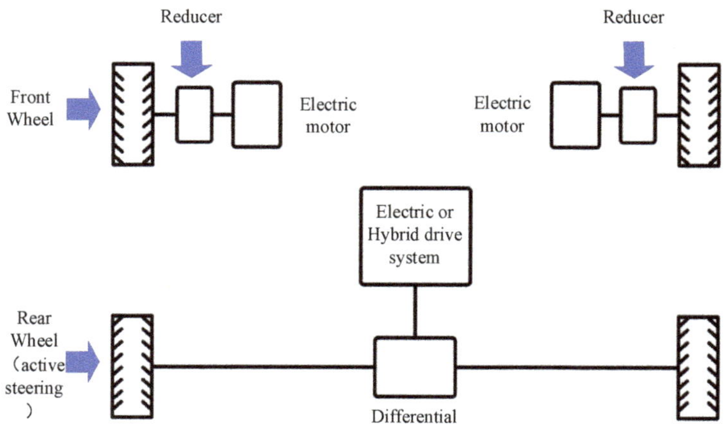

Figure 7. The layout structure of the power system of an all-wheel-steer distributed-drive new energy vehicle.

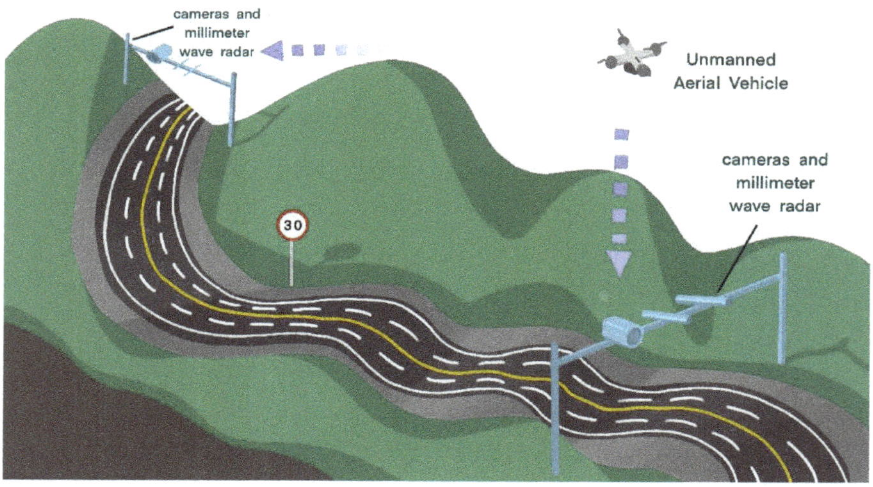

Figure 8. Use traffic sensors to collect changes in traffic parameters.

Once we understand the evolution of traffic flow and can predict changes in traffic parameters, we then need to understand the dynamics of the vehicles we are riding in. In order to improve traffic efficiency, we assume that the minimum speed of the vehicle is the traffic velocity at this time. However, to get to their destination earlier, passengers sometimes need to travel faster than the traffic speed. At this time, vehicle dynamics plays a key role in the decision-making and speed prediction of intelligent vehicles. Because at this time, whether in the curve or during an overtaking lane change condition, vehicle tire force is closer to its adhesion limit. As shown in Figure 9, when the host vehicle (HV3) is in these conditions, it needs to comprehensively consider the distance (s13 means the relative distance between HV3 and SV1, s23 means the relative distance between HV3 and SV2, s12 means the relative distance between SV1 and SV2), relative speed (V1, V2, V3 indicate the speed of the SV1, SV2, and HV3, respectively) with surrounding vehicles, and the speed limit on the curve. The vehicle needs to assess traffic risks and predict changes in its speed before performing these actions. At this time, the speed prediction is no longer passive, but an active planning process. Thus, one gets a more accurate prediction of speed when traffic flow and vehicle dynamics are taken into account.

Figure 9. An example of the influence of vehicle lateral dynamics on lane change decisions and speed prediction.

For the problem of vehicle routing, most of the existing literature adopt the optimization of single objective as the criterion. This paper proposes a multi-objective optimization method for hybrid vehicle path planning as follows:

4.1. Vehicle Handling Stability Criterion Model by Neural Network

The vehicle handling stability criterion model is investigated with test data in this part. In this approach, previous measurements of traffic flow velocity and traffic density are mainly considered. An artificial neural network is represented in Figure 10. As the figure depicts, the inputs of the network include traffic flow speed and traffic density in Equation (1).

$$\delta_{\mathrm{TFSC}} = f[\rho(1), \rho(2), \rho(3) \ldots \rho(n); V(\rho(1)), V(\rho(2)), V(\rho(3)) \ldots V(\rho(n))], . \tag{1}$$

where $\rho(1), \rho(2), \rho(3), \ldots, \rho(n)$ are traffic density,$V(\rho(1)), V(\rho(2)), V(\rho(3)), \ldots, V(\rho(n))$ are traffic speed. The δ_{TFSC} is the tire force saturation coefficient represented by Equations (2) and (3):

$$\delta_{\mathrm{TFSCb}} = (\frac{F_{\mathrm{xk}}}{\mu F_{\mathrm{zk}}})^2 + (\frac{F_{\mathrm{yk}}}{F_{\mathrm{ymaxk}}})^2, . \tag{2}$$

$$\delta_{\mathrm{TFSCmax}} = max((\frac{F_{\mathrm{xk}}}{\mu F_{\mathrm{zk}}})^2 + (\frac{F_{\mathrm{yk}}}{F_{\mathrm{ymaxk}}})^2), . \tag{3}$$

where: $F_{\mathrm{xk}}, F_{\mathrm{yk}}, F_{\mathrm{zk}}$ are the longitudinal, lateral, and vertical tire forces of the vehicle tires, respectively; μ is the friction factor between the tire and the road; k is the number of the tire (specifically refers to four different tires); F_{ymaxk} is the maximum lateral force of the kth tire.

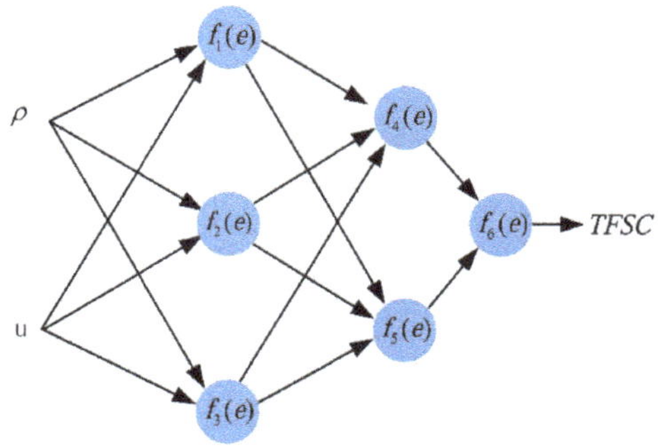

Figure 10. Artificial neural network structure.

The idea of stability criteria model is that we first build the lateral dynamics model of the vehicle (please refer to our previous work [68]), and then we build the traffic environment shown in Figure 11. A driver-in-the-loop simulation environment for the presented scenario is shown in Figure 12. The traffic environment in the figure mainly shows two main traffic parameters: Distance and speed of SV and HV. Distance represents traffic density; at the same time, the speed of the SV indicates the speed of the traffic. We use this simulation environment to simulate the change in the saturation coefficient of the vehicle's tires when the driver faces different traffic conditions and makes a lane change. For a front-wheel-drive ordinary vehicle, Figure 13 shows the change in the tire's tire saturation coefficient as a function of traffic flow speed and density. As can be seen from Figure 13, the δ_{TFSC} increases as the vehicle speed and density increase when the driver performs a lane change.

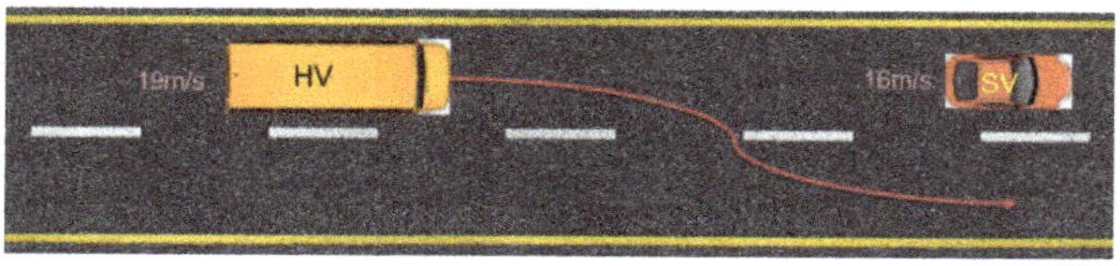

Figure 11. One of the working conditions for collecting driver data.

Figure 12. Driver-in-the-loop experiment to collect drivers' reactions to different working conditions.

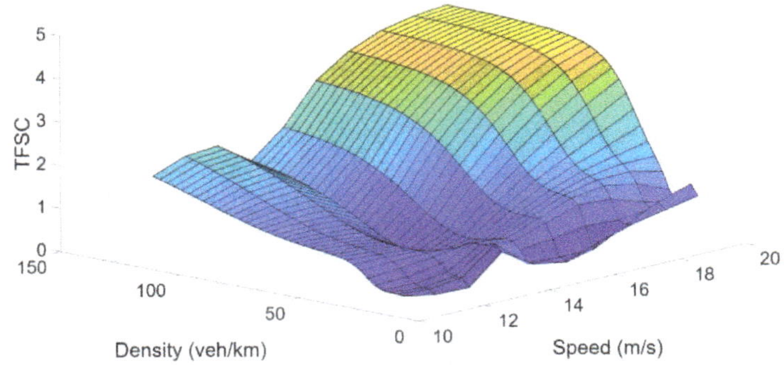

Figure 13. Neural network fitting results of TFSC for traffic speed and density.

4.2. Multi-Objective Optimization Path Planning for Hybrid Vehicles

In this article, the speed prediction method based on the traffic flow model, speed prediction method based on traffic historical data and real-time data, and speed prediction method based on the vehicle lateral dynamics are reviewed. The relationship between the energy management method and speed prediction of HEVs is briefly summarized. The existing problems of speed prediction methods are presented and a new system structure of a hybrid electric vehicle is constructed. Based on the new system structure of the hybrid electric vehicle and the proposed stability criterion model, a vehicle path planning method and its application case based on the vehicle speed prediction method and vehicle lateral risk assessment are given below. According to the various traffic flow prediction methods reviewed above, this paper assumes that future traffic flow parameters, such as traffic density ρ and traffic velocity V, can be accurately predicted. Furthermore, the vehicle tire force saturation factor δ_{TFSC} corresponding to each traffic density ρ and traffic velocity V was calculated by the vehicle stability criterion model established above. The constructed vehicle stability criterion model based on the use of tire force saturation factors can reflect the result of vehicle safety during driving. Next, the stability criterion model shown in Figure 13 will be used as the basis of the multi-objective optimization path planning to make the vehicle reach its destination safely, quickly, and efficiently. Driving safety, driving time, and energy consumption are trade-offs to achieve the best overall performance. The problem of vehicle path planning and selection is described as an optimization problem. The optimized performance indicators are described below.

4.2.1. Composite Index

The composite index $\{J_{\mathrm{com}}\}$ is used to evaluate the overall performance of vehicle driving safety, time, and energy consumption, and the composite index is represented by Equation (4):

$$\min J_{\mathrm{com}} = \lambda J_{\mathrm{s}_i} + \beta J_{\mathrm{t}_i} + \gamma J_{\mathrm{e}_i}, \tag{4}$$

where λ, β, γ represent the weighting factors of vehicle safety, travel time, and energy consumption, respectively.

4.2.2. Driving Safety Index

Using the steering stability criterion of the vehicle, the tire force saturation factor can be used to evaluate this basis, and the driving safety index can be represented by $\{J_{\mathrm{s}_i}\}$. Driving safety is often the most important factor when the driver is driving a vehicle on the road. The larger the driving safety index $\{J_{\mathrm{s}_i}\}$, the greater the influence of path security on the planning result.

$$J_{\mathrm{s}_i} = e^{\delta_{\mathrm{TFSC}(i)}}, \tag{5}$$

4.2.3. Travel Time Index

The driver expects that the driving time of the vehicle be as short as possible. However, most of the time, we have an expected period of time to arrive at our destination, and the weight function should reflect penalties that are greater than the expected time and rewards that are less than the expected time. The travel time index of the vehicle path can be represented by $\{J_{t_i}\}$:

$$J_{t_i} = \begin{cases} 0.8 + (t_i - t^{lb}) & t_i < t^{lb} \\ 0.8 & t^{lb} \leq t_i \leq t^{ub} \\ 0.8 + (t_i - t^{ub}) & t_i > t^{ub} \end{cases}, \tag{6}$$

where $\{t_i\}$ is the travel time of the vehicle on the ith path, and t^{lb} and t^{ub} are the lower and upper bounds of the expected vehicle travel time, respectively.

4.2.4. Energy Expenditure Index

According to the longitudinal vehicle dynamics model, it can be seen that the vehicle is driving on a flat road, which means that there is no up and down gradient during the driving process [65]. At this time, the longitudinal dynamics model is expressed from Equations (7)–(10). The energy consumed is represented by Equation (11):

$$F_\omega = F_{air} + F_{roll} + F_{inertia}, \tag{7}$$

$$F_{air} = \frac{C_D A_f}{21.15} V^2, \tag{8}$$

$$F_{roll} = mgf, \tag{9}$$

$$F_{inertia} = \sigma m \frac{du}{dt}, \tag{10}$$

$$J_{e_i} = V t F_\omega, \tag{11}$$

where C_D is the coefficient of air resistance; A_f is the front area of the vehicle; m is the vehicle quality; g is the acceleration of gravity; f is the rolling resistance coefficient; t is the time interval; σ is the rotating mass correction coefficient; F_ω is the total resistance of the wheel when driving; F_{air} is the air resistance; F_{roll} is the rolling resistance; $F_{inertia}$ is the acceleration resistance; J_{e_i} is the energy consumed by the vehicle.

4.2.5. Determination of Weighting Factors

To compare the above three indexes, there is a need to weigh each of them by a coefficient. For driving safety index $\{J_{s_i}\}$, the weight factor $\{\lambda\}$ is determined as follows (Equation (12)):

$$\lambda = \begin{cases} 1, & \text{nomally} \\ \eta, & \text{other characteristics} \end{cases}, \tag{12}$$

where η is adjusted according to the driver's aggressiveness level, road friction coefficient, vehicle stability characteristics, etc. The travel time weighting factor β is determined as follows (Equation (13)):

$$\beta = \begin{cases} 1, & \text{nomally} \\ \kappa, & \text{other characteristics} \end{cases}, \tag{13}$$

where κ is determined according to the driver anxious degree, the congestion of the road network, rush hour, etc. Note that κ is used to avoid excessively congested roads.

Generally, we can find the shortest path from the starting point to the destination. Because the driver expects that the vehicle consumes as little energy as possible, we choose the energy consumed by the vehicle in the shortest route under a smooth traffic environment as a reference. The energy weight factor $\{\gamma\}$ is determined as follows (Equation (14)):

$$\gamma = \frac{1}{J_{\text{normal}}},. \tag{14}$$

where J_{normal} is the normally expended energy or the average value of energy consumption from the beginning to the end.

The control framework proposed in this section aims to find a trade-off between driving safety, driving time, and energy consumption during the vehicle driving process under the premise of ensuring the safety of the vehicle, so as to help the vehicle rationally plan and select the path. As shown in Figure 14, consider there are i paths available for vehicles to choose, and it is assumed that the parameters of traffic flow can be accurately predicted. Then, the vehicle stability criterion model is used to evaluate the stability of the future parameters of each path, and the saturation factor of the tire force δ_{TFSC} is obtained when the vehicle is traveling at a predicted traffic density and speed in the future. To evaluate the tire force saturation factor of each path, if the tire force saturation factor of each path $\delta_{\text{TFSC}_i} < 1$, vehicles can adopt driving behavior strategies such as lateral overtaking and changing lanes, at this time, calculate the driving safety index of each path J_{s_i}. Then we calculate travel time index J_{t_i} and energy consumption index J_{e_i} of each path, respectively, by substituting the driving index of each path and the weighting factor corresponding to each index into the comprehensive index model to obtain the best path. If the tire force saturation factor of each path is $\delta_{\text{TFSC}_i} > 1$, drivers on such roads should avoid overtaking and changing lanes, and choose conservative driving behaviors, such as following the car in front. Under the premise of ensuring the safety of the vehicle, we then get the travel time index J_{t_i} and energy consumption index J_{e_i} of each path, and obtain the comprehensive index of each path and the weighting factor corresponding to each index. We then analyze the weight function "composite index" to get the best path. If some sections of a path have a criterion $\delta_{\text{TFSC}_i} < 1$ and others sections have $\delta_{\text{TFSC}_i} > 1$, then it needs to be discussed separately. For the sections of $\delta_{\text{TFSC}_i} < 1$, vehicles are free to make overtaking lane changes. In the sections of $\delta_{\text{TFSC}_i} > 1$, in order to ensure the stability of the vehicle, the driver needs to follow the traffic flow instead of lane-changing and overtaking. At this time, we recalculate the driving safety index J_{s_i}, the travel time index J_{t_i}, and energy consumption index J_{e_i} of each path, and obtain the weighting factor corresponding to each index, using the composite index model to finally obtain the best choice path.

4.3. Path Planning Application

The process of achieving the path planning goals can be expressed as follows: The traffic parameter set $P(x,t)$ (vehicle velocity, traffic density, etc.) is predicted for each sub-path, and it is provided to the vehicles in the traffic system so as to arrange the journey according to their respective characteristics. Then, through machine learning, the stability criterion model $S(\rho,u)$ of all-wheel-drive (AWD) vehicles and front-wheel-drive (FWD) vehicles is established facing different traffic parameter sets $P(x,t)$. Then the optimal path is chosen according to $S(\rho,u)$. In the end, the optimal path for the current vehicle, among multiple paths, is obtained by using the multi-objective optimal path planning method $J(\rho,u)$ proposed in this paper.

When the expected vehicle speed is higher than the traffic flow speed, it means that the vehicle needs to change lanes frequently. At a certain vehicle flow speed and density, the maximum front-wheel angle of a successful lane change can be predicted. Furthermore, we can calculate the lateral tire force of the vehicle through the vehicle lateral dynamics model. Similar to energy and time, the vehicle's tire force saturation coefficient δ_{TFSC} can be used as a basis for the vehicle's path selection. Finally, depending on the travel time, energy consumption, environment (road friction coefficient), and vehicle characteristics (vehicle handling stability), one can choose a safe, efficient, and fast road for the driver.

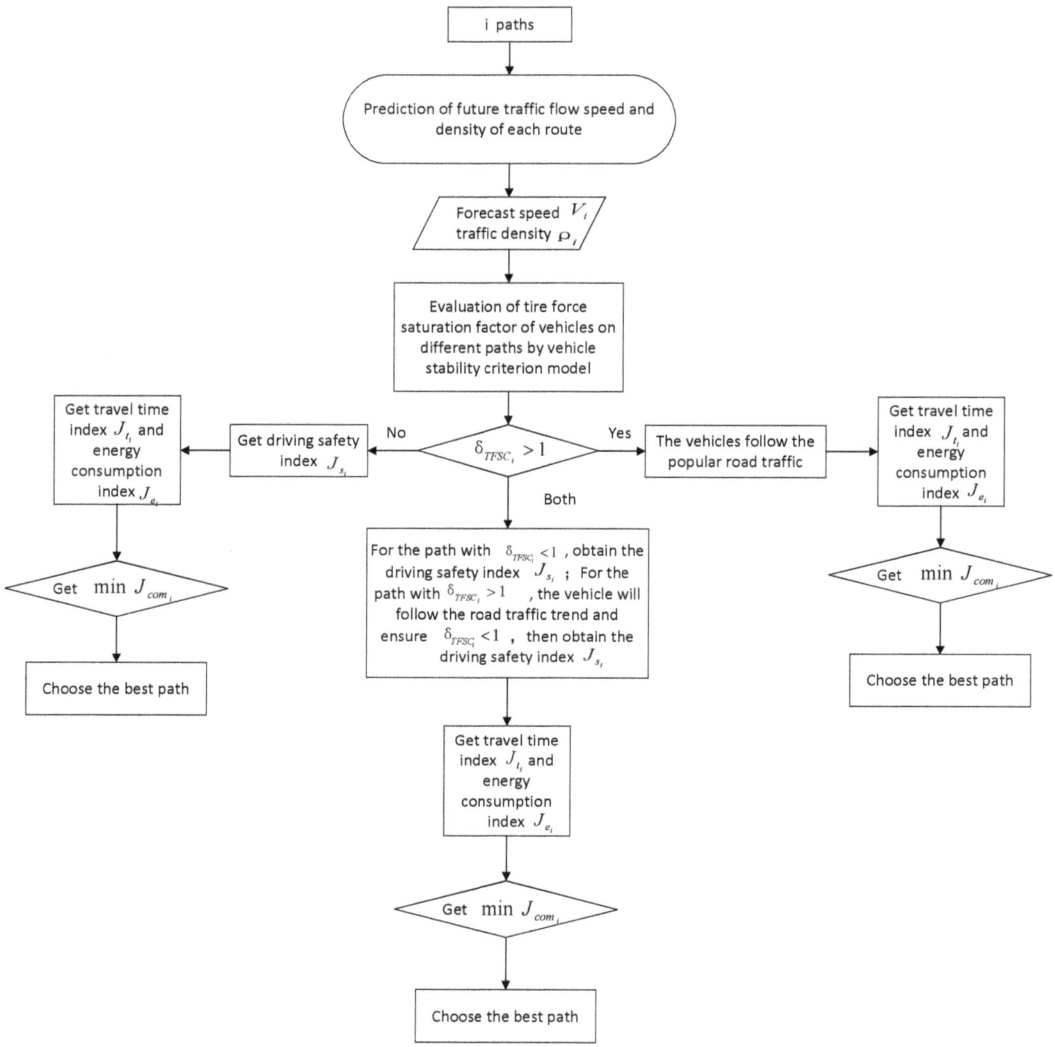

Figure 14. Path planning/selection logic diagram based on multi-objective optimization.

We consider route planning of a traffic system using traffic flow model and vehicle stability criterion as an example in Figure 15. Assume that Elbert intends to drive his vehicle from point A, seen on the left side of the figure, to point B, shown on the right side of the figure. There are three routes for Elbert to choose, namely Route 1, Route 2, and Route 3. Route 1 is closer, but the traffic flow is dense and the speed is slow. The distance of Route 2 is longer, but the traffic density is small and vehicle speed can be fast while traveling. The length and traffic flow of Route 3 are moderate, between Route 1 and Route 2, but road friction is poor. At this time, the predicted traffic flow parameters, the stability criterion model, and the multi-objective optimization route planning method can help Elbert select a safe, efficient, and fast road that matches the characteristics of his vehicle.

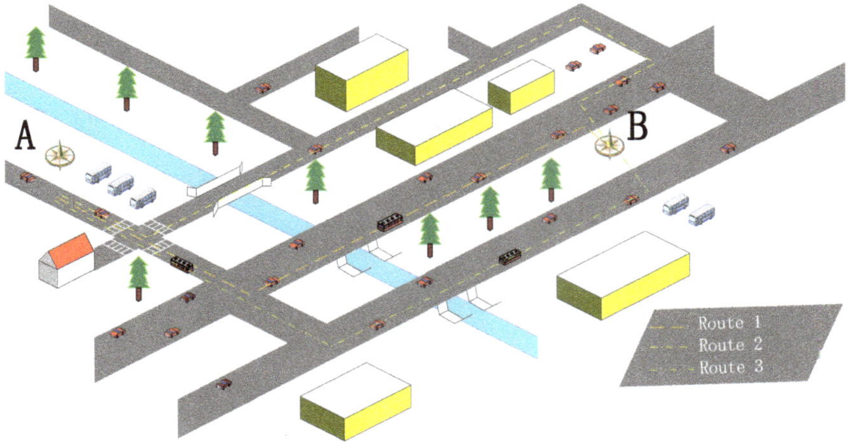

Figure 15. Example of route planning for a traffic system using traffic flow model and vehicle stability criterion.

In addition, for the problem shown in Figure 15, we consider that Elbert encounters a rainy and snowy day (Route 1's and Route 2's tire-road friction coefficient is 0.4, Route 3's friction coefficient is 0.1 due to road icing). According to the statement above, he will have different options to drive All-Wheel-Drive (AWD) and Front-Wheel-Drive (FWD) vehicles. Figure 11 shows a lane change and overtaking condition when the driver is close to the preceding vehicle, the target vehicle (HV)'s speed is large, and the front vehicle (SV)'s speed is small. Furthermore, Figure 16 shows the variation in δ_{TFSC} for AWD and FWD vehicles at the same steering input in an overtaking and lane change condition from Route 1 [100]. We can observe that the FWD vehicle has been destabilized, but the AWD vehicle can keep the vehicle stable under such extreme conditions. For Route 3, where the road is icy, both AWD and FWD cannot stabilize the vehicle during a lane change and overtaking condition. Therefore, although the path 1 traffic flow speed is slow, when the driver drives the AWD vehicle, the destination can be reached faster by continuously overtaking and changing lanes. However, when driving a FWD vehicle, Path 2 would be a better choice.

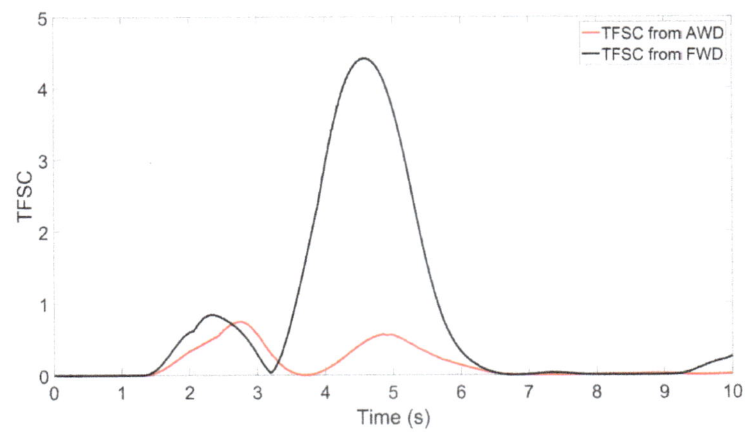

Figure 16. TFSC comparison results between an AWD vehicle with stability control and a FWD vehicle without stability control for the same steering input.

5. Conclusions

Energy management strategies of new energy vehicles (NEVs) highly depend on accurate prediction of future velocity. In this paper, we present a review study on various vehicle speed prediction methods for NEVs. In this regard, various macroscopic traffic flow models, data-based traffic flow models, and the influence of vehicle lateral dynamics are introduced.

Through a detailed review and comparison of each method, it is clear that each approach is suitable for different application scenarios. Macroscopic and data-based traffic flow models are introduced and compared in terms of their pros and cons, potentially leading to better identification of development trends for prospective designers. Questions regarding the error in the macro traffic flow model, the magnitude of the error, and how to reduce the error with delay are to be answered. Moreover, prediction error between data-based and macro traffic models as well as how to combine these to reduce error remains an open headline.

Since the core issue of vehicle energy management is accurate speed forecast and planning, an emerging field, namely, vehicle stability control methods on speed prediction are investigated. Key questions on establishing a mixed traffic flow model and safety risk assessments are investigated from a traffic flow model and influence on vehicle lateral dynamics viewpoints.

The link between vehicle stability and energy efficiency is demonstrated by the application field of speed prediction methods. Benefitting from the fast development of vehicular technologies; software developers in the field of artificial intelligence; and sensors, cameras, and radars, potential future developments for velocity prediction methods could guide and inspire prospective researchers.

Lastly, examples of driving safety, traffic efficiency, and energy management are used to demonstrate the applications of speed prediction method based on vehicle handling dynamics and driving environment in path planning.

Funding: This research was funded by Northeast Forestry University No. 520-60201418.

Conflicts of Interest: The authors declare no conflict of interest.

References

1. Li, K.Q.; Dai, Y.F.; Li, S.B.; Bian, M.Y. Development status and trend of intelligent connected vehicle (ICV) technology. *J. Automot. Saf. Energy* **2017**, *8*, 1–14.
2. Pei, J.; Su, Y.; Zhang, D.; Qi, Y.; Leng, Z. Velocity forecasts using a combined deep learning model in hybrid electric vehicles with V2V and V2I communication. *Sci. China Technol. Sci.* **2020**, *63*, 55–64. [CrossRef]
3. Hu, X.; Yuan, H.; Zou, C.; Li, Z.; Zhang, L. Co-estimation of state of charge and state of health for lithium-ion batteries based on fractional-order calculus. *IEEE Trans. Veh. Technol.* **2018**, *67*, 10319–10329. [CrossRef]
4. Zhang, F.; Hu, X.; Langari, R.; Wang, L.; Cui, Y.; Pang, H. Adaptive energy management in automated hybrid electric vehicles with flexible torque request. *Energy* **2021**, *214*, 118873. [CrossRef]
5. Zhang, F.; Hu, X.; Liu, T.; Duan, Z.; Xu, K.; Pang, H. Computationally Efficient Energy Management for Hybrid Electric Vehicles Using Model Predictive Control and Vehicle-to-Vehicle Communication. *IEEE Trans. Veh. Technol.* **2020**, *70*, 237–250. [CrossRef]
6. Hu, X.; Zou, C.; Tang, X.; Liu, T.; Hu, L. Cost-optimal energy management of hybrid electric vehicles using fuel cell/battery health-aware predictive control. *IEEE Trans. Power Electron.* **2019**, *35*, 382–392. [CrossRef]
7. Yang, S.; Wang, J.; Zhang, F.; Xi, J. Self-adaptive Equivalent Consumption Minimization Strategy for Hybrid Electric Vehicles. *IEEE Trans. Veh. Technol.* **2020**, *70*, 189–202. [CrossRef]
8. China Association of Automobile Manufacturers. 2021. Available online: http://en.caam.org.cn/ (accessed on 1 June 2021).
9. Taha, A.E.; AbuAli, N. Route planning considerations for autonomous vehicles. *IEEE Commun. Mag.* **2018**, *56*, 78–84. [CrossRef]
10. Lu, J.; Cheng, Z.Y. Research progress in road traffic network security risk identification. *J. Southeast Univ.* **2019**, *49*.
11. Huang, H.L.; Jiang, M.X.; Han, C.Y.; Xu, G.M. Multi-class user traffic assignment model based on safety and reliability. *China J. Highw. Transp.* **2018**, *31*, 312–321.
12. Zhang, F.; Wang, L.; Coskun, S.; Pang, H.; Cui, Y.; Xi, J. Energy management strategies for hybrid electric vehicles: Review, classification, comparison, and outlook. *Energies* **2020**, *13*, 3352. [CrossRef]
13. Zhang, F.; Hu, X.; Langari, R.; Cao, D. Energy management strategies of connected HEVs and PHEVs: Recent progress and outlook. *Prog. Energy Combust. Sci.* **2019**, *73*, 235–256. [CrossRef]

14. Tran, D.D.; Vafaeipour, M.; El Baghdadi, M.; Barrero, R.; Van Mierlo, J.; Hegazy, O. Thorough state-of-the-art analysis of electric and hybrid vehicle powertrains: Topologies and integrated energy management strategies. *Renew. Sustain. Energy Rev.* **2020**, *119*, 109596. [CrossRef]
15. Zhou, Y.; Ravey, A.; Péra, M.C. A survey on driving prediction techniques for predictive energy management of plug-in hybrid electric vehicles. *J. Power Sources* **2019**, *412*, 480–495. [CrossRef]
16. Djahel, S.; Doolan, R.; Muntean, G.M.; Murphy, J. A communications-oriented perspective on traffic management systems for smart cities: Challenges and innovative approaches. *IEEE Commun. Surv. Tutor.* **2014**, *17*, 125–151. [CrossRef]
17. Papamichail, I.; Bekiaris-Liberis, N.; Delis, A.I.; Manolis, D.; Mountakis, K.S.; Nikolos, I.K.; Papageorgiou, M. Motorway traffic flow modelling, estimation and control with vehicle automation and communication systems. *Annu. Rev. Control* **2019**, *48*, 325–346. [CrossRef]
18. Seo, T.; Bayen, A.M.; Kusakabe, T.; Asakura, Y. Traffic state estimation on highway: A comprehensive survey. *Annu. Rev. Control* **2017**, *43*, 128–151. [CrossRef]
19. Van Wageningen-Kessels, F.; Van Lint, H.; Vuik, K.; Hoogendoorn, S. Genealogy of traffic flow models. *EURO J. Transp. Logist.* **2015**, *4*, 445–473. [CrossRef]
20. Hoogendoorn, S.P.; Bovy, P.H. State-of-the-art of vehicular traffic flow modelling. *Proc. Inst. Mech. Eng. Part I J. Syst. Control Eng.* **2001**, *215*, 283–303. [CrossRef]
21. Lighthill, M.J.; Whitham, G.B. On kinematic waves II. A theory of traffic flow on long crowded roads. *Proc. R. Soc. London. Ser. A Math. Phys. Sci.* **1955**, *229*, 317–345.
22. Jin, W.L.; Zhang, H.M. The formation and structure of vehicle clusters in the Payne–Whitham traffic flow model. *Transp. Res. Part B Methodol.* **2003**, *37*, 207–223. [CrossRef]
23. Papageorgiou, M.; Blosseville, J.M.; Hadj-Salem, H. Macroscopic modelling of traffic flow on the Boulevard Périphérique in Paris. *Transp. Res. Part B Methodol.* **1989**, *23*, 29–47. [CrossRef]
24. Zhang, H.M. A non-equilibrium traffic model devoid of gas-like behavior. *Transp. Res. Part B Methodol.* **2002**, *36*, 275–290. [CrossRef]
25. Lebacque, J.P.; Mammar, S.; Haj-Salem, H. The Aw–Rascle and Zhang's model: Vacuum problems, existence and regularity of the solutions of the Riemann problem. *Transp. Res. Part B Methodol.* **2007**, *41*, 710–721. [CrossRef]
26. Blandin, S.; Work, D.; Goatin, P.; Piccoli, B.; Bayen, A. A general phase transition model for vehicular traffic. *SIAM J. Appl. Math.* **2011**, *71*, 107–127. [CrossRef]
27. Delis, A.I.; Nikolos, I.K.; Papageorgiou, M. High-resolution numerical relaxation approximations to second-order macroscopic traffic flow models. *Transp. Res. Part C Emerg. Technol.* **2014**, *44*, 318–349. [CrossRef]
28. Crandall, M.G.; Ishii, H.; Lions, P.L. User's guide to viscosity solutions of second order partial differential equations. *Bull. Am. Math. Soc.* **1992**, *27*, 1–67. [CrossRef]
29. Claudel, C.G.; Bayen, A.M. Lax–Hopf based incorporation of internal boundary conditions into Hamilton–Jacobi equation. Part I: Theory. *IEEE Trans. Autom. Control* **2010**, *55*, 1142–1157. [CrossRef]
30. Laval, J.A.; Leclercq, L. The Hamilton–Jacobi partial differential equation and the three representations of traffic flow. *Transp. Res. Part B Methodol.* **2013**, *52*, 17–30. [CrossRef]
31. Jin, W.L. On the equivalence between continuum and car-following models of traffic flow. *Transp. Res. Part B Methodol.* **2016**, *93*, 543–559. [CrossRef]
32. Madaan, N.; Sharma, S. A lattice model accounting for multi-lane traffic system. *Phys. A Stat. Mech. Appl.* **2021**, *564*, 125446. [CrossRef]
33. Costeseque, G.; Duret, A. Mesoscopic multiclass traffic flow modeling on multi-lane sections. In Proceedings of the 95th Annual Meeting Trans-portation Research Board-TRB, Transportation Research Board, Washington, DC, USA, 10–14 January 2016; p. 27.
34. Sumalee, A.; Zhong, R.X.; Pan, T.L.; Szeto, W.Y. Stochastic cell transmission model (SCTM): A stochastic dynamic traffic model for traffic state surveillance and assignment. *Transp. Res. Part B Methodol.* **2011**, *45*, 507–533. [CrossRef]
35. Jabari, S.E.; Zheng, J.; Liu, H.X. A probabilistic stationary speed–density relation based on Newell's simplified car-following model. *Transp. Res. Part B Methodol.* **2014**, *68*, 205–223. [CrossRef]
36. Jabari, S.E.; Liu, H.X. A stochastic model of traffic flow: Theoretical foundations. *Transp. Res. Part B Methodol.* **2012**, *46*, 156–174. [CrossRef]
37. Wada, K.; Usui, K.; Takigawa, T.; Kuwahara, M. An optimization modeling of coordinated traffic signal control based on the variational theory and its stochastic extension. *Transp. Res. Proedria* **2017**, *23*, 624–644. [CrossRef]
38. Liu, Z.; Liu, Y.; Meng, Q.; Cheng, Q. A tailored machine learning approach for urban transport network flow estimation. *Transp. Res. Part C Emerg. Technol.* **2019**, *108*, 130–150. [CrossRef]
39. Mackenzie, J.; Roddick, J.F.; Zito, R. An evaluation of HTM and LSTM for short-term arterial traffic flow prediction. *IEEE Trans. Intell. Transp. Syst.* **2018**, *20*, 1847–1857. [CrossRef]
40. Zhu, L.; Yu, F.R.; Wang, Y.; Ning, B.; Tang, T. Big data analytics in intelligent transportation systems: A survey. *IEEE Trans. Intell. Transp. Syst.* **2018**, *20*, 383–398. [CrossRef]
41. Malek, Y.N.; Najib, M.; Bakhouya, M.; Essaaidi, M. Multivariate deep learning approach for electric vehicle speed forecasting. *Big Data Min. Anal.* **2020**, *4*, 56–64. [CrossRef]

42. Lin, X.; Wang, Z.; Wu, J. Energy management strategy based on velocity prediction using back propagation neural network for a plug-in fuel cell electric vehicle. *Int. J. Energy Res.* **2021**, *45*, 2629–2643. [CrossRef]
43. Nallaperuma, D.; Nawaratne, R.; Bandaragoda, T.; Adikari, A.; Nguyen, S.; Kempitiya, T.; Pothuhera, D. Online incremental machine learning platform for big data-driven smart traffic management. *IEEE Trans. Intell. Transp. Syst.* **2019**, *20*, 4679–4690. [CrossRef]
44. Kong, F.; Li, J.; Jiang, B.; Song, H. Short-term traffic flow prediction in smart multimedia system for Internet of Vehicles based on deep belief network. *Future Gener. Comput. Syst.* **2019**, *93*, 460–472. [CrossRef]
45. Zang, D.; Ling, J.; Wei, Z.; Tang, K.; Cheng, J. Long-term traffic speed prediction based on multiscale spatio-temporal feature learning network. *IEEE Trans. Intell. Transp. Syst.* **2018**, *20*, 3700–3709. [CrossRef]
46. Chen, M.; Yu, X.; Liu, Y. PCNN: Deep convolutional networks for short-term traffic congestion prediction. *IEEE Trans. Intell. Transp. Syst.* **2018**, *19*, 3550–3559. [CrossRef]
47. Zhang, Y.; Cheng, T.; Ren, Y.; Xie, K. A novel residual graph convolution deep learning model for short-term network-based traffic forecasting. *Int. J. Geogr. Inf. Sci.* **2020**, *34*, 969–995. [CrossRef]
48. Chen, W.; An, J.; Li, R.; Fu, L.; Xie, G.; Bhuiyan, M.Z.A.; Li, K. A novel fuzzy deep-learning approach to traffic flow prediction with uncertain spatial–temporal data features. *Future Gener. Comput. Syst.* **2018**, *89*, 78–88. [CrossRef]
49. Tang, J.; Liu, F.; Zou, Y.; Zhang, W.; Wang, Y. An improved fuzzy neural network for traffic speed prediction considering periodic characteristic. *IEEE Trans. Intell. Transp. Syst.* **2017**, *18*, 2340–2350. [CrossRef]
50. Qu, L.; Li, W.; Li, W.; Ma, D.; Wang, Y. Daily long-term traffic flow forecasting based on a deep neural network. *Expert Syst. Appl.* **2019**, *121*, 304–312. [CrossRef]
51. Wu, Y.; Tan, H.; Qin, L.; Ran, B.; Jiang, Z. A hybrid deep learning based traffic flow prediction method and its understanding. *Transp. Res. Part C Emerg. Technol.* **2018**, *90*, 166–180. [CrossRef]
52. Han, L.; Huang, Y.S. Short-term traffic flow prediction of road network based on deep learning. *IET Intell. Transp. Syst.* **2020**, *14*, 495–503. [CrossRef]
53. Li, L.; Qin, L.; Qu, X.; Zhang, J.; Wang, Y.; Ran, B. Day-ahead traffic flow forecasting based on a deep belief network optimized by the multi-objective particle swarm algorithm. *Knowl. Based Syst.* **2019**, *172*, 1–14. [CrossRef]
54. Shin, J.; Yeon, K.; Kim, S.; Sunwoo, M.; Han, M. Comparative Study of Markov Chain with Recurrent Neural Network for Short Term Velocity Prediction Implemented on an Embedded System. *IEEE Access* **2021**, *9*, 24755–24767. [CrossRef]
55. Guo, Y.; Li, Z.; Liu, P.; Wu, Y. Modeling correlation and heterogeneity in crash rates by collision types using full Bayesian random parameters multivariate Tobit model. *Accid. Anal. Prev.* **2019**, *128*, 164–174. [CrossRef] [PubMed]
56. Rahman, M.S.; Abdel-Aty, M. Longitudinal safety evaluation of connected vehicles' platooning on expressways. *Accid. Anal. Prev.* **2018**, *117*, 381–391. [CrossRef]
57. Wen-Xing, Z.; Li-Dong, Z. A new car-following model for autonomous vehicles flow with mean expected velocity field. *Phys. A Stat. Mech. Its Appl.* **2018**, *492*, 2154–2165. [CrossRef]
58. Zhu, W.X.; Zhang, H.M. Analysis of mixed traffic flow with human-driving and autonomous cars based on car-following model. *Phys. A Stat. Mech. Appl.* **2018**, *496*, 274–285. [CrossRef]
59. Li, G.; Yang, Y.; Zhang, T.; Qu, X.; Cao, D.; Cheng, B.; Li, K. Risk assessment based collision avoidance decision-making for autonomous vehicles in multi-scenarios. *Transp. Res. Part C Emerg. Technol.* **2021**, *122*, 102820. [CrossRef]
60. Zhang, L.; Wang, S.; Chen, C.; Yang, M.; Zhe, X. Lane-changing risk model for expressway exit zone based on natural driving data. *J. Tongji Univ.* **2019**, *47*, 1446–1453.
61. Peng, Z.; Gao, S.; Li, Z.; Xiao, B.; Qian, Y. Vehicle safety improvement through deep learning and mobile sensing. *IEEE Netw.* **2018**, *32*, 28–33. [CrossRef]
62. Yu, Z.; Feng, Y.; Xiong, L. Review on vehicle dynamics control of distributed drive electric vehicle. *Jixie Gongcheng Xuebao (Chin. J. Mech. Eng.)* **2013**, *49*, 105–114. [CrossRef]
63. Jiang, K.; Yang, D.; Xie, S.; Xiao, Z.; Victorino, A.; Charara, A. Real-time estimation and prediction of tire forces using digital map for driving risk assessment. *Transp. Res. Part C Emerg. Technol.* **2019**, *107*, 463–489. [CrossRef]
64. Lin, Y.; Nguyen, H.; Balas, V.; Lin, T.; Kuo, I. Adaptive prediction-based control for an ecological cruise control system on curved and hilly roads. *J. Intell. Fuzzy Syst.* **2020**, *38*, 6129–6144. [CrossRef]
65. Chu, D.; Li, Z.; Wang, J.; Wu, C.; Hu, Z. Rollover speed prediction on curves for heavy vehicles using mobile smartphone. *Measurement* **2018**, *130*, 404–411. [CrossRef]
66. Imine, H.; Benallegue, A.; Madani, T.; Srairi, S. Rollover Risk Prediction of Heavy Vehicle Using High-Order Sliding-Mode Observer: Experimental Results. *IEEE Trans. Veh. Technol.* **2014**, *63*, 2533–2543. [CrossRef]
67. Liu, K.; Gong, J.; Chen, S.; Zhang, Y.; Chen, H. Model Predictive Stabilization Control of High-Speed Autonomous Ground Vehicles Considering the Effect of Road Topography. *Appl. Sci.* **2018**, *8*, 822. [CrossRef]
68. Li, L.; Coskun, S.; Zhang, F.; Langari, R.; Xi, J. Energy Management of Hybrid Electric Vehicle Using Vehicle Lateral Dynamic in Velocity Prediction. *IEEE Trans. Veh. Technol.* **2019**, *68*, 3279–3293. [CrossRef]
69. Song, P.; Gao, B.; Xie, S.; Fang, R. Optimal Predictive Control for Path Following of a Full Drive-by-Wire Vehicle at Varying Speeds. *Chin. J. Mech. Eng.* **2017**, *30*, 711–721. [CrossRef]
70. Krid, M.; Benamar, F.; Zamzami, Z. Design of an active device for controlling lateral stability of fast mobile robot. *Robotica* **2016**, *34*, 2629–2651. [CrossRef]

71. Ren, H.; Chen, S.; Yang, L.; Zhao, Y. Optimal Path Planning and Speed Control Integration Strategy for UGVs in Static and Dynamic Environments. *IEEE Trans. Veh. Technol.* **2020**, *69*, 10619–10629. [CrossRef]
72. Jin, X.J.; Yin, G.; Chen, N. Gain-scheduled robust control for lateral stability of four-wheel-independent-drive electric vehicles via linear parameter-varying technique. *Mechatronics* **2015**, *30*, 286–296. [CrossRef]
73. Coskun, S.; Li, L. Vehicle lateral motion control via robust delay-dependent Takagi-Sugeno strategy. *Trans. Inst. Meas. Control* **2021**, *43*, 1430–1444. [CrossRef]
74. Coskun, S. Autonomous overtaking in highways: A receding horizon trajectory generator with embedded safety feature. *Eng. Sci. Technol. Int. J.* **2021**. [CrossRef]
75. Wang, R.; Hu, C.; Yan, F.; Chadli, M. Composite nonlinear feedback control for path following of four-wheel independently actuated autonomous ground vehicles. *IEEE Trans. Intell. Transp. Syst.* **2016**, *17*, 2063–2074. [CrossRef]
76. Guo, J.; Luo, Y.; Li, K. Dynamic coordinated control for over-actuated autonomous electric vehicles with nonholonomic constraints via nonsingular terminal sliding mode technique. *Nonlinear Dyn.* **2016**, *85*, 583–597. [CrossRef]
77. Abzi, I.; Kabbaj, M.N.; Benbrahim, M. Fault tolerant control of vehicle lateral dynamic using a new pneumatic forces multiple model. *Actuators* **2020**, *9*, 120. [CrossRef]
78. Yang, X.; Wang, Z.; Peng, W. Coordinated control of AFS and DYC for vehicle handling and stability based on optimal guaranteed cost theory. *Veh. Syst. Dyn.* **2009**, *47*, 57–79. [CrossRef]
79. Zhang, H.; Zhang, X.; Wang, J. Robust gain-scheduling energy-to-peak control of vehicle lateral dynamics stabilisation. *Veh. Syst. Dyn.* **2014**, *52*, 309–340. [CrossRef]
80. Zhang, H.; Wang, J. Vehicle lateral dynamics control through AFS/DYC and robust gain-scheduling approach. *IEEE Trans. Veh. Technol.* **2015**, *65*, 489–494. [CrossRef]
81. Li, B.; Goodarzi, A.; Khajepour, A.; Chen, S.K.; Litkouhi, B. An optimal torque distribution control strategy for four-independent wheel drive electric vehicles. *Veh. Syst. Dyn.* **2015**, *53*, 1172–1189. [CrossRef]
82. Yang, W.; Zheng, L.; Li, Y.; Ren, Y.; Xiong, Z. Automated highway driving decision considering driver characteristics. *IEEE Trans. Intell. Transp. Syst.* **2019**, *21*, 2350–2359. [CrossRef]
83. Conti, J.; Holtberg, P.; Diefenderfer, J.; LaRose, A.; Turnure, J.T.; Westfall, L. *International Energy Outlook 2016 with Projections to 2040*; No. DOE/EIA-0484 (2016); Office of Energy Analysis, USDOE Energy Information Administration (EIA): Washington, DC, USA, 2016.
84. Ramachandran, S.; Stimming, U. Well to wheel analysis of low carbon alternatives for road traffic. *Energy Environ. Sci.* **2015**, *8*, 3313–3324. [CrossRef]
85. Huang, Y.; Wang, H.; Khajepour, A.; Li, B.; Ji, J.; Zhao, K.; Hu, C. A review of power management strategies and component sizing methods for hybrid vehicles. *Renew. Sustain. Energy Rev.* **2018**, *96*, 132–144. [CrossRef]
86. Li, J.; Zhou, Q.; He, Y.; Shuai, B.; Li, Z.; Williams, H.; Xu, H. Dual-loop online intelligent programming for driver-oriented predict energy management of plug-in hybrid electric vehicles. *Appl. Energy* **2019**, *253*, 113617. [CrossRef]
87. Zhou, Y.; Li, H.; Ravey, A.; Péra, M.C. An integrated predictive energy management for light-duty range-extended plug-in fuel cell electric vehicle. *J. Power Sources* **2020**, *451*, 227780. [CrossRef]
88. Jinquan, G.; Hongwen, H.; Jiankun, P.; Nana, Z. A novel MPC-based adaptive energy management strategy in plug-in hybrid electric vehicles. *Energy* **2019**, *175*, 378–392. [CrossRef]
89. Li, G.; Görges, D. Energy management strategy for parallel hybrid electric vehicles based on approximate dynamic programming and velocity forecast. *J. Frankl. Inst.* **2019**, *356*, 9502–9523. [CrossRef]
90. Guo, J.; He, H.; Sun, C. ARIMA-based road gradient and vehicle velocity prediction for hybrid electric vehicle energy management. *IEEE Trans. Veh. Technol.* **2019**, *68*, 5309–5320. [CrossRef]
91. Chen, Z.; Hu, H.; Wu, Y.; Zhang, Y.; Li, G.; Liu, Y. Stochastic model predictive control for energy management of power-split plug-in hybrid electric vehicles based on reinforcement learning. *Energy* **2020**, *211*, 118931. [CrossRef]
92. Lian, R.; Tan, H.; Peng, J.; Li, Q.; Wu, Y. Cross-type transfer for deep reinforcement learning based hybrid electric vehicle energy management. *IEEE Trans. Veh. Technol.* **2020**, *69*, 8367–8380. [CrossRef]
93. Zhai, C.; Luo, F.; Liu, Y. A novel predictive energy management strategy for electric vehicles based on velocity prediction. *IEEE Trans. Veh. Technol.* **2020**, *69*, 12559–12569. [CrossRef]
94. Chen, H.; Guo, L.; Ding, H.; Li, Y.; Gao, B. Real-time predictive cruise control for eco-driving taking into account traffic constraints. *IEEE Trans. Intell. Transp. Syst.* **2018**, *20*, 2858–2868. [CrossRef]
95. Xie, S.; Qi, S.; Lang, K.; Tang, X.; Lin, X. Coordinated management of connected plug-in hybrid electric buses for energy saving, inter-vehicle safety, and battery health. *Appl. Energy* **2020**, *268*, 115028. [CrossRef]
96. Li, Y.; Zhang, J.; Lv, C.; Yuan, Y. Coordinated control of the steering system and the distributed motors for comprehensive optimization of the dynamics performance and the energy consumption of an electric vehicle. *Proc. Inst. Mech. Eng. Part D J. Automob. Eng.* **2017**, *231*, 1605–1626. [CrossRef]
97. Ahmed, A.A.; Ramadan, H.S. Prototype implementation of advanced electric vehicles drivetrain system: Verification and validation. *Appl. Energy* **2020**, *266*, 114807. [CrossRef]
98. Guo, L.; Gao, B.; Gao, Y.; Chen, H. Optimal energy management for HEVs in eco-driving applications using bi-level MPC. *IEEE Trans. Intell. Transp. Syst.* **2016**, *18*, 2153–2162. [CrossRef]

99. Alcantar, J.V.; Assadian, F. Vehicle dynamics control of an electric-all-wheel-drive hybrid electric vehicle using tyre force optimization and allocation. *Veh. Syst. Dyn.* **2019**, *57*, 1897–1923. [CrossRef]
100. Li, L.; Pei, Y.; Yin, L.; Zhou, L. Vehicle Path Planning Based on Stability and Macroscopic Traffic Flow Model. *China J. Highw. Transp.* **2020**, *33*, 71–80.

energies

Article

Active Disturbance Rejection Control of Differential Drive Assist Steering for Electric Vehicles

Junnian Wang [1,*], Xiandong Wang [1], Zheng Luo [2] and Francis Assadian [3]

1 State Key Laboratory of Automotive Simulation and Control, Jilin University, Changchun 130022, China; wangxd18@mails.jlu.edu.cn
2 Motor Technical Center, Shanghai Automotive Industry Corporation, Shanghai 201804, China; LuoZheng01@saicmotor.com
3 Department of Mechanical and Aerospace Engineering, University of California Davis, Davis, CA 95616, USA; fassadian@ucdavis.edu
* Correspondence: wjn@jlu.edu.cn; Tel.: +86-431-85095443

Received: 4 May 2020; Accepted: 20 May 2020; Published: 22 May 2020

Abstract: The differential drive assist steering (DDAS) system makes full use of the advantages of independent control of wheel torque of electric vehicle driven by front in-wheel motors to achieve steering assistance and reduce the steering effort of the driver, as the electric power steering (EPS) system does. However, as an indirect steering assist technology that applies steering system assistance via differential drive, its linear control algorithm, like existing proportion integration differentiation (PID) controllers, cannot take the nonlinear characteristics of the tires' dynamics into account which results in poor performance in road feeling and tracking accuracy. This paper introduces an active disturbance rejection control (ADRC) method into the control issue of the DDAS. First, the third-order ADRC controller of the DDAS is designed, and the simulated annealing algorithm is used to optimize the parameters of ADRC controller offline considering that the parameters of ADRC controller are too many and the parameter tuning is complex. Finally, the 11-DOF model of the electric vehicle driven by in-wheel motors is built, and the standard working conditions are selected for simulation and experimental verification. The results show that the ADRC controller designed in this paper can not only obviously reduce the steering wheel effort of the driver like PID controller, but also have better nonlinear control performance in tracking accuracy and smooth road feeling of the driver than the traditional PID controller.

Keywords: independent-wheel drive; steering assistance; nonlinear system; active disturbance rejection control; smooth road feeling

1. Introduction

With the intensification of interest in environmental protection and energy issues, electric vehicles have ushered in significant development opportunities. Compared with traditional centralized drive electric vehicles, the torque of each wheel of an electric vehicle driven by several in-wheel motors, also commonly called independent-wheel-drive electric vehicle (IWDEV), can be independently controlled. Differential drive assist steering (DDAS) is a novel power steering technology based on the unique advantages of independent-drive of the electric vehicle driven by several in-wheel motors [1]. It uses the different driving force of two-side front wheels to generate steering assistance, which can substitute the traditional power steering system, such as hydraulic power steering (HPS) system or electric power steering (EPS) system. The reason is that the DDAS technology has the advantages of more compact structure and lower cost. Specifically, on the one hand, the DDAS system does not need an add-on actuator, like the steering motor of the EPS system. On the other hand, due to having the same actuators, the in-wheel motors, with the driving system the controller of the DDAS system can be integrated into the vehicle driving controller. Hence, the DDAS technology appeals to many researchers' attention and interest.

Although the concept of DDAS has only been proposed in recent years, many scholars have conducted much research and achieved some progress. Wang [1] first proposed the concept of DDAS technology, conducted a theoretical analysis and proposed a steering-wheel-torque direct control strategy based on an anti-windup PID algorithm. The steering assist feasibility, steering return ability and driving torque coordination of the DDAS were studied in depth and verified by software simulations and real vehicle tests. Zhao [2] further studied a coupling control strategy of force and displacement for a DDAS system to improve the steering maneuverability and handling stability of EVs with motorized wheels. By analyzing the key factors that affect the interaction between vehicle and driver, the optimum hand wheel torque of a DDAS system is designed and achieved by the torque difference between two front wheels based on $H_2/H\infty$ control method, and its effectiveness was verified by a MatLab/Simulink software (MathWorks, Natick, Massachusetts, USA) simulation. Hu [3] studied the lane keeping control for four-wheel independently actuated autonomous vehicles only when the active-steering motor entirely fails and designed an adaptive multivariable super-twisting control strategy and verified its effectiveness by CarSim (Mechanical Simulation Corporation, Ann Arbor, Michigan, USA) and Simulink co-simulations. Kuslits [4] proposed a full state feedback control system for scenarios at higher speeds, whereas a simple angle tracking controller can be used for a DDAS system in scenarios at lower speed. The effectiveness of the strategy was verified through simulations. Peng [5] developed a coordinated steering control strategy with a hierarchical structure for a multi-axle independent-drive electric vehicle, which is steered simultaneously by traditional mechanical steering and differential drive steering and verified the effectiveness of the strategy through simulations. Wang [6] also designed a hierarchical coordinated controller for the DDAS and vehicle stability control based on the phase plane theory. Various typical simulations on roads with different adhesion characteristics showed the effects on expanding the working range of DDAS systems and simultaneously mitigating the additional influence of the DDAS on vehicle stability. Römer [7] studied the potential of independent-wheel-drive influencing the driver's steering torque using a control technique based on classical EPS control plans, and compared the energy saving potential of DDAS system with the conventional EPS system. The energy saving potential was proved through realistic driving cycles experiments which included the Karlsruhe motorway, the Herzogenaurach highway and a trip through the city of Karlsruhe. It was concluded that the DDAS system can save up to 121.93 Wh/100 km of energy, or approximately 0.95% in lateral acceleration ranges below 4.5 m/s^2, and about 0.43% (55.2 Wh/100 km) in mean value compared to a conventional EPS system. Interestingly, if the tractive energy of an independent-wheel-drive electric vehicle is considered, the comprehensive energy conservation generated by the optimization of the two-side differential drive torque, as DDAS does, can reach up to 4% without any loss of vehicle stability [8].

A review of these references shows that a DDAS system can be used to substitute for a traditional power steering system, such as EPS and as a result, the energy consumption of these traditional add-on assistance steering systems is removed. Although the aspect of energy conservation is not the core purpose of our research, it can be concluded that the existing studies of DDAS system have proved the apparent engineering application value of this novel technology and its feasibility for providing steering assistance and the coordination issue with other chassis control system have been well studied in the above existing literature. However, other important aspects, such as the steering assistance quality issue of DDAS system, are also important factors determining whether this technology can be finally applied in practice. Unfortunately, the steering assistance quality of the DDAS system, which reflects smooth steering hand force with less interference and good tracking performance with ideal steering force characteristic, has not been widely studied until now. To better understand this requirement of the technology, the basic principles and assistance characteristics of a DDAS system and some previous experimental results have to be reviewed and discussed first. Figure 1 shows the working principle diagram of a DDAS system.

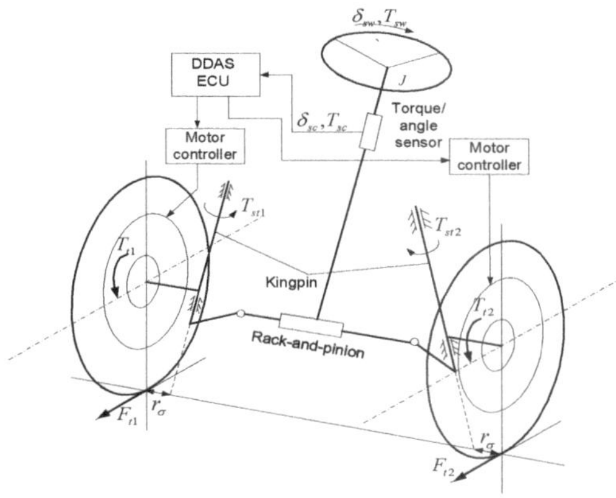

Figure 1. Working principle diagram of a differential drive assist steering (DDAS) system.

As shown in Figure 1, the DDAS system maintains the traditional mechanical steering mechanism but removes the power assist steering actuators, such as a hydraulic cylinder or an electric motor. Since the longitudinal driving force of the left and right steering wheels of the IWDEV is independent and controllable, the torque generated by the front wheels around the respective kingpin can be unequal. Here we define this torque as the driving steering torque. At the same time, because the two steering wheels are connected by the steering trapezium and have a fixed geometric motion relationship, therefore, the driving steering torque will drive the two steering wheels to turn to the side with a small driving force. In theory, the controller of the DDAS system controls the outer steering wheel to increase the driving torque properly and the inner steering wheel to reduce the driving torque equally, which can ensure that the driving steering torque generated by the inner steering wheel and the outer steering wheel on the steering rack are exactly equal to the required steering power torque. It is obvious that the DDAS system can realize the power steering function without changing the total driving torque. Compared with a traditional EPS system, the DDAS system can achieve the same power steering effect without needing a power assist steering actuator. The DDAS system saves the part of energy used to drive the power assist steering actuator, so the DDAS system must be more energy-saving compared with a traditional EPS system. Since the energy consumption of the power assist steering actuator is small, the energy saving of the DDAS system is limited, but it still plays an important role in improving the driving range of pure electric vehicles. DDAS system has the same actuator, two front in-wheel motors, as the driving system of IWDEV, and the DDAS electric control unit (ECU) is also integrated into the driving controller. Consequently, a DDAS system has advantages over other power assist steering systems in layout and cost.

However, it should be noted that the DDAS system is an indirect power steering system, that is, the steering assistance provided by the system is achieved by indirectly acting on the mechanical steering rack through changing the tire forces of two-side steerable driving wheels. The steering assistance generated by DDAS system can be expressed as follows:

$$F_{ast} = (T_1 - T_3)\frac{r_\sigma}{r_w}N_L + I_w\left(\frac{\mathrm{d}\omega_1}{dt} - \frac{\mathrm{d}\omega_3}{dt}\right)\frac{r_\sigma}{r_w}N_L \tag{1}$$

where T_1 and T_3 are the driving torques of the left and right front wheels, r_σ is the scrub radius, r_w the tire rolling radius, N_L is the transmission ratio of the rack translation to the knuckle arm angular displacement, I_w is the moment of inertia of the wheel about its central axis, ω_1 and ω_3 are the rotational velocity of left and right front wheels. According to Equation (1), it can be seen that the assistance provided by the DDAS system is related to the wheel rotational dynamic characteristics and suspension parameters. During the operation of the vehicle, tires may work in a nonlinear range, and the scrub radius of the wheels is also constantly changing. In addition, the steering wheel torque/angle sensor noise may also have a great impact on the control of a DDAS system.

Based on the review of the characteristics of DDAS technology, its control issue has also been studied in many published references. Most of the researchers applied conventional control algorithms, such as classical open-loop look-up table control plan [7] like EPS does, anti-windup PID control plan [6] and fuzzy adaptive PID control plan [1] to the control issue of DDAS system and the their control effects on steering assistance and returnability performance look good in the corresponding simulations, but the control effects of these classic linear control methods on the steering assistance qualities, such as road feeling, the steering wheel torque control stability and robustness against system parameters variation and sensor noise in real applications, are considered to be unacceptable. To better understand the lack of competence of the traditional linear PID control plan with fixed control parameters, Figure 2 shows a real world double-lane-change road test result of the steering wheel torque of an IWDEV that is controlled by a conventional anti-windup PID controller based on a DDAS system published in reference [1]. It can be seen that the PID controller has poor tracking performance to the reference steering wheel torque though the steering assistance function is achieved. This means the smooth road feeling and accurate hand force feedback cannot be fully achieved in the real application of PID controllers for DDAS system. In addition, because the nonlinear mathematical models of tire dynamics, steering system and suspension system are difficult to establish accurately, the changing laws of these interferences are difficult to identify. Thus, despite having better robustness and optimality, some advanced controllers that depend on the accurate model of the controlled system with interference observer or estimator, such as H infinite control, linear quadratic regulator (LQR) control, etc. may be not easy or suitable to apply to the DDAS system, too.

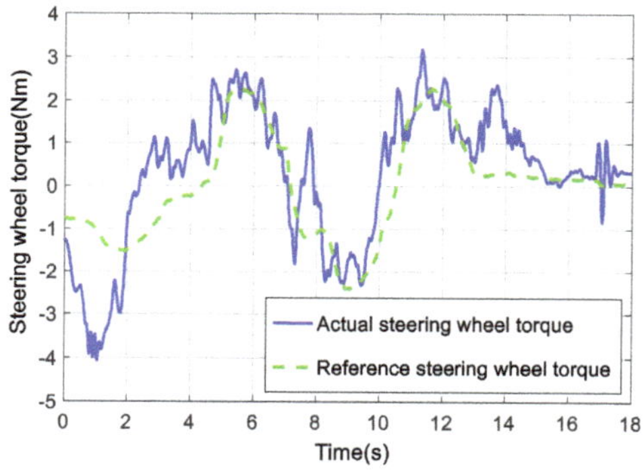

Figure 2. Steering wheel torque.

In summary, though the driver's steering effort can be obviously reduced by the DDAS system, it can be seen from the above analysis that the selected control strategies and control algorithms may highly impact its effect on steering assistance quality. This performance will ultimately decide whether this novel power assistance steering technology can be actually applied in a real car. In this paper, having good robustness in nonlinear control issues, the use of the active disturbance rejection control (ADRC) method is attempted for this purpose. As an improved form of PID controller, the ADRC approach combines the advantages of the PID controller and some robust algorithms. It is relatively easy to implement, robust against possible system interferences and one does not need to know an accurate controlled system model [9].

Compared with the existing literature, the main purpose or main contribution of this paper is that we try to pay more attention on the improvement of the steering assistance quality of the DDAS system before its real application, and firstly attempt to apply the ADRC control approach to improve the steering assistance quality of the DDAS system, in order to make the driver have a better road feeling, and achieve a smooth steering force with less interference caused by possible sensor noise and model parameter changes.

The structure of this paper is as follows: Firstly, the independent-wheel-drive electric vehicle model with four degrees of freedom mechanical steering system is established, and then the ADRC controller model of DDAS is designed for the steering-wheel-torque direct control strategy. Secondly, aiming at solving the problem that the parameters of the ADRC controller are numerous and difficult to set, a simulated annealing algorithm is used to optimize the parameters offline. Finally, typical driving conditions are selected for simulation and experimental verification, which verify the effectiveness of the control method proposed in this paper.

2. Independent-Wheel-Drive Electric Vehicle Dynamic Model

Figure 3 shows the overall framework of the independent-wheel-drive electric vehicle model, which is composed of a vehicle body model, mechanical steering system model, DDAS controller model, wheel rotation dynamics model, tire model and in-wheel-motor model, etc. The specific modelling processes of the core parts are shown in the following subsections, the DDAS controller model is introduced in Section 3, and for the others readers may refer to [1,10]. The descriptions of all the symbols of the variables can be found in the nomenclature section.

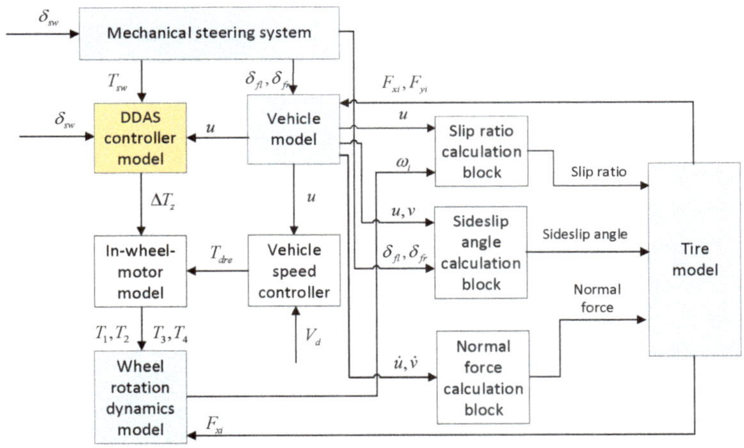

Figure 3. The whole block diagram of independent-wheel-drive electric vehicle model.

2.1. Vehicle Body Dynamic Model

The vehicle body dynamic model established in this paper includes three degrees of freedom: longitudinal motion, lateral motion and yaw motion [1,10]. As shown in Figure 4, the following equations can be established:

$$m(\dot{u}-v\omega_r)=F_{x1}\cos\delta_{fl}+F_{x2}\cos\delta_{fr}+F_{x3}+F_{x4}-F_{y1}\sin\delta_{fl}-F_{y2}\sin\delta_{fr}-\frac{1}{2}C_D\rho Au^2 \quad (2)$$

$$m(\dot{v}+u\omega_r)=F_{y1}\cos\delta_{fl}+F_{y2}\cos\delta_{fr}+F_{y3}+F_{y4}+F_{x1}\sin\delta_{fl}+F_{x2}\sin\delta_{fr} \quad (3)$$

$$\begin{aligned}I_z\dot{\omega}_r=&(F_{x2}\cos\delta_{fr}-F_{y2}\sin\delta_{fr}-F_{x1}\cos\delta_{fl}+F_{y1}\sin\delta_{fl})\tfrac{B}{2}+(F_{x4}-F_{x3})\tfrac{B}{2}\\&+(F_{x2}\sin\delta_{fr}+F_{y2}\cos\delta_{fr}+F_{x1}\sin\delta_{fl}+F_{y1}\cos\delta_{fl})L_f-(F_{y4}+F_{y3})L_r\end{aligned} \quad (4)$$

where F_{xi} and $F_{yi}(i=1,2,3,4)$ are the longitudinal and lateral forces of the left front wheel, the right front wheel, the left rear wheel, and the right rear wheel respectively; m is the mass of the whole vehicle, u and v are the longitudinal and lateral speed of the body centroid respectively, ω_r is the yaw rate of the vehicle body, I_z is the inertia of the body around the vertical axis, C_D is the air resistance coefficient, ρ is the air density, A is the frontal area, δ_{fl} and δ_{fr} are the steering angles of the left front wheel and the right front wheel, L_f is the distance from the body centroid to the front axle, L_r is the distance from the body centroid to the rear axle, B is the wheelbase.

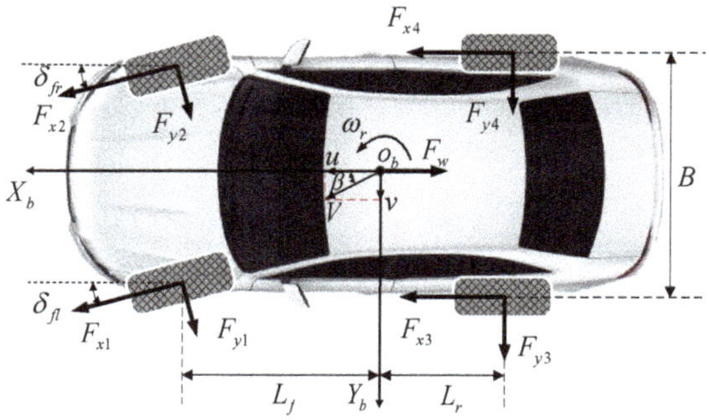

Figure 4. Vehicle body dynamic model.

2.2. Four-DoF Mechanical Steering System Sub-Model

As mentioned above, the DDAS system maintains the mechanical steering system. The four degrees of freedom model of the steering system is shown in Figure 5. The corresponding dynamic equations of the steering system are shown as follows [11,12]:

$$J_C\ddot{\delta}_{sw} + B_C\dot{\delta}_{sw} + K_C\left(\delta_{sw} - \frac{Y_R}{r_P}\right) = T_{sw} \tag{5}$$

$$M_R\ddot{Y}_R + B_R\dot{Y}_R + \eta_F\frac{K_C}{r_P}\left(\frac{Y_R}{r_P} - \delta_{sw}\right) + CF_R + \eta_B\left(\frac{T_{KL1}}{N_{L1}} + \frac{T_{KL2}}{N_{L2}}\right) = 0 \tag{6}$$

$$J_{FW1}\ddot{\delta}_{FW1} + B_{FW1}\dot{\delta}_{FW1} + CF_{FW1} + AT_1 = T_{KL1} \tag{7}$$

$$J_{FW2}\ddot{\delta}_{FW2} + B_{FW2}\dot{\delta}_{FW2} + CF_{FW2} + AT_2 = T_{KL2} \tag{8}$$

$$T_{KL1} = K_{SL1}\left(\frac{Y_R}{N_{L1}} - \delta_{FW1}\right) \tag{9}$$

$$T_{KL2} = K_{SL2}\left(\frac{Y_R}{N_{L2}} - \delta_{FW2}\right) \tag{10}$$

$$T_{SC} = K_C\left(\delta_{sw} - \frac{Y_R}{r_P}\right) \tag{11}$$

where T_{sw} is the steering wheel torque, B_C and J_C are the damping of the steering column and equivalent inertia of steering wheel and column, δ_{sw} is the steering wheel angle, M_R and B_R are the mass and damping of the rack, Y_R is the displacement of the rack, r_p is the radius of the pinion, η_f and η_B are the forward transmitting efficiency and backward transmitting efficiency of the steering gear respectively, K_C is the torsional stiffness of the torsion bar, CF_R is the Coulomb friction of the gear and rack [13], N_{Li} is the ratio of the rack transfer displacement to knuckle angular displacement, J_{FWi} and $B_{FWi}(i = 1, 2)$ are the inertia of the road wheels round their kingpin and damping of kingpin, CF_{FWi} is the coulomb friction caused by the left front wheel and the right front wheel rotating the kingpin, and the specific calculation formula is shown in the literature [14], T_{KLi} is the total torque from the kingpins of the left front wheel and the right front wheel, δ_{FWi} is the steering angle of front wheels, K_{SLi} is the torsional stiffness of the kingpin of the front wheels.

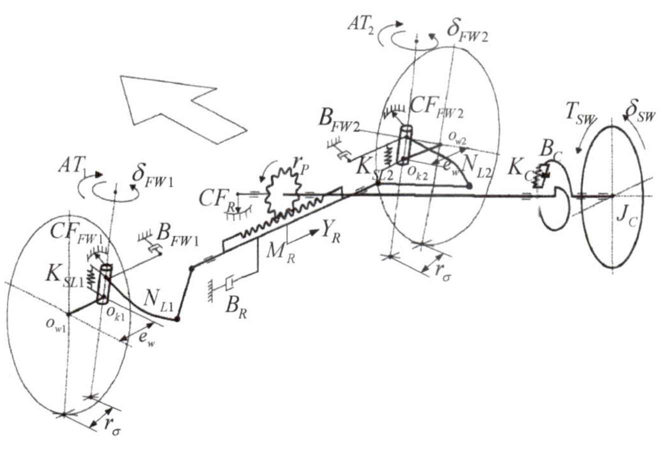

Figure 5. 4-DoF mechanical steering system model.

AT_i represents the alignment torque of each front wheel around the kingpin, which is mainly composed of the following four parts [15,16]:

$$M_{sy} = F_y \cdot r_w \sin\tau \cos\sigma \tag{12}$$

$$M_{sx} = F_x \cdot r_\sigma \cos\tau \cos\sigma \tag{13}$$

$$M_{zz} = M_{tz} \cdot \cos\tau \cos\sigma \tag{14}$$

$$M_{sz} = F_z \cdot \cos\tau \sin\sigma \sin\delta_{FW} \cos\sigma (r_\sigma + r_w \tan\sigma) \tag{15}$$

where M_{sy} is the alignment torque generated by the lateral force of the tire, M_{sx} is the alignment torque generated by the longitudinal force of the tire, M_{zz} is the component of the self-alignment torque around the kingpin, M_{sz} is the alignment torque generated by the normal force of the front axle, M_{tz} is the self-alignment torque of the tire, F_x is the longitudinal force of the wheels, F_y is the lateral force of the wheels, F_z is the normal force of the wheels, τ is the kingpin caster angle, σ is the kingpin inclination angle.

2.3. Wheel Rotation Dynamic Sub-Model

The wheel rotation dynamic model is shown in Figure 6.

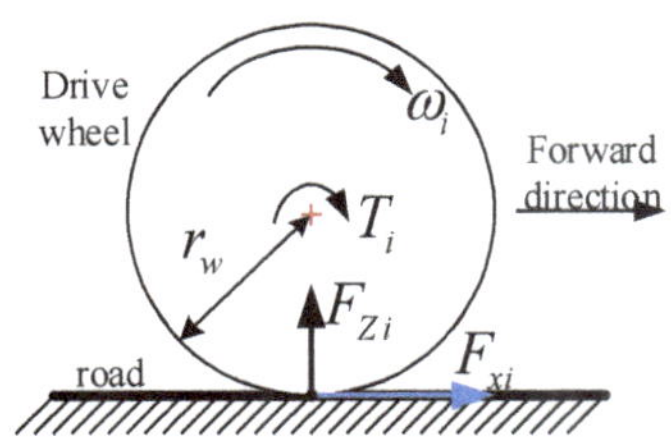

Figure 6. Wheel rotation dynamic model.

The rotation dynamic equation of each driving wheel can be established as follows:

$$I_w \dot{\omega}_i = T_i - F_{xi} r_w \tag{16}$$

where I_w is the moment of inertia of the wheels, $\omega_i(i = 1,2,3,4)$ is the angular velocity of the wheels, $T_i(i = 1,2,3,4)$ is the driving torque of the wheels.

2.4. Tire Sub-Model

The tire model is an important part of the vehicle dynamic model. This paper selects the magic formula tire model which is widely used in the study of handling dynamics. The specific definition and parameters of the model are detailed in the literature [17].

3. Design of DDAS Controller

3.1. Steering-Wheel-Torque Direct Control Strategy

The existing DDAS controller mostly adopts an open-loop strategy based on an assist characteristic curve look-up table inherited from the EPS control plan [7,18]. Different from the working principle of an EPS system, the steering assistance generated by a DDAS system through the ground traction force difference is indirectly applied on the steering system. Given the complexity of real road conditions and nonlinear characteristics of vehicle tires as well as inconstant scrub radius, acquiring the actual ground traction force accurately is difficult. Hence, a traditional control strategy like EPS which generates the steering assistance torque command directly based on a look-up table of control current of the steering electric motor is hard to implement. In another words, it is difficult to directly determine the torque commands of the two front in-wheel motors, since we do not know the accurate law that defines how much torque difference can generate required steering assistance torque for the steering system. Therefore, a reference steering wheel torque following control law called as steering-wheel-torque direct control strategy has to be proposed to avoid the embarrassment that the steering assistance torque is hard to know in DDAS system. It is proved that this control strategy is suitable to solve this problem and it is not hard to be carried out in real applications. The architecture of the steering-wheel-torque direct control strategy is shown in Figure 7.

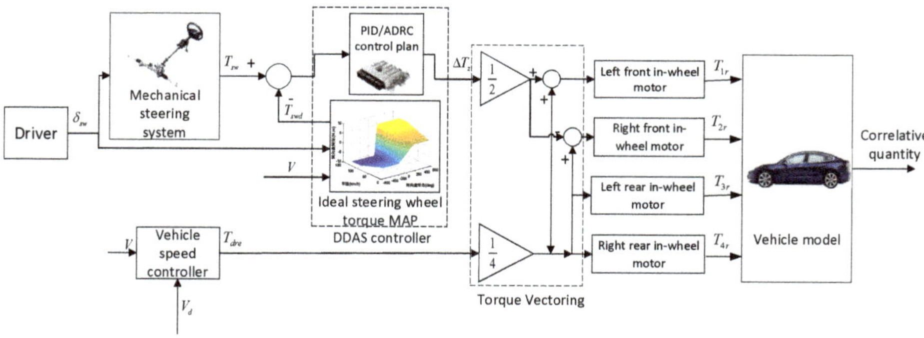

Figure 7. Steering-wheel-torque direct control strategy.

As Figure 7 shows, the actual steering wheel torque T_{sw} is measured through a steering wheel torque sensor and the steering wheel angle δ_{sw} is measured through a steering wheel angle sensor, and the vehicle speed signal V is obtained from the CAN bus. Then δ_{sw} and V are delivered to the ideal steering wheel torque map to obtain the target steering wheel torque T_{swd}. The difference of the ideal steering wheel torque T_{swd} and the actual steering wheel torque T_{sw} is delivered to the DDAS controller to obtain the front wheels torque difference ΔT_z. The front wheels torque difference ΔT_z is distributed by the torque vectoring block, as shown in the Equation (17), which distributes half of the torque difference to the two side wheels with equal absolute value but opposite sign. Then the differential of ground traction force of the front wheels generated by torque vectoring makes the actual steering wheel torque track the ideal steering wheel torque, which consequently reduces the steering wheel hand force of the driver:

$$\begin{cases} T_1, T_2 = \frac{T_{dre}}{4} \pm \frac{\Delta T_z}{2} \\ T_3, T_4 = \frac{T_{dre}}{4} \end{cases} \tag{17}$$

where T_{dre} is the total demand driving torque determined by the longitudinal driver model, which can be calculated by the following formula:

$$T_{dre} = k_p(V_d - V) + k_i \int_0^t (V_d - V)\mathrm{d}t + k_d \frac{\mathrm{d}(V_d - V)}{\mathrm{d}t} \tag{18}$$

where V_d is the target speed, k_p is the proportional coefficient of the PID controller, k_i is the integral coefficient of the PID controller, k_d is the differential coefficient of the PID controller.

As for the driver's ideal steering wheel torque, many research institutes have conducted a lot of researches very early, mainly through real vehicle test or driving simulator measurement. According to the previous research conclusion, the preference steering wheel torque characteristic of many drivers is closely related to the vehicle speed and the steering wheel angle [19–22]. Hence, as an example, a kind of driver's preference steering wheel torque map derived from other's experimental results is illustrated in Figure 8.

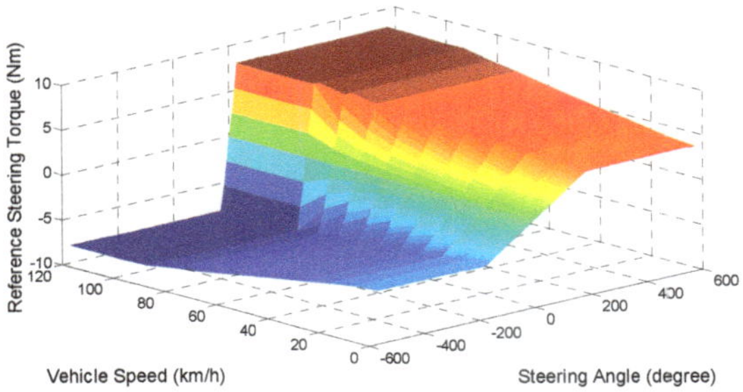

Figure 8. The ideal steering wheel torque map.

3.2. Design of the ADRC Controller of DDAS System

As discussed in Section 1, on one hand, the control effect of a conventional controller may be influenced by several interferences, such as inevitable steering wheel torque sensor noise and frequently changing suspension parameters as well as continuous road unevenness, while on the other hand, the change laws of these interferences are hardy to identify and accurately model. Therefore, PID control plans, especially the ones with fixed control parameters, as well as some robust and optimization control plan may not be competent for the DDAS control problem. As a result, the ADRC control method which has good robustness against inner parameter changes and outer sensor noise without knowing the accurate mathematical model of the controlled system in advance is proposed for the DDAS control system. Actually, as an improved PID controller, the ADRC controller treats the inner and outer interferences as a whole interference and the influence of the interference on the control effect is compensated by the disturbance estimation compensator. Therefore, compared with a PID controller, the ADRC controller has better anti-interference ability.

The ADRC controller is mainly composed of four parts: tracking differentiator, extended state observer, nonlinear state error feedback law and disturbance estimation compensator [23]. ADRC controller solves the problem that the differential signal of the error is difficult to extract in the traditional PID controller by using the tracking differentiator and the extended state observer. The extended state observer of ADRC controller obtains the state of the system, the differential signal of the state and the disturbance acting on the system by observing the input and output of the system. The nonlinear state error feedback of ADRC makes the controller more adaptive to nonlinear system by introducing nonlinear function *fal*. The disturbance estimation compensator of the ADRC controller can compensate the control result of nonlinear state error feedback by choosing appropriate compensation coefficient, so as to effectively reduce the influence of disturbance on control effect.

According to the characteristics of the DDAS system, the specific structure of the third-order ADRC controller of DDAS system designed in this paper is shown in Figure 9.

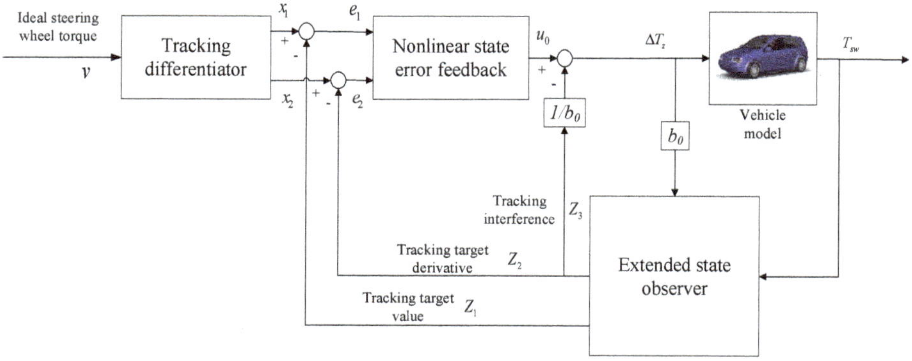

Figure 9. The structure of ADRC controller of DDAS system.

The specific design process of the ADRC controller of DDAS system is introduced as follows:

(a) Tracking differentiator.

In classical control theory, the differential value of a given signal is solved by the following equation:

$$y = w(s)v = \frac{s}{Ts+1}v = \frac{1}{T}(1 - \frac{1}{Ts+1})v \tag{19}$$

where T is the time constant of the controlled system. The smaller the time constant T is, the closer the output of the system is to the real differential value of the signal. Therefore, the time constant is usually a small value [9]. The structure which tracks the dynamic characteristics of the signal as fast as possible through the first order inertia link and obtains the approximate differential signal by solving the differential equation is called a tracking differentiator.

To track the dynamics of the input signal fast, generally, the following nonlinear tracking differentiator can be selected:

$$\begin{cases} \dot{x}_1 = x_2 \\ \dot{x}_2 = -R\text{sign}(x_1 - v_0 + \frac{x_2|x_2|}{2R}) \end{cases} \tag{20}$$

where v_0 is the ideal steering wheel torque T_{swd}, R is the speed factor. The tracking performance is better when the R is bigger, which means that x_1 and x_2 are closer to v_0 and the differential of v_0, respectively.

Although the nonlinear tracking differentiator can track the target steering wheel torque T_{swd} well, it is easy to generate high frequency oscillation due to the bang-bang characteristic of the selected sign function in Equation (20). In order to prevent the occurrence of high frequency oscillations when the system comes into a steady state, this paper uses the time-optimal control synthesis function $fhan(x_1, x_2, R, h)$ to design the tracking differentiator. The specific formula of this function is as follows [24]:

$$\begin{cases} d = R \cdot h \\ d_0 = h \cdot d \\ y = x_1 + hx_2 \\ a_0 = \sqrt{d^2 + 8r|y|} \\ a = \begin{cases} x_2 + 0.5(a_0 - d)\text{sign}(y), & |y| > d_0 \\ x_2 + \frac{y}{h}, & |y| \leq d_0 \end{cases} \\ fhan = -\begin{cases} R\text{sign}(a), & |a| > d \\ R\frac{a}{d}, & |a| \leq d \end{cases} \end{cases} \tag{21}$$

where h is the tracking step and R is the speed factor.

Therefore, the final designed tracking differentiator in this paper is as follows:

$$\begin{cases} \dot{x}_1 = x_2 \\ \dot{x}_2 = fhan(x_1 - T_{swd}, x_2, R, h) \end{cases} \tag{22}$$

(b) Extended state observer.

In the running process of the system, signal interaction with the external environment is constantly carried out. Therefore, the internal state information of the system can be determined by monitoring the system input and output. The device for determining the internal state information of the system is called the state observer [9].

For a general nonlinear system as shown in the following equations:

$$\begin{cases} \dot{x}_1 = x_2 \\ \dot{x}_2 = f(x_1, x_2) + bu \\ y = x_1 \end{cases} \tag{23}$$

Then the state observer of this system can be established as follows by selecting a nonlinear feedback form:

$$\begin{cases} e = z_1 - y \\ \dot{z}_1 = z_2 - \gamma_{01} g_1(e) \\ \dot{z}_2 = -\gamma_{02} g_2(e) + bu \end{cases} \tag{24}$$

where γ_{01}, γ_{02} are control parameters, e is the error term, u is the external input, $g_i(e)$ is nonlinear function that satisfies the following conditions:

$$eg_i(e) \geq 0 \tag{25}$$

As long as the appropriate γ_{01}, γ_{02} and nonlinear function $g_i(e)$ are chosen for the state observer, the state variables can be well estimated in a wide range of system. Let $x_3(t) = f(x_1(t), x_2(t))$, and denote $\dot{x}_3(t) = w_t$. Then, the system can be expanded into a new linear control system as follows:

$$\begin{cases} \dot{x}_1 = x_2 \\ \dot{x}_2 = f(x_1, x_2) + bu \\ \dot{x}_3 = w(t) \\ y = x_1 \end{cases} \tag{26}$$

The state observer established for this new expanded control system is as follows:

$$\begin{cases} e = z_1 - y \\ \dot{z}_1 = z_2 - \gamma_{01} g_1(e) \\ \dot{z}_2 = z_3 - \gamma_{02} g_2(e) \\ \dot{z}_3 = -\gamma_{03} g_3(e) + bu \end{cases} \tag{27}$$

This new state observer is known as the extended state observer of the new system. Among them, $x_3(t)$ is called the expanded state. According to the research needs of this paper, the third-order extended state observer established in this paper is rewritten as follows:

$$\begin{cases} e = z_1 - T_{sw} \\ \dot{z}_1 = z_2 - \gamma_{01}e \\ \dot{z}_2 = z_3 - \gamma_{02}|e|^{\alpha_1}\text{sign}(e) + b\Delta T_z \\ \dot{z}_3 = -\gamma_{03}|e|^{\alpha_2}\text{sign}(e) \end{cases} \tag{28}$$

In order to prevent the phenomenon of high frequency flutter in the control process, the function $|e|^{\alpha}\text{sign}(e)$ in the extended state observer is replaced by the power function with linear segment at the origin as follows:

$$fal(e,\alpha,\phi) = \begin{cases} \frac{e}{\phi^{\alpha-1}} & |e| \le \phi \\ |e|^{\alpha}\text{sign}(e) & |e| > \phi \end{cases} \tag{29}$$

where ϕ is the length of the linear segment and is an important parameter. The final third-order extended state observer is as follows:

$$\begin{cases} e = z_1 - T_{sw} \\ \dot{z}_1 = z_2 - \gamma_{01}e \\ \dot{z}_2 = z_3 - \gamma_{02}fal(e,\alpha_1,\phi_1) + b_0\Delta T_z \\ \dot{z}_3 = -\gamma_{03}fal(e,\alpha_2,\phi_2) \end{cases} \tag{30}$$

where α_1, α_2 are the nonlinear factors in the fal function, z_1 tracks the target value of the actual steering wheel torque T_{sw}, z_2 tracks the target signal of the changing speed of steering wheel torque, and z_3 tracks the total disturbance term of the system.

(c) Nonlinear state error feedback.

The nonlinear state error feedback is an important part of the ADRC controller. This part can quickly adjust the deviation and make the system balance between response fastness and overshoot. Hence, the following nonlinear state error feedback is chosen:

$$\begin{cases} e_1 = x_1 - z_1 \\ e_2 = x_2 - z_2 \\ u_0 = \gamma_1 fal(e_1,\alpha_3,\phi_3) + \gamma_2 fal(e_2,\alpha_4,\phi_4) \end{cases} \tag{31}$$

where e_1 is the steering wheel torque error term, e_2 is the steering wheel torque change rate error term, γ_1 and γ_2 are nonlinear combination coefficients, u_0 is the control output of the nonlinear state error feedback.

After completing the design of above three parts, the final control value can be obtained based on the nonlinear state error feedback control value plus the compensation of the disturbance estimation value. This part is called disturbance estimation compensator mentioned above, which is expressed as follows:

$$\Delta T_z = u_0 - z_3/b_0 \tag{32}$$

where ΔT_z is the front wheels torque difference required by the final decision of the ADRC controller, b_0 is the compensation factor, which determines the strength of the compensation and is an important parameter of the ADRC controller, directly affecting the ADRC controller performance [25].

4. Controller Parameter Optimization Based on Simulated Annealing Algorithm

Compared with the PID controller, though the ADRC controller has the advantages of better robustness, simple structure and easy implementation without knowing the accurate mathematical model of the controlled system, it also has the disadvantages of needing more control parameters and complicated parameter tuning [26], which severely limits its further industrial application. At the same time, these parameters have a great impact on the performance of the controller, so appropriate method selection to set the values of each parameter has to be done first.

According to the theory of ADRC, some of the parameters are determined empirically, and once these parameters are determined, no correction is needed. For example, α_1, α_2, α_3, α_4 and ϕ_1, ϕ_2, ϕ_3, ϕ_4 are the parameters of the nonlinear function *fal*, which affect the change trend of the nonlinear function, but they usually do not change with the change of the controlled system. Therefore, the ranges of α_3 and α_4 in the nonlinear state error feedback are generally $0 < \alpha_3 < 1$, $\alpha_4 > 1$, so in this paper, α_3 and α_4 are chosen as fixed values, 0.95 and 1.25, respectively, α_1 and α_2 in the third-order extended state observer are chosen as fixed values, 0.5 and 0.25, respectively. The values of ϕ_1, ϕ_2, ϕ_3 and ϕ_4 have a great influence on the nonlinearity of the controller. After multiple simulations, the value of ϕ_1, ϕ_2, ϕ_3 and ϕ_4 are chosen as 0.01 which is ten times of the sampling step. The value of speed factor R in this paper is 10.

In summary, in addition to the empirically determined parameters, the other parameters which need to be specifically set are the following six parameters γ_{01}, γ_{02}, γ_{03}, γ_1, γ_2 and b_0. Generally, there is no relationship between the six parameters γ_{01}, γ_{02}, γ_{03}, γ_1, γ_2, b_0 and the parameters α_1, α_2, α_3, α_4, ϕ_1, ϕ_2, ϕ_3, ϕ_4 mentioned above. At the same time, γ_{01}, γ_{02} and γ_{03} in the extended state observer are mainly related to sampling step size [23], which can be designed separately. In addition, γ_1 and γ_2 in nonlinear state error feedback are also important parameters of the controller and b_0 is an important parameter to characterize the difference of different systems. Due to the fact that there is a certain mutual influence between these parameters, and manual adjustment is too complicated, offline optimization to set the values of these six parameters is implemented. In this optimization process, three parameters γ_{01}, γ_{02}, γ_{03} are optimized first, and then the rest three parameters are optimized.

There are many existing optimization algorithms, such as genetic algorithm, simulated annealing algorithm and particle swarm optimization, etc. Among them, the simulated annealing algorithm has the advantages of simple description, flexible use, high operational efficiency and less constraint on initial conditions [27]. Therefore, the simulated annealing algorithm is chosen as the optimization algorithm.

In order to implement the optimization, the objective function of the optimization problem according to the needs of this paper should be determined first. The target of this paper is to design a better DDAS controller, which is to control the actual steering wheel torque to follow the ideal steering wheel torque in real time by controlling the front wheels driving torque difference. Therefore, the objective function is defined as follows:

$$J = \int_0^\infty |T_{swd} - T_{sw}| \mathrm{d}t \tag{33}$$

In order to speed up the optimization process, the initial values of each parameter are determined at first by multiple simulations as shown in Table 1. Then the relevant optimization program is coded in MatLab software, in which the *sim* function is used to call the simulation model. The simulation condition selects the sinusoidal steering angle input at 30 km/h vehicle speed, and the road surface adhesion coefficient is high adhesion, which is 0.8. As an example, the iterative optimization process of the three parameters γ_{01}, γ_{02} and γ_{03} in extended state observer is shown in Figure 10.

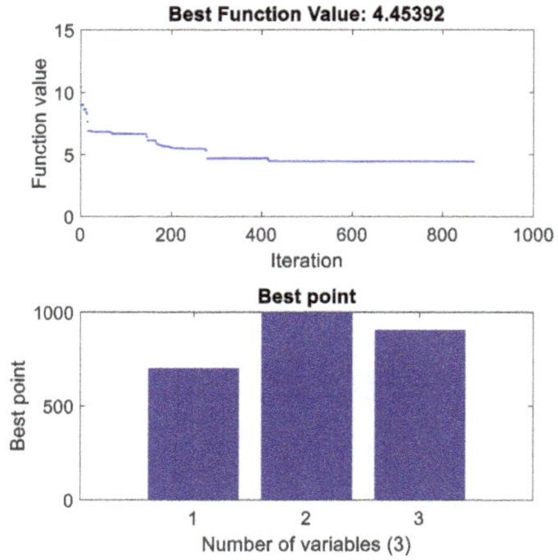

Figure 10. Optimization process.

As shown in Figure 10, after around 870 generations, the fitness function basically reaches the optimal value. The final six optimized parameters are shown in Table 1 below:

Table 1. ADRC controller parameter optimization

Parameters	γ_{01}	γ_{02}	γ_{03}	γ_1	γ_2	b_0
Initial value	450	600	1000	−350	30	1
Optimal value	700.955	997.714	903.922	−425.893	0.372	2.801

5. Simulation Analysis

5.1. Sinusoidal Steering Wheel Angle Input Simulation

The simulation condition is selected as the sinusoidal steering wheel angle input without the driver's steering model. First the vehicle gradually accelerates to 50 km/h and maintains this speed. As shown in Figure 11a, the vehicle is input a sinusoidal steering wheel angle at the 5th second and its amplitude and the frequency are 45 degrees and 0.2 Hz, respectively, and the road adhesion coefficient is set to 0.8. In order to verify the anti-interference performance of the proposed DDAS controller, the white noise model is used to imitate the sensor noise of the steering wheel torque, of which power and frequency are 0.01 and 27.5 Hz, respectively. During this procedure, two different controllers—one a PID controller, and the other the proposed ADRC controller—are used as two comparison simulation cases to control the DDAS system to assist the driver to steer the car. For better comparing their performance, the three parameters in the PID controller are also repeatedly calibrated after multiple groups of simulations with same sinusoidal steering angle input. Figure 11 shows all the comparison simulation results.

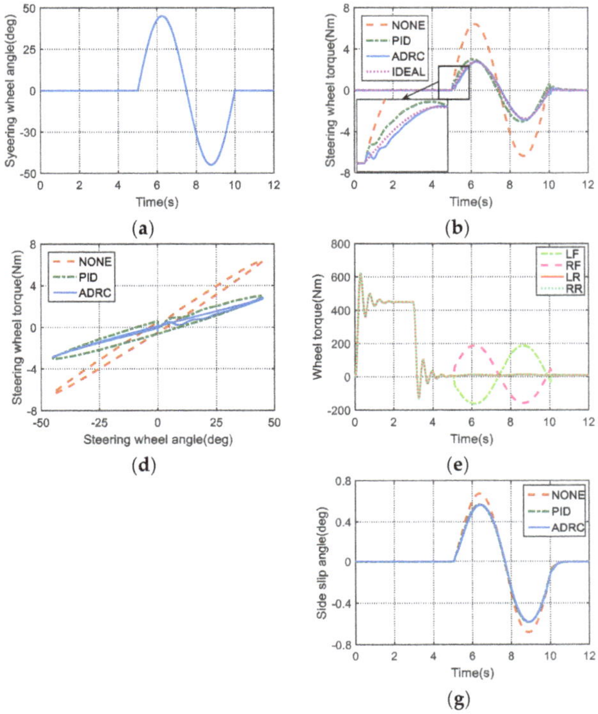

Figure 11. Sinusoidal input simulation: (**a**) Steering wheel angle; (**b**) Steering wheel torque; (**c**) Steering wheel torque with white noise; (**d**) Cross-plot of steering wheel angle and torque; (**e**) Wheel torque of ADRC controller; (**f**) Yaw rate; (**g**) Side slip angle.

It can be seen from Figure 11b that both PID controller and ADRC controller have achieved good performance in power assistance. The maximum torque of the steering wheel dropped from 6.3 Nm to about 2.7 Nm, a reduction by 56%, so the effect is remarkable. Moreover, it is obvious that the steering wheel torque not only has a certain leading response, but also has a little big peak value with respect to the ideal steering wheel torque when the PID controller is adopted. The trend looks good, but the tracking accuracy of PID controller is a little poor. However, as a contrast, when the ADRC controller is used, the steering wheel torque can better track the ideal steering wheel torque, and almost has no difference in peak value with respect to the ideal value. In addition, Figure 11c shows the actual steering wheel torque curve after adding white noise to the steering system. It can be seen from Figure 11c that the chatter of the actual steering wheel torque when the PID controller and ADRC controller are adopted, respectively, is similar during straight line driving conditions, while during steering conditions, it is obvious that the chatter of the actual steering wheel torque controlled by the ADRC controller is smaller than with the PID controller. During the steering stage, the function of the ADRC controller of the DDAS system is to generate steering assistance for the driver and simultaneously mitigate the system noise interference on the steering wheel torque control performance. In contrast, the linear PID controller shows difficulties in dealing with the interference. The result proves that the ADRC controller has better anti-interference ability than the PID controller.

Besides that, it can be seen from the relationship between the steering angle and the torque of the steering wheel shown in Figure 11d that the DDAS system using the PID controller and the ADRC controller can effectively reduce the steering efforts of the driver. However, the ADRC controller shows better assistance performance with smaller fluctuation and more stable steering wheel torque, indicating that the ADRC controller is significantly better than the PID controller. In addition, the wheel torque curves of the ADRC controller in Figure 11e clearly show that the right and left front wheels begin to generate differential torque, and the maximum and minimum values are 190 Nm and −180 Nm, respectively, which are within the normal range. There is almost no difference between the two controllers in yaw rate and side slip angle of center of mass as shown in Figure 11f,g.

Table 2 summarizes the root mean square (RMS) values of the steering wheel torque tracking error of these control methods. It is obvious that the tracking error RMS of the ADRC controller is much smaller than that of the PID controller, which proves that the assistance performance of the ADRC controller is significantly better than that of the PID controller.

Table 2. The RMS comparison of the steering wheel torque following error in sinusoidal input simulation.

Performance	Uncontrolled	PID	ADRC
$T_{sw} - T_{swd}$(Nm)	1.674	0.2795	0.0867

5.2. Closed-Loop Lemniscate Simulation

A driver-vehicle-road closed-loop simulation is also conducted to verify the control effect of the DDAS control strategy designed in this paper. The closed-loop simulation condition selects the standard lemniscate trajectory driving condition. The trajectory function is established according to the relevant standard, and the minimum curvature radius is set to $R_0 = 6$ m. In order to have a smooth steering process during the whole simulation, the standard lemniscate is rotated by 45 degrees, and a straight line segment is set before entering and after exiting the standard lemniscate road. The vehicle speed is chosen to be 10.8 km/h, and the road adhesion coefficient is 0.8. The white noise model with same amplitude and frequency as above simulation condition is also embedded into the comparing simulation. The simulation results are shown in Figure 12.

As seen in Figure 12a, the vehicle in each case with different controller can track the target trajectory better. As shown in Figure 12b, DDAS has achieved good power assisting effect with PID controller and ADRC controller, and the maximum torque of the steering wheel dropped from about 11.5 Nm to about 5.1 Nm, reduced by 50%. The effect of DDAS system on steering hand force reduction is remarkable. Furthermore, comparing to the case of PID controller, the steering wheel torque in the case of the ADRC controller can better track the ideal steering wheel torque with no advance response phenomenon, and has smaller chatter while reaching the peak value. In addition, Figure 12c shows the actual steering wheel torque result after adding white noise to the steering system. It also can be seen from Figure 12c that when the car is cornering, the influence of the sensor noise to the road feeling of the driver is greater than the car is in the straight line driving condition, especially for the case of the PID controller. Moreover, when the vehicle enters into the steering condition, the chatter of the actual steering wheel torque when using the PID controller is significantly greater than that when using the ADRC controller. The result fully proves that when the interference of the steering wheel torque sensor noise or other external or internal interference comes out, such as the changing suspension parameters as well as the nonlinear characteristics of the tires, the anti-interference ability of ADRC controller is significantly better than that of PID controller. Both Figures 12b and 12c prove that the ADRC controller shows better assistance performance with smaller fluctuations and smoother steering wheel torque when compared with the PID controller. Similar results can also be seen from the cross-plot of the steering wheel angle versus steering wheel torque as in Figure 12e. By observing the time history of yaw rate shown in Figure 12f and side slip angle shown in Figure 12g, so it also can be concluded that the DDAS system does not obviously influence the stability of the vehicle which is driving all the time at constant speed on a high-adhesion road.

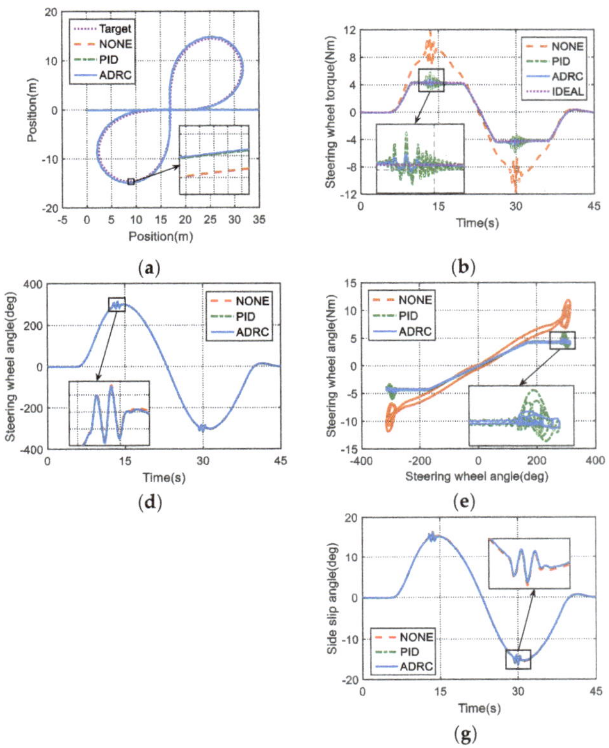

Figure 12. Closed-loop lemniscate simulation: (**a**) Vehicle trajectory; (**b**) Steering wheel torque; (**c**) Steering wheel torque with white noise; (**d**) Steering wheel angle; (**e**) Cross-plot of steering wheel angle and torque; (**f**) Yaw rate; (**g**) Side slip angle.

Table 3 is the RMS value of the steering wheel torque tracking error with respect to the ideal value of each method under lemniscate simulation. It can be concluded that both the PID controller and the ADRC controller can make the RMS of the deviation of the steering wheel torque from ideal value be much smaller compared to without control, indicating that DDAS system has a good power- assisting effect. When the ADRC controller is adopted, the RMS of the deviation of the steering wheel torque from the ideal value is 69% smaller than that of the PID controller, indicating that the power assisting effect of the ADRC controller is significantly better than that of the PID controller.

Table 3. The RMS comparison of the steering wheel torque tracking errors in lemniscate simulation.

Performance	Uncontrolled	PID	ADRC
$T_{sw} - T_{swd}$(Nm)	2.176	0.2106	0.07809

6. Hardware-in-the-Loop Experimental Validation

Taking the need for more facilities, time requirements and test yard support into account as well as the risk of losing stability while carrying out the double-lane-change test in actual vehicle experiments, a hardware-in-the-loop simulation experiment is selected to replace real vehicle testing and validate the control effect of the DDAS control strategy proposed in this paper. The testing platform of the driving simulator as shown in Figure 13 is composed of a host computer, G29 driving simulator with steering wheel angle/torque sensor and dSPACE 1103 hardware running the vehicle model.

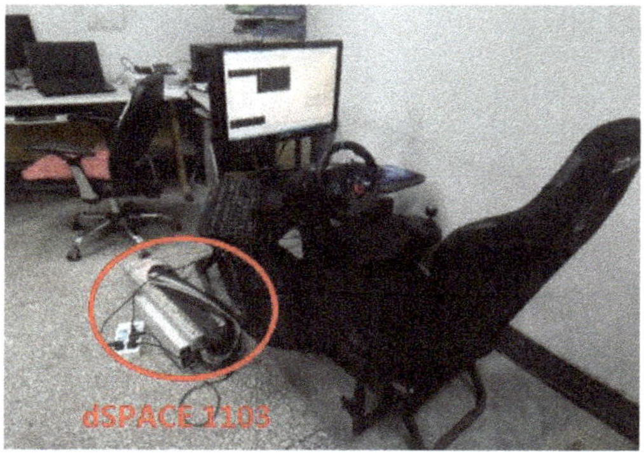

Figure 13. Testing simulator platform based on dSPACE hardware.

The driver-in-the-loop experiment selects a double-lane-change test case, with a target constant-speed 40 km/h and a uniform road adhesion coefficient 0.85. The experimental results are shown in Figure 14.

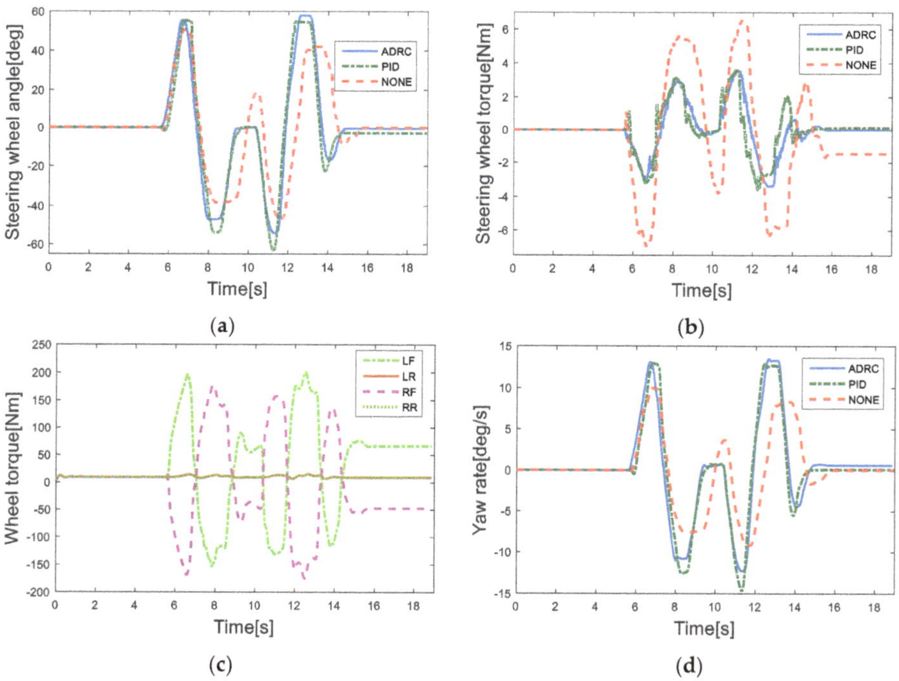

Figure 14. Simulator experiment results: (**a**) Steering wheel angle; (**b**) Steering wheel torque; (**c**) Wheel torque of ADRC controller; (**d**) Yaw rate.

It can be seen from Figure 14a that the steering wheel angle input of the drivers in three experimental cases with different controller of DDAS system or without any power steering control are similar. Figure 14b shows that DDAS has achieved very good power steering assistance effect using both the PID controller and the ADRC controller. The peak value of the steering wheel torque is reduced from 7 Nm in the case of having no any power steering to 4 Nm in the case of having the

DDAS system, a reduction rate of about 43%, and the assistance effect is obvious. Moreover, when the ADRC controller is used, the steering wheel torque fluctuation is much smaller and smoother. The disturbance estimation compensator of the ADRC controller provides a compensation function for the possible disturbance from inside or outside of the steering system, thereby effectively improving the control effect of the differential drive assist steering system. The comparison result indicates that the ADRC controller is more robust when facing possible steering wheel torque or angular sensor noise, thus achieving a better road feeling for the driver than the PID controller. As shown in Figure 14d, when DDAS is applied, it is interesting that the yaw rate of the vehicle is increased, but the steering wheel angle input of the driver in this process also increases.

7. Conclusions

In this paper, a DDAS control strategy based on the steering-wheel-torque direct control is developed for the first time. Considering the disadvantages of the traditional PID controller and the particularities of the DDAS system, the ADRC technology is introduced to design the DDAS controller to reduce the steering effort of the driver and improve the driver's road feeling simultaneously. As for the problem that the parameters of the ADRC controller are difficult to set, the simulated annealing algorithm is used to optimize the controller parameters offline. Finally, a variety of working conditions are selected to verify the developed strategy by using the vehicle model established in this paper. All the simulation and experiment results show that compared with the PID controller commonly used in DDAS, the proposed ADRC controller developed in this paper can not only reduce the steering effort of the driver obviously like previous conventional control method, but also have better control performance in tracking accuracy and smooth road feeling of the driver.

Author Contributions: J.W. administrated the whole research and revised the manuscript as the scientific project leader. X.W. implemented the experiments and wrote the initial manuscript. Z.L. designed the control algorithm model and performed the simulation. F.A. supervised the writing and critically reviewed. All authors have read and agreed to the published version of the manuscript.

Funding: This research was funded by the National Natural Science Foundation of China under Grant 51875235, and in part by the Jilin Province Science and Technology Development Project under Grant 20180414011GH and 20190802015ZG.

Acknowledgments: The authors are grateful to the National Natural Science Foundation of China and all the reviewers for their constructive comments.

Conflicts of Interest: The authors declare no conflict of interest.

Nomenclature

r_σ	scrub radius
r_w	tire rolling radius
N_L	transmission ratio of the rack translation to the knuckle arm angular displacement
I_w	moment of inertia of the wheel about its central axis
$F_{xi}, F_{yi}(i = 1,2,3,4)$	the longitudinal and lateral forces of the left front wheel, the right front wheel, the left rear wheel, and the right rear wheel respectively
m	the mass of the whole vehicle
u	the longitudinal speed of the body centroid
v	the longitudinal speed of the body centroid
ω_r	the yaw rate of the vehicle body
I_z	the inertia of the body around the Z axis
C_D	the air resistance coefficient
ρ	the air density
A	the frontal area
δ_{fl}, δ_{fr}	steering angles of the left front wheel and the right front wheel

L_f	the distance from the body centroid to the front axle
L_r	the distance from the body centroid to the rear axle
B	the wheelbase
T_{sw}	the steering wheel torque
B_C	the damping of the steering column
J_C	equivalent inertia of steering wheel and column
δ_{sw}	steering wheel angle
M_R	the mass of the rack
B_R	the damping of the rack
Y_R	the displacement of the rack
r_p	the radius of the pinion
η_f	the forward transmitting efficiency of the steering gear
η_B	the backward transmitting efficiency of the steering gear
K_C	the torsional stiffness of the torsion bar
CF_R	the coulomb friction of the gear and rack
J_{FWi}	the inertia of the road wheel round their kingpin
B_{FWi}	the damping of kingpin
CF_{FWi}	the coulomb friction caused by the steerable wheels rotating the kingpin
T_{KLi}	the total torque from the kingpins of the left front wheel and the right front wheels
δ_{FWi}	the steering angle of front wheels
K_{SLi}	the torsional stiffness of the kingpin of the front wheels
M_{sy}	the alignment torque generated by the lateral force of the tire
M_{sx}	the alignment torque generated by the longitudinal force of the tire
M_{zz}	the component of the tire self-alignment torque about the kingpin
M_{sz}	the alignment torque generated by the normal force of the front axle
M_{tz}	the self-alignment torque of the tire
F_z	the normal force of the wheels
τ	the kingpin caster angle
σ	the kingpin inclination angle
$\omega_i(i = 1,2,3,4)$	the angular velocity of the wheels
$T_i(i = 1,2,3,4)$	the driving torque of the wheels
T_m^*	the target electromagnetic torque
ξ	the motor characteristic parameter
T_{swd}	the target steering wheel torque
ΔT_z	the front wheels torque difference
T_{dre}	the total demand driving torque determined by the longitudinal driver model
V_d	the target speed
k_p	the proportional coefficient of the PID controller
k_i	the integral coefficient of the PID controller
k_d	the differential coefficient of the PID controller
R	the speed factor
h	the tracking step
$\gamma_{0i}(i = 1,2,3)$	the control parameters
e	the error term
ϕ	the length of the linear segment
α_1, α_2	the nonlinear factor in the *fal* function
e_1	the steering wheel torque error term
e_2	the steering wheel torque change rate error term
γ_1, γ_2	the nonlinear combination coefficients
u_0	the control output of the nonlinear state error feedback
R_0	the minimum curvature radius

References

1. Wang, J.; Wang, Q.; Jin, L.; Song, C. Independent wheel torque control of 4WD electric vehicle for differential drive assisted steering. *Mechatronics* **2011**, *21*, 63–76. [CrossRef]
2. Zhao, W.Z.; Zhang, H. Coupling Control Strategy of Force and Displacement for Electric Differential Power Steering System of Electric Vehicle with Motorized Wheels. *IEEE Trans. Veh. Technol.* **2018**, *10*, 8118–8128. [CrossRef]
3. Hu, C.; Qin, Y.C.; Cao, H.T. Lane keeping of autonomous vehicles based on differential steering with adaptive multivariable super-twisting control. *Mech. Syst. Signal Process.* **2019**, *125*, 330–346. [CrossRef]
4. Kuslits, M.; Bestle, D. Modelling and control of a new differential steering concept. *Veh. Syst. Dyn.* **2019**, *57*, 520–542. [CrossRef]
5. Peng, B.; Li, J.Q.; Sun, F.C. Research on Differential Steering Control Strategy of Distributed Electric Drive Vehicle. In Proceedings of the International Conference on Civil Engineering, Mechanics and Materials Science, Bangkok, Thailand, 17–19 May 2019.
6. Wang, J.; Luo, Z.; Wang, Y.; Yang, B.; Assadian, F. Coordination control of differential drive assist steering and vehicle stability control for four-wheel-independent-drive EV. *IEEE Trans. Veh. Technol.* **2018**, *67*, 11453–11467. [CrossRef]
7. Römer, J.; Kautzmann, P.; Frey, M.; Gauterin, F. Reducing Energy Demand Using Wheel-Individual Electric Drives to Substitute EPS-Systems. *Energies* **2018**, *11*, 247. [CrossRef]
8. Sun, W.; Wang, J.; Wang, Q.; Assadian, F.; Fu, B. Simulation investigation of tractive energy conservation for a cornering rear-wheel-independent-drive electric vehicle through torque vectoring. *Sci. China Technol. Sci.* **2018**, *61*, 257–272. [CrossRef]
9. Han, J. *Auto Disturbances Rejection Control Technique*; National Defense Industry Press: Beijing, China, 2008.
10. Esmailzadeh, E.; Vossoughi, G.R.; Goodarzi, A. Dynamic modeling and analysis of a four motorized wheels electric vehicle. *Veh. Syst. Dyn.* **2001**, *35*, 163–194. [CrossRef]
11. Pang, D.Y.; Jang, B.C.; Lee, S.C. *Steering Wheel Torque Control of Electric Power Steering by PD-Control*; ICCAS: Kintex, Korea, 2005.
12. Badawy, A.; Zuraski, J.; Bolourchi, F.; Chandy, A. *Modeling and Analysis of an Electric Power Steering System*; International Congress and Exposition: Detroit, MI, USA, 1999.
13. Salaani, M.K.; Heydinger, G.J.; Grygier, P.A. *Closed Loop Steering System Model for the National Advanced Driving Simulator*; 2004 SAE World Congress: Detroit, MI, USA, 2004.
14. Guo, K. *Vehicle Handling Dynamics*; Jilin Science Press: Changchun, China, 1991.
15. Jin, L.Q.; Song, C.X.; Peng, Y.H. Theory for Optimization Design of Steer Wheel Alignment Based on Returnability and Handiness. *Trans. Chin. Soc. Agric. Mach.* **2006**, *37*, 20–23.
16. Wang, Q.; Wang, J.; Yan, L. Differential assisted steering applied on electric vehicle with electric motored wheels. *J. Jilin Univ. (Eng. Technol. Ed.)* **2009**, *39*, 1–6.
17. Bakker, E.; Nyborg, L.; Pacejka, H.B. *Tyre Modeling for Use in Vehicle Dynamics Studies*; Society of Automotive Engineers: Warrendale, PA, USA, 1987.
18. Shen, R.; Lin, Y.; Tai, X. Research on Modeling and Compensation Control Strategy of Electric Power Steering System. *Trans. Chin. Soc. Agric. Mach.* **2007**, *38*, 5–9.
19. Bertollini, G.P.; Hogan, R.M. *Applying Driving Simulation to Quantify Steering Effort Preference as a Function of Vehicle Speed*; International Congress and Exposition: Detroit, MI, USA, 1999.
20. Green, P.; Gillespie, T.; Reifeis, S. *Subjective Evaluation of Steering Effort Levels*; The University of Michigan Transportation Research Institute: Ann Arbor, MI, USA, 1984.
21. Toffin, D.; Reymond, G.; Kemeny, A.; Droulez, J. Role of steering wheel feedback on driver performance: Driving simulator and modeling analysis. *Veh. Syst. Dyn.* **2007**, *45*, 375–388. [CrossRef]
22. Shen, R.; Lin, Y.; Tai, X. Steering torque direct control strategy for vehicle electric power steering system. *J. Jilin Univ. (Eng. Technol. Ed.)* **2007**, *37*, 504–508.
23. Ahi, B.; Haeri, M. Linear Active Disturbance Rejection Control from the Practical Aspects. *IEEE/ASME Trans. Mechatron.* **2018**, *23*, 2909–2919. [CrossRef]
24. Han, J. Auto Disturbances Rejection Control Technique. *Front. Sci.* **2007**, *1*, 24–31.
25. Zheng, Q.; Dong, L.; Lee, D.H. Active Disturbance Rejection Control for MEMS Gyroscopes. *IEEE Trans. Control Syst. Technol.* **2009**, *17*, 1432–1438. [CrossRef]

26. Wang, K. *Research on Active Disturbance Rejection Control of Full Vehicle Active Suspension Based on Adaptive Genetic Algorithm*; Jilin University: Changchun, China, 2017.
27. Zhao, J. *Optimization Technology and MATLAB Optimization Toolbox*; Mechanical Industry Press: Beijing, China, 2011.

energies

MDPI

Article

Virtual Simulation of Electric Bus Fleets for City Bus Transport Electrification Planning

Jakov Topić *, Jure Soldo, Filip Maletić, Branimir Škugor and Joško Deur

Faculty of Mechanical Engineering and Naval Architecture, University of Zagreb, 10002 Zagreb, Croatia; jsoldo@fsb.hr (J.S.); fmaletic@fsb.hr (F.M.); bskugor@fsb.hr (B.Š.); jdeur@fsb.hr (J.D.)
* Correspondence: jtopic@fsb.hr; Tel.: +385-1-6168-555

Received: 20 May 2020; Accepted: 26 June 2020; Published: 2 July 2020

Abstract: City bus transport electrification has a strong potential of improving city air quality, reducing noise pollution and increasing passenger satisfaction. Since the city bus operation is rather deterministic and intermittent, the driving range- and charging-related concerns may be effectively overcome by means of fast charging at end stations and/or slow charging in depot. In order to support decision making processes, a simulation tool for planning of city bus transport electrification has been developed and it is presented in this paper. The tool is designed to use real/recorded driving cycles and techno-economic data, in order to calculate the optimal type and number of e-buses and chargers, and predict the total cost of ownership including investment and exploitation cost. The paper focuses on computationally efficient e-bus fleet simulation including powertrain control and charging management aspects, which is illustrated through main results of a pilot study of bus transport electrification planning for the city of Dubrovnik.

Keywords: city bus transport; electric vehicles; electrification; software tool; planning; control; charging management; simulation; analysis

1. Introduction

Due to environmental concerns, there is a strong tendency of electrifying road transport systems by means of introducing different types of electric vehicles [1]. Apart from reducing pollutant and CO_2 emissions, electric vehicles (EV) are characterised by substantially reduced noise pollution, lower operating cost (including energy and maintenance cost) and generally better driving characteristics. On the other hand, higher investment cost, slow battery charging and limited driving range inhibit their faster proliferation [2]. This is why the transition to fully electric vehicles (FEV or BEV) is characterised by application of hybrid electric vehicles (HEV) and plug-in hybrid electric vehicles (PHEV) [3].

City bus transport is a natural candidate for electrification, aiming at improvement in city air quality and reduction of noise. Since the city bus routes are known in advance and the operation is intermittent, the range- and charging-related issues are of lesser significance than with passenger cars. These issues can be tackled by the following two basic approaches [4]: (i) the buses are equipped with large enough battery packs to sustain half a day or even full day of operation, and the buses are efficiently recharged by using slow charging (typically overnight); and (ii) the battery size is minimised and superfast charging is employed at bus stops (typically end stations). Therefore, it is generally of interest to find optimal locations of charging stations, as considered in [5–11] with a focus on passenger cars and urban areas.

In [12], e-buses and corresponding charging systems are analysed from the total cost perspective by using data related to route specifications, timetables and other local conditions. Additionally, the authors have developed a user-friendly tool which enables the user to investigate and quantify trade-offs between EV battery size, charging infrastructure cost and vehicle fleet operational costs. However, all routes are modelled as straight lines between the bus stops and elevation is collected only at stops, thus hiding some relevant topographical features along the routes and affecting the analysis accuracy. In [13], a stochastic integer program was developed to jointly optimise charging station locations and bus fleet size, while considering stochastic bus charging demand and time-of-use electricity tariffs for the case of real-world network in Melbourne. However, this study does not consider charging scheduling for individual buses. The aforementioned city bus transport optimisation study [4] relies on an artificial urban bus driving cycle when conducting bus fleet simulations, thus neglecting influential effects of city traffic congestion, road slopes, often stops and so on. The authors of [14] indicated that lowering the fuel cost for electric buses can balance the high investment costs related to building charging infrastructure, while additionally achieving a significant reduction in pollutant emissions. To that extent, detailed techno-economic analyses comparing the total cost of ownership (TCO) of conventional and EV fleets should be conducted [15].

To the best of the authors' knowledge, studies dealing with extensive virtual simulations of different e-bus-type fleets based on real-life driving cycles and concerning spatially-distributed charging management and related TCO analyses have not been considered in the literature thus far. To fill the gap, a unique simulation tool for planning of city bus transport electrification, which contains all of the above functionalities, has been developed, which is described in this paper, including its application to a pilot study for the City of Dubrovnik. The tool consists of four modules aimed at (i) post-processing and statistical analysis of a large set of recorded driving cycles, (ii) simulation of conventional (CONV) and different types of e-buses (HEV, PHEV and BEV), (iii) virtual simulation of e-bus fleets over recorded driving cycles including user-defined setting of charging station locations and charging management itself and (iv) techno-economic analyses. The main contributions of the paper include: (i) creating a unique and flexible/transferable simulation tool resulting in realistic, data-driven transport electrification analyses; (ii) building a static map-based form of HEV/PHEV-type bus model including its control strategy, which drastically boost computational efficiency of large-scale fleet simulations; (iii) performing a detailed techno-economic analysis based on realistic virtual bus fleet simulation and actual technical data provided by city bus transport companies.

The paper is organised as follows. Section 2 describes the methodology of recording driving cycle data and overviews the structure of developed simulation tool. Section 3 outlines the Data Post-Processing Module (DPPM) and presents corresponding results of conventional city-bus transport system characterisation. Section 4 deals with E-Bus Simulation Module (EBSM) and discusses belonging simulation results for the four considered types of city buses. Section 5 describes the Charging Optimisation Module (COM) and overviews the results related to obtaining near-optimal charging infrastructure configuration for the cases of PHEV- and BEV-type bus fleets. The Techno-Economic Analysis Module (TEAM) is briefly described in Section 6, and the corresponding TCO results are discussed for various e-bus fleet scenarios. Concluding remarks are given in Section 7.

2. Pilot Data and Simulation Tool Structure

2.1. Recording of Driving Cycle Data

The driving cycle data have been collected on a sub-fleet of 10 MAN Lyon's City NL323 buses operating in the city of Dubrovnik. They are considered a good representative of the overall fleet, as they cover all major bus routes (Figure 1) and represent around 1/3 of regularly used city buses. The driving data recording was performed by utilising a commercial GPS/GPRS vehicle tracking device installed in the selected buses for the purpose of this study. The data, collected from a built-in GPS device and vehicle CAN bus, are summarised in Table 1. Recording was conducted continuously for a period of five months, and the data sampling time was 1 second.

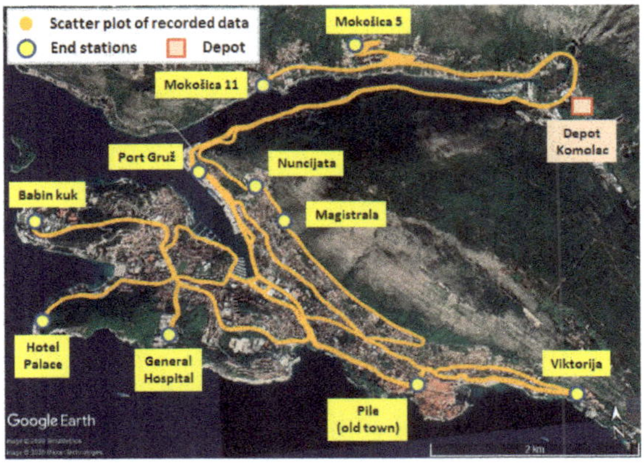

Figure 1. City bus routes in Dubrovnik along with end-station and depot locations.

Table 1. List of available city bus tracking data.

Measured Data	Resolution
GPS coordinates (latitude, longitude), (°)	1.0×10^{-7}
Elevation height (altitude), (m)	0.1
Vehicle speed, (km/h)	0.1
Travelled distance (from odometer), (km)	1.0
Accelerator pedal position, (%)	1.0
Cumulative fuel consumption, (L)	0.5
Fuel level in reservoir, (%)	1.0
Engine speed, (rpm)	1.0
Engine load, (%)	1.0

2.2. *Organisational Structure of Simulation Tool*

Figure 2 illustrates the structure of the developed tool, including the connections between the main modules. The tool is driven by recorded driving cycle data, and as the main output, it delivers the TCO over the projected fleet operational period (12 years, herein). The DPPM transforms the recorded driving data into individual driving cycles, and it also calculates various statistical features characterising the conventional city bus transport behaviours. The EBSM provides computer simulations of different types of city buses (CONV, HEV, PHEV, BEV) over the driving cycles extracted by the DPPM. The module outputs include the individual bus energy consumption (fuel and/or electricity) and various features of powertrain response (e.g., engine/e-motor operating points, gear ratio trajectories, etc.). The COM utilises the outputs of DPPM and EBSM to simulate the overall city bus fleet over the recorded driving cycles and optimise the PHEV- and BEV-type bus charging configuration and management. This module provides the number, location and type of chargers, the bus battery capacity and the number of reserve buses in the BEV case, which are required to fulfil the driving routes with sufficient battery charge. The COM also outputs the total fuel and/or electricity consumption over the considered period of operation. The TEAM uses the output data from the COM module, as well as the data on bus transport investment and exploitation/maintenance cost, in order to calculate the TCO.

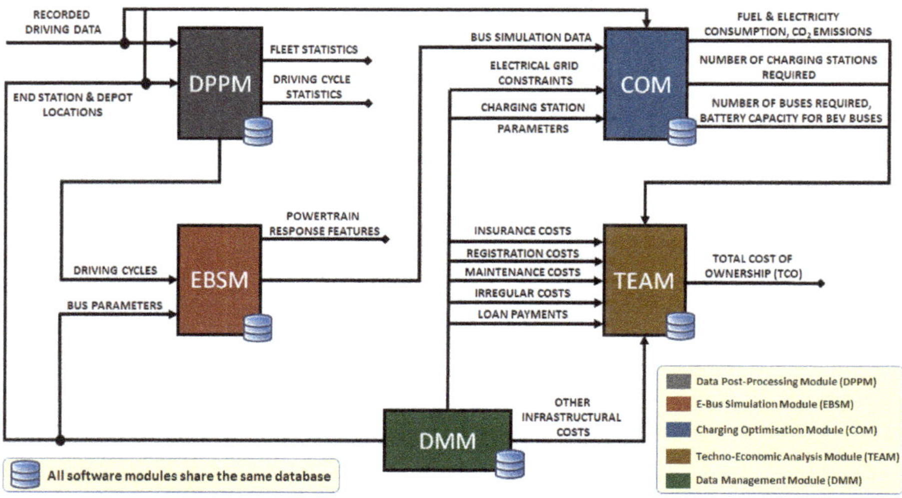

Figure 2. Organisational structure of simulation tool.

The simulation tool is written in Python object-oriented programming language, with computationally demanded routines coded in C language. The Python language has been selected because it is available in open-source domain, it supports multiple platforms, has a vast number of available libraries, and is dynamically typed language (i.e., does not require declaring of data type of a variable) convenient for rapid prototyping. The simulation tool is designed in a user-friendly way (based on a graphical user interface (GUI) including windows, tabs, I/O data interfaces, etc.) and has in mind the transferability to other cities using a common/shared database. The database serves as a main storage for recorded driving cycle data, and plays the role of an intermediary between the main tool modules. In addition, the simulation tool includes the Data Management Module (DMM), which provides greater flexibility and adaptability to different cities' transport system configurations. DMM enables the user to define all static data (system parameters) required by the simulation tool, e.g., those related to vehicle model parameters, end-station and depot locations, charging station parameters and techno-economic data.

3. Data Post-Processing Module (DPPM)

3.1. General Description

According to the DPPM flowchart shown in Figure 3, the recorded driving data are first processed and stored in a database. Next, the driving cycles are extracted by using the DMM-based data related to geographical coordinates of end stations and depot (Section 3.2). At the same time, the fleet statistics are calculated for the entire fleet and individual buses (Section 3.3). Finally, the module outputs including driving cycles and statistical features can be plotted in different formats, and they are saved into the database.

3.2. Extraction of Driving Cycles

A single driving cycle is defined by the velocity vs. time and road slope vs. travelled distance profiles between two consecutive end stations, including depot (see Figure 1). The corresponding time profiles of cumulative fuel consumption are also extracted.

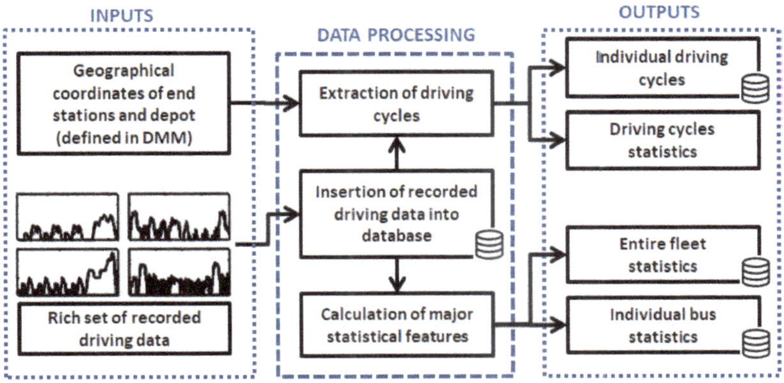

Figure 3. Flowchart of Data Post-Processing Module (DPPM).

The driving cycle segmentation process resulted in a total of 122,727 extracted driving cycles. An example of recorded driving cycle is shown in Figure 4. Note that the vehicle velocity is directly measured, while the road grade is reconstructed from the rate of change of altitude when expressed with respect to the travelled distance. In order to reduce the noise in the reconstructed road slope profile, the altitude signal is pre-filtered by a low-pass double-sided Butterworth filter [16].

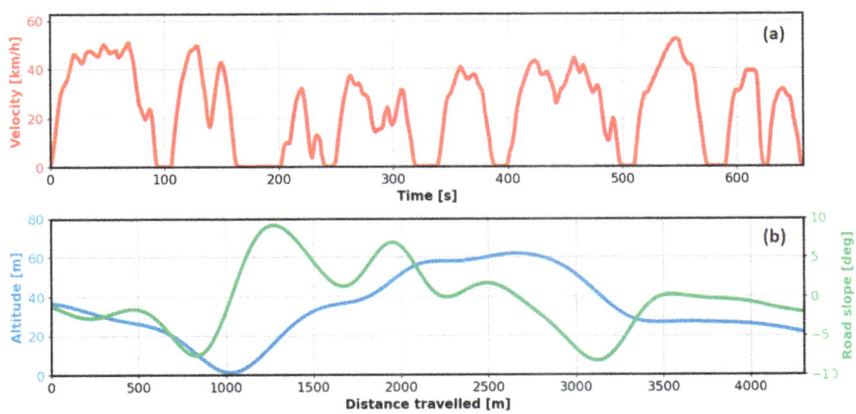

Figure 4. Sample of reconstructed driving cycle for city bus route Hotel Palace–Pile and afternoon hours: velocity vs. time profile (**a**) and road slope vs. distance travelled given along altitude source profile (**b**).

3.3. Calculation of Vehicle Fleet Statistics

A rich set of statistically significant driving features is calculated for the purpose of actual/conventional city-bus transport system characterisation and in support of transport electrification (e.g., locating charging stations, Section 5). The features related to individual buses, all given per-day-basis, include: the total fuel consumption and distance travelled; average fuel consumption in L/100 km; the total time the bus is dwelling at depot, individual end stations or any other locations (typically bus stops); total driving time; mean velocity; number of bus stops per kilometre; number of bus visits to depot and end stations. The results are stored in a two-dimensional (2D) matrix (one per bus), whose rows and columns represent individual days and the statistical features, respectively.

Once the individual statistics are stored, they are further used to calculate the same features for the entire fleet on the basis of individual day, week, month or year. The results can be presented in different ways, e.g., instead of individual dwelling times, one can get the information about percentage share of time the buses or entire fleet are resting at depot, end stations and other locations (see Figure 5). Specific fleet features requiring denser sampling (30 min, herein) are also calculated. An example of the daily average fleet velocity profile is shown in Figure 6. Other features related to entire fleet include: average number of buses being parked at depot on 24-hour time basis (Figure 6), clusters of buses parking durations in relation to geographical coordinates, count of transitions between individual end stations/depot, etc.

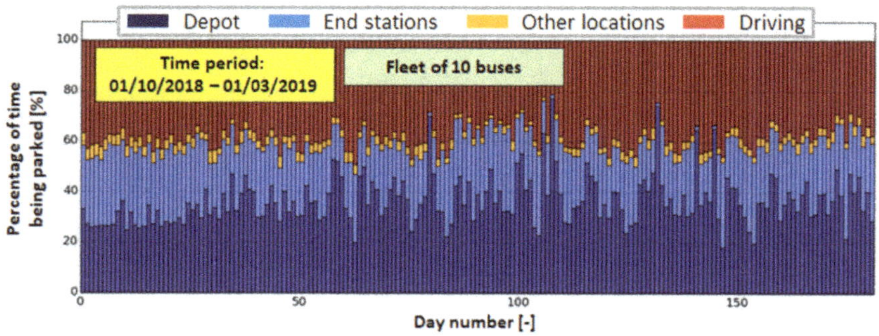

Figure 5. Percentage time share of buses being operated and parked in depot, end stations and other locations.

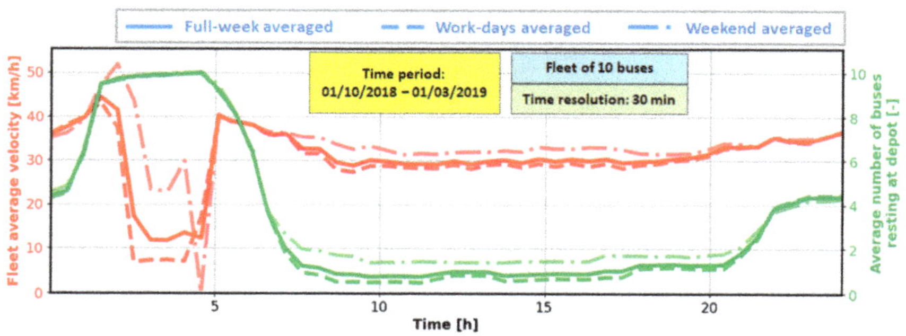

Figure 6. Time profiles of fleet average velocity and average number of buses resting at the depot, both given on daily basis and averaged over the considered five-month period.

Figure 6 indicates that the average bus velocity when operating is around 30 km/h and it is higher in early morning and night hours, as well as over weekends. The low-velocity gap between 2 a.m. and 5 a.m. corresponds to the interval when most of the buses are parked in the depot. The buses rarely visit the depot in other time intervals, particularly over the work days (when the average number of buses is lower than 1).

Figure 5 confirms that the share of total time of buses being parked in the depot is relatively small (approx. 30%) and comparable to the share of end-station parking time (approx. 25%). The rest of the time the buses spend in driving (40%), while only for a small portion of time (approx. 5%), they rest elsewhere, typically at bus stops.

Figure 7a indicates that there are significant differences in bus resting time at different end stations. The average resting durations for most pronounced end stations are between 10 and 20 min, thus making them good candidates for installation of fast chargers. When selecting the best candidates, a charging station utilisation factor should also be considered (Figure 7b). The final end stations targeted for electrification are then obtained by taking the cross-section of these two criteria (end stations underlined in red in Figure 7).

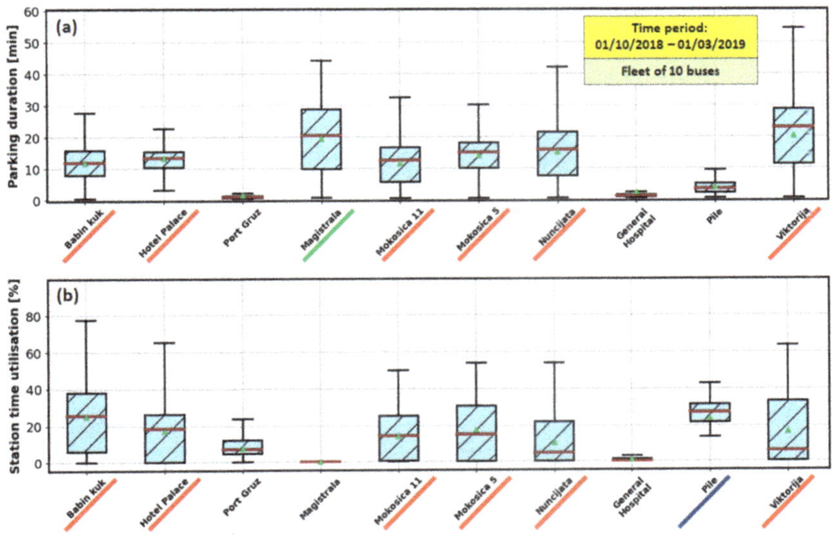

Figure 7. Boxplots of time duration of buses being parked at different end stations (**a**) and end stations charging utilisation (**b**).

4. E-Bus Simulation Module (EBSM)

4.1. General Description

As illustrated in flowchart in Figure 8, the EBSM simulates different types of city buses (CONV, HEV, PHEV and BEV) over the recorded driving cycles extracted by the DPPM (Section 3). The simulation first involves loading of vehicle-related parameters from the database, which need to be previously defined in the DMM. Next, the vehicle is simulated over the selected driving cycles by using the numerically-efficient backward-looking model (Section 4.2). Note that the vehicle model includes a control strategy that manages the gear ratio in the CONV and BEV cases, and the internal combustion engine (ICE) torque in the HEV and PHEV cases (Section 4.3). The emphasis has been on transforming the previously developed control strategy [17,18] to a form of off-line optimised maps, instead of using an on-line optimisation algorithm. The EBSM outputs time responses of key powertrain variables, such as cumulative fuel and electricity consumption, CO_2 emissions and transmission gear ratio.

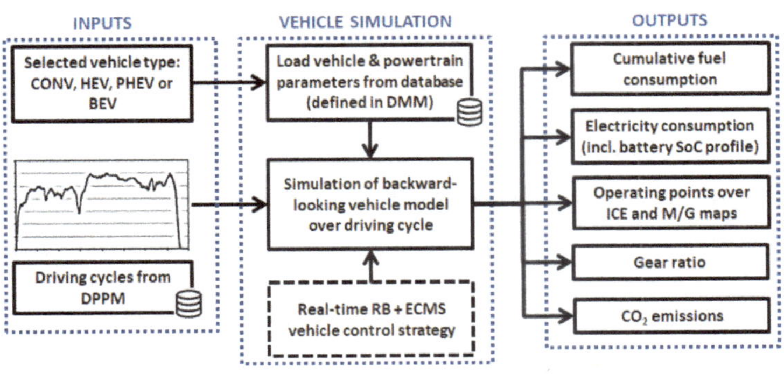

Figure 8. Flowchart of E-Bus Simulation Module (EBSM).

4.2. Vehicle Modelling

4.2.1. Considered City Buses

The MAN Lion's City buses with the length of 12 m and the capacity of up to 126 passengers are represented in virtual simulation by the Volvo 7900 bus model. The Volvo 7900 platform was chosen because it includes all three e-bus variants considered (HEV, PHEV and BEV; Table 2).

Table 2. Basic parameters of modelled 12 m city buses [19].

Parameter	CONV	HEV	PHEV	BEV
Model label	Volvo 7900 (Diesel)	Volvo 7900 Hybrid	Volvo 7900 Electric Hybrid	Volvo 7900 Electric
Maximum ICE power	228 kW	161 kW	173 kW	N/A
Maximum e-motor power	N/A	120 kW	150 kW	160 kW
Battery capacity	N/A	4.8 kWh	19 kWh	76 kWh
Transmission model (type)	ZF 6AP 400B (AT)	Volvo AT2412 I-Shift (AMT)		Volvo 2-speed (AMT)
Number of gears	6	12		2
Maximum fast charging power	N/A	N/A	150 kW	300 kW

Volvo e-buses use lithium iron phosphate (LFP) battery due to its high specific power required for propulsion and fast charging. The battery packs of Volvo 7900 HEV, PHEV and BEV bus variants have energy capacities of 4.8 kWh, 19 kWh and 76 kWh, respectively (Table 2).

4.2.2. Modelling

In the backward-looking models, the powertrain variables are calculated in the direction from the wheels towards the engine and/or e-motor, starting from the wheel speed and torque being defined by the driving cycles [20]. In order to boost the computational efficiency, the powertrain dynamics is neglected, except for the battery state-of-charge (SoC) dynamics that are represented by a first-order model.

The considered parallel configuration of a HEV/PHEV-type bus is illustrated in Figure 9a [17]. The battery is represented by the equivalent battery circuit model shown in Figure 9b, which is described by the following state equation [20,21]:

$$\dot{SoC}(t) = -\frac{I_{batt}(t)}{Q_{\max}} = \frac{\sqrt{U_{oc}^2(SoC) - 4R(SoC)P_{batt}(t)} - U_{oc}(SoC)}{2Q_{max}R(SoC)}, \quad (1)$$

where U_{oc} is the open-circuit voltage, R is the internal resistance, I_{batt} is the battery current, $Q_{\max}$ is the maximal battery charge capacity and the SoC is defined as $SoC = Q/Q_{\max}$, with Q denoting the actual charge.

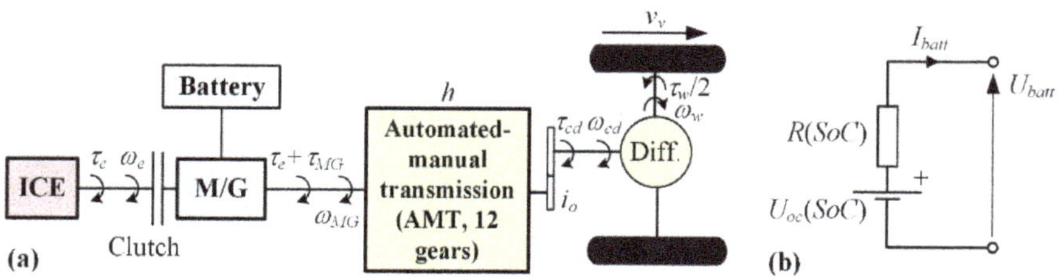

Figure 9. Functional scheme of considered parallel HEV/PHEV powertrain (**a**) and battery equivalent circuit model (**b**).

The model input P_{batt} represents the battery output power defined as:

$$P_{batt} = \eta_{MG}^{k} \cdot \tau_{MG} \cdot \omega_{MG} \,, \tag{2}$$

where τ_{MG} and ω_{MG} are the motor/generator (M/G) torque and speed, respectively, η_{MG} is the M/G machine efficiency (represented by $\eta_{MG}(\omega_{MG},\tau_{MG})$ map; see Figure 10c and [17]) and the coefficient k is equal to 1 or −1 depending on whether the M/G machine operates as a generator or motor, respectively. The M/G machine speed and torque are given by the following kinematic equations:

$$\omega_{MG} = i_o h \omega_w = i_o h \frac{v_v}{r_w} \,, \tag{3}$$

$$\tau_{MG} = \frac{\frac{\tau_w}{\eta_{tr}(\tau_w)} + \frac{P_0(\omega_w)}{\omega_w}}{i_o h} - \tau_e \,, \tag{4}$$

where v_v is the vehicle velocity, r_w is the tire effective radius, h and i_0 are the transmission and final drive ratios, respectively, τ_w and w_w are the total wheels torque and speed, respectively, $\eta_{tr}(\tau_w)$ and $P_0(\omega_w)$ are drivetrain efficiency and idle power loss maps [17] and τ_e is the engine torque considered as a control variable (in addition to h). The wheel torque is determined according to vehicle longitudinal dynamics equation covering the vehicle acceleration torque and aerodynamic, road grade and rolling resistances [20].

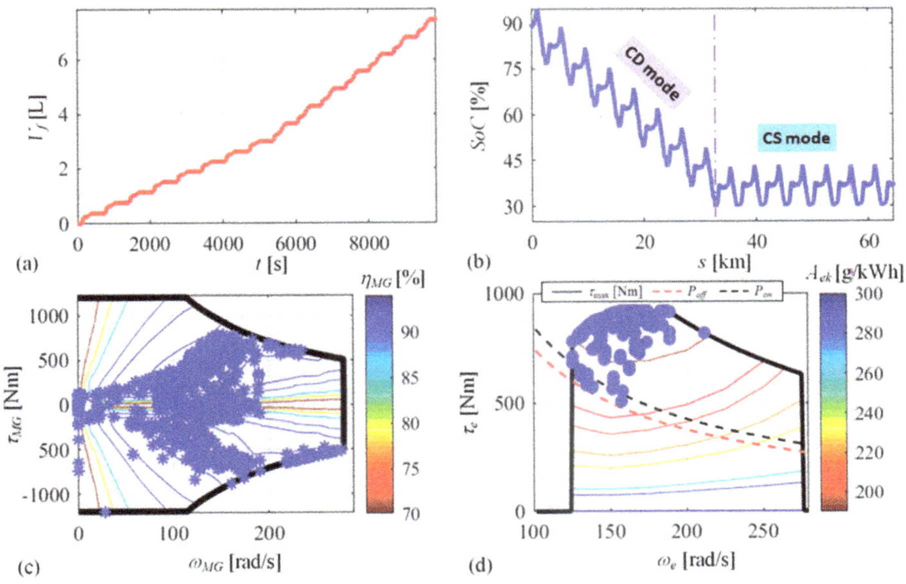

Figure 10. Simulation results for PHEV bus over the driving cycle given in Figure 4 repeated 15 times, including: cumulative fuel consumption time response (**a**), battery SoC vs. distance travelled (**b**), M/G machine operating points (**c**), and engine operating points (**d**).

The fuel consumption at the driving cycle end time t_f is determined as:

$$V_f = \frac{1}{\rho_{fuel}} \int_0^{t_f} \dot{m}_f dt = \frac{1}{\rho_{fuel}} \int_0^{t_f} \left(A_{ek}(\tau_e, \omega_e) \frac{\tau_e \omega_e}{3.6 \cdot 10^6} \right) dt \,, \tag{5}$$

where A_{ek} is the engine specific fuel consumption given by the map shown in Figure 11, ρ_{fuel} is the diesel fuel density (ρ_{fuel} = 845 g/L) and ω_e equals ω_{MG} or 0 when the engine is switched on or off, respectively. Note that the integral Equation (5) is realised by using the Euler integration method with the common sample time of the backward model equal to 1 s.

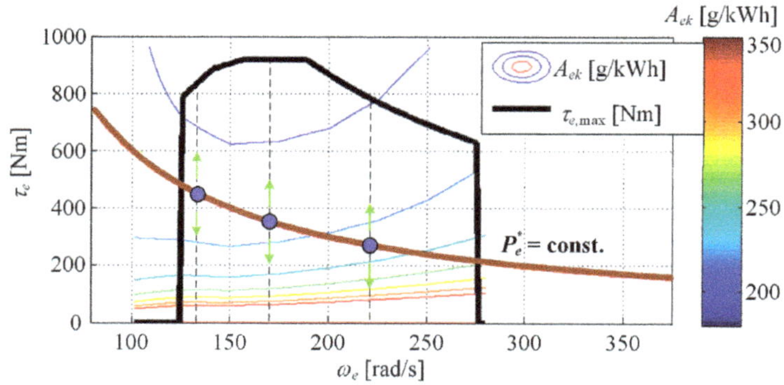

Figure 11. Engine specific fuel consumption map including illustration of ECMS-based operating point search.

In the case of a conventional (CONV) bus, the battery and M/G machine are omitted from the functional scheme in Figure 9, while the AMT is replaced by a torque converter AT. The torque converter is represented by a backward-looking map ω_i (ω_t,τ_t) derived offline from the well-known static torque converter model (see [22] and references therein) and the map $\tau_i = \tau_t / R_\tau(\omega_t/\omega_i)$, where the subscripts i and t denote impeller/engine and turbine/transmission input variables, and R_τ (.) is the static model torque ratio map. For the BEV-type bus, the engine is omitted and a two-speed AMT is used.

Vehicle auxiliary devices (HVAC system, servo steering, air compressor, engine cooling fan and alternator) are modelled based on the nominal power of each device and a binary power-modulating signal, whose duty cycle is made dependent on the driving and atmospheric conditions (urban driving conditions and ambient temperature dependence are assumed) [23].

4.3. Control Strategy

In the BEV case, the aim of control strategy is to find an optimal gear ratio h^* in each sampling instant, which minimises the battery discharging power and maximises the battery recharging/ regenerative power:

$$h^* = \underset{h}{\mathrm{argmin}} \begin{cases} \eta_{batt}^{k} P_{batt}(P_d, \omega_w, h), & \text{for } P_{batt} < 0 \text{ (charging)} \\ \eta_{batt}^{k} P_{batt}(P_d, \omega_w, h), & \text{for } P_{batt} \geq 0 \text{ (discharging)} \end{cases}, \tag{6}$$

where η_{batt} is the battery efficiency and P_d is the transmission input power demand calculated from v_v and τ_w as shown in Figure 12. The optimal gear ratio h^* is calculated offline and mapped as h^* (P_d, ω_w). The SoC dependence of P_{batt} has a minor effect on h^* and is neglected in Equation (6), and further on.

Similarly, in the CONV case, the control strategy finds an optimal gear ratio h^* that minimises the fuel mass flow $\dot{m}_f$:

$$h^* = \underset{h}{\mathrm{argmin}} \dot{m}_f(P_d, \omega_w, h) . \tag{7}$$

The off-line obtained optimal solutions are mapped as $h^*(P_d, \omega_w)$.

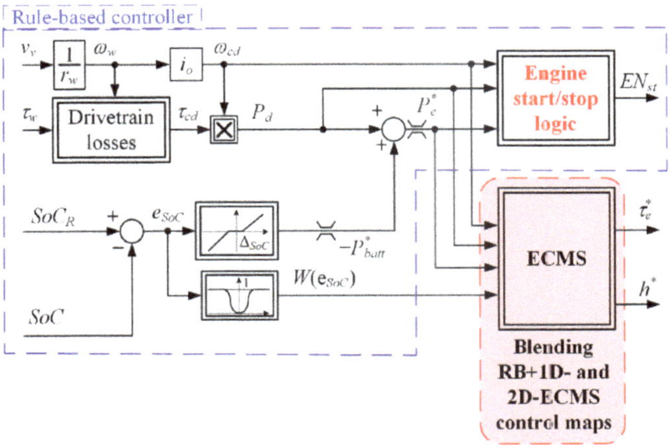

Figure 12. Block diagram map-based RB+ECMS control strategy.

The HEV and PHEV control strategy determines the transmission gear ratio h and the engine torque τ_e based on combining a rule-based (RB) controller and an equivalent consumption minimisation strategy (ECMS) [17,24]. The original RB+ECMS strategy is modified here to allow for computationally-efficient map-based realization, which is shown in Figure 12. A proportional-like battery SoC controller commands the battery recharging power—P_{batt}, which is added to the transmission input power demand P_d to obtain the engine power demand P^*_e. The demanded engine power P^*_e is compared with engine on and off thresholds P_{on} and $P_{off} < P_{on}$, respectively, in order to determine the engine on/off state EN_{st}. The engine is exceptionally kept switched on in the case $P^*_e < P_{off}$ when the speed-dependent M/G machine power limit is not high enough to satisfy the driver power demand P_d in the fully electric driving mode. If the engine is switched on, the signal P^*_e is fed to the ECMS to find the optimal values of h and τ_e. Otherwise, the electric driving mode is activated and the gear ratio is determined according to Equation (6).

In the original RB+ECMS strategy [17], the equivalent fuel consumption $\dot{m}_{eq}(P_e, P_d, \omega_w, h, SoC)$ is minimised instantaneously and on-line with respect to both control variables h and τ_e. In the simplified map-based RB+ECMS version considered here, the equivalent fuel consumption is minimised in two stages. In the first stage, the ECMS is applied to discrete operating points along the constant power curve P^*_e (denoted in Figure 11 by blue circles) to determine the optimal gear ratio:

$$h^* = \underset{h}{\operatorname{argmin}}\, \dot{m}_{eq}(P^*_e, P_d, \omega_w, h)\,, \tag{8}$$

where $\dot{m}_{eq}$ is the equivalent fuel consumption rate containing the actual fuel consumption rate $\dot{m}_f$ and a battery power-equivalent fuel rate (see [17,24] for details). The off-line obtained optimal solutions are stored in a three-dimensional (3D) map $h^*(P^*_e, P_d, \omega_w)$ representing the RB+1D-ECMS control map. In the second stage, the ECMS is applied along the engine torque axis (see green arrows in Figure 11):

$$\tau^*_{e,2D} = \underset{\tau_e}{\operatorname{argmin}}\, \dot{m}_{eq}(P_d, \omega_e, \tau_e)\,. \tag{9}$$

The off-line optimisation results are stored in a 2D map $\tau^*_{e,2D}(P_d, \omega_e)$ representing the 2D-ECMS control map. Finally, the engine torque obtained by the RB+1D-ECMS (as $\tau^*_{e,RB} = P^*_e/(h^* i_o \omega_w)$) and the one obtained by the 2D-ECMS are combined/blended on-line using the SoC control error (e_{SoC})-dependent weighting factor $W(e_{SoC})$ [17,24]:

$$\tau^*_e = \tau^*_{e,RB} W(e_{SoC}) + \tau^*_{e,2D}(1 - W(e_{SoC}))\,, \tag{10}$$

where $W(.)$ is an SoC control error-dependent weighting function illustrated in Figure 12. For small SoC control errors, the 2D-ECMS is dominant, while for large SoC control errors, the RB+1D-ECMS is preferred to satisfy the engine power demand P^*_e (and, thus, the battery power demand P^*_{batt}).

Finally, gear shift delay logic is implemented to prevent frequent gear switching [17]. The goal is to prevent gear shift occurrence in k^{th} time step, i.e., rather use the gear ratio from the previous $(k-1)^{th}$ step, h_{k-1}, if the time elapsed since the last gear shift t_{sh} is lower than an arbitrarily set shift delay threshold t_{th} and if h_{k-1} gives feasible set $\mathbf{u}_k$ of engine and M/G machine operating points in the kth step (denoted by Π):

$$h_k^* = \begin{cases} h_{k-1}^*, & \text{for } t_{sh} < t_{th} \text{ and } \mathbf{u}_k\left(P_e^*, P_{d,k}, \omega_{w,k}, h_{k-1}^*\right) \in \Pi \\ h^*\left(P_{e,k}^*, P_{d,k}, \omega_{w,k}\right), & \text{otherwise} \end{cases} \tag{11}$$

The same gear shifting delay logic is applied in the CONV and BEV cases. The shift delay threshold is set here to $t_{th} = 2$ s.

The above described simplified, map-based control strategy has been found to result in a negligible model response deviation when compared to the use of original strategy, which was proven to be close to the dynamic programming-based global optimum [17]. On the other hand, the execution time is reduced by around 200 times. The achieved execution time, expressed as the amount of microseconds needed to simulate one second of real time (for a workstation having 16 GB RAM and Intel® Xeon® Processor E5-1620 v3 @ 3.50GHz) falls in the range from 50 to 87 µs/s depending on vehicle type. This results in approximate yearly 10-bus fleet simulation time ranging from approximately 4.5 h to 7.5 h, which is deemed acceptable for such a large-scale fleet simulation. Note that the execution time could further be reduced by using parallel computing.

The PHEV can operate in two characteristic modes [20]: (i) charge depleting (CD) followed by charge sustaining (CS), where the former involves the engine only when absolutely needed and the latter correspond to hybrid operation at the target SoC of 30%; (ii) blended mode where engine is regularly used all over the driving cycle for additional energy savings. For the sake of simplicity, the CD/CS mode is considered in this paper.

4.4. Simulation Results

The results related to relative fuel and/or electricity consumptions for different city bus types are given in Table 3 for the full recording period. The relative difference between the simulated (Sim) and recorded (Rec) fuel consumptions for the CONV bus is equal to only 1.4%. Therefore, the CONV simulation model used as a basis for e-bus modelling can be considered accurate. Note that although the real and simulated buses are different (MAN Lion City and Volvo 7900), the validation is considered fair, as the two buses are rather comparable in terms of size, mass, engine power, number of passengers and other similar factors.

The simulated electricity consumptions of PHEV- and BEV-type buses are close to recorded ones documented in the ZeEUS project report [25] for Volvo 7900 bus series (Table 3). In the PHEV case, the simulated fuel consumption is by 30% higher than the ZeEUS recorded one, but this discrepancy is compensated for by 26% higher recorded electricity consumption when compared to the simulated one. In the BEV case, the relative difference in electricity consumption equals 6%. The simulated HEV fuel consumption is reduced by 50% when compared to CONV simulation results, while the manufacturer states the fuel consumption reduction from 39% to 45% reported by operators [19]. The observed, relatively modest discrepancies in fuel/electricity consumption may be related to difference in considered bus weights (passenger weight is fixed to 1250 kg), road slope and traffic congestion conditions, as well as regenerative braking capacity (set to the maximum amount of 100% in simulation).

Table 3. Recorded and simulated relative fuel and electricity consumptions for different bus types and full recording period.

	CONV		HEV		PHEV		BEV	
	Rec *	Sim	Est **	Sim	Rec *	Sim	Rec *	Sim
Fuel consumption, (L/100 km)	42.9	43.5	24.8	21.6	10.2	13.3	N/A	
Electricity consumption, (kWh/100 km)	N/A		N/A		53	42.4	83	77.9

* Recorded PHEV and BEV fuel and electricity consumptions are taken from the ZeEUS project report [25]. ** Estimated based on information on fuel consumption reduction from 39% to 45% for Volvo 7900 Hybrid vs. Volvo 7900 according to [19].

Figure 10 shows the PHEV-case simulation results for the recorded driving cycle shown in Figure 4 repeated 15 times and the initial battery SoC equal to 90%. When the CD mode is active, the engine is used only when needed and the cumulative fuel consumption is often constant (a stepwise-like response, Figure 10a). After entering the CS mode, the engine is more active to sustain the battery SoC (Figure 10b). The control strategy deploys the operating points of engine (when switched on) and M/G machine in the high efficiency areas of corresponding maps (Figure 10c–d), thus minimising the energy consumption.

5. Charging Optimisation Module (COM)

5.1. General Description

A generic framework for virtual simulation of an e-bus fleet over the recorded driving cycles is represented by the flowchart shown in Figure 13. The model of city bus of any type (Section 4) is initialised based on the data stored in DMM and virtually run over the recorded driving cycles for the specified period of time, thus resulting in fuel and/or electricity consumption output data. In the case of a PHEV- or BEV-type bus the user needs to specify locations and types of charging stations, the nominal vehicle battery capacity and grid power constraints to execute the simulation. The simulation is repeatedly run for a peak day or peak week for different charging infrastructure and battery capacity specifications, in order to find a nearly-optimal configuration, which would be finally re-run for a given, longer period of time to calculate the fuel/electricity consumption and charging station utilisation statistics. In the case of BEV-type bus, the COM automatically adds reserve bus(es) if needed, and calculate their final number and related statistics. The bus intervening algorithm monitors each BEV's battery SoC, and when it drops below a predefined minimum value (0.3, herein), the immediate replacement with reserve e-bus is conducted. At the same time, it is taken into account that the depleted bus needs some constant time to reach the nearest charging station, where it is to be charged (as any other bus), and once it is fully charged, it will be waiting for the next replacement/intervention.

5.2. Charging Management Algorithm

Charging management is described by the flowchart shown in Figure 14. First, it is checked if a PHEV- or BEV-type bus has arrived to an end station/depot and if that station has a charger installed. If the charger is not occupied or if the bus has a lower battery SoC of any of the buses already being charged, the bus is put on charge; otherwise, it remains in the charging queue. Note that each station can be set to have an arbitrary number of chargers, as described with Figure 13.

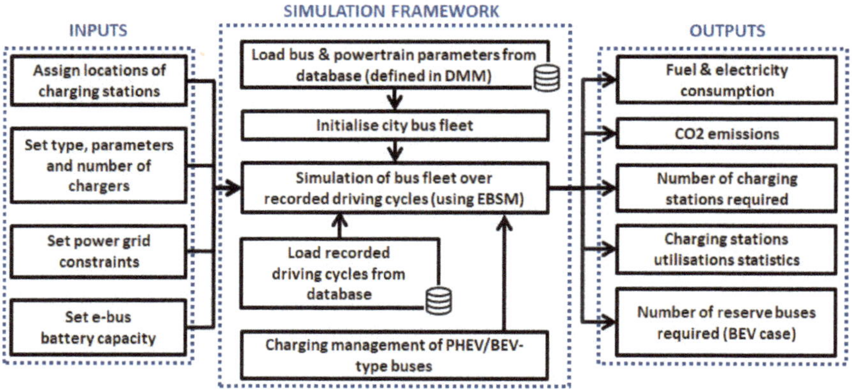

Figure 13. Flowchart of Charging Optimisation Module (COM) with embedded EBSM functionality.

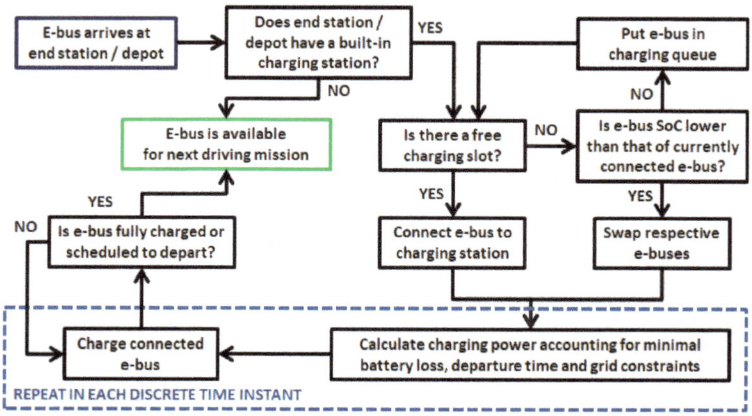

Figure 14. Flowchart of rule-based charging management algorithm (applied to PHEV- and BEV-type buses).

The charging process is managed by taking into account the requirements on satisfying the departure schedule, minimising battery power loss and respecting the grid power constraints. According to [26,27], the battery energy loss is minimised by demanding a linear change of SoC all over the remaining charging interval $\Delta T_{ch} = t_f - t_k$. Therefore, the SoC rate is updated in each sampling instant k according to:

$$\frac{dSoC}{dt} = \frac{SoC_f - SoC_k}{t_f - t_k}, \tag{12}$$

where SoC_k is the current SoC and SoC_f is the target SoC at departure. Inserting Equation (12) into the battery state Equation (1) yields the charging power $P_{batt} < 0$ to be applied in the kth sampling instant:

$$P_{batt} = \frac{U_{OC}^2(SoC_k) - \left[2Q_{max}R_{int}(SoC_k)\frac{SoC_f - SoC_k}{t_f - t_k} + U_{OC}(SoC_k)\right]^2}{4R_{int}(SoC_k)}. \tag{13}$$

If the charging power $-P_{batt}$ calculated from Equation (13) is greater/less than the maximum/minimum allowable power (defined by the charger selected), the charging power is limited to the maximum/minimum power, respectively. Note that $\Delta T_{ch} = t_f - t_k$ is saturated in Equation (13) to its lower limit of 30 s to avoid division by zero.

Once the charging power profile is obtained for each sampling step k and for i^{th} vehicle from the total number of N_v vehicles connected to chargers at the same grid sections (e.g., depot), it is checked if the total charging power is greater than the maximum grid power $P_{grid,max}$. If this applies, the charging power is scaled down to satisfy the grid power constraint:

$$P_{batt,k,i,corr} = \frac{P_{grid,\ max}}{\sum_{i=1}^{N_v} P_{batt,k,i}} P_{batt,k,i}, \text{ if } \sum_{i=1}^{N_v} P_{batt,k,i} > P_{grid,\ max} \,. \tag{14}$$

5.3. Obtaining of Near-Optimal Charging System Configurations

According to the city bus transport characterisation results from Figure 7, there is a number of end stations with relatively long bus resting durations and potentially high utilization of charger units. Additionally, the end station resting time share approaches that of depot (Figure 5), and there are no other emphasised stop locations. Therefore, fast charging stations and belonging transformer substations can be installed at end stations to provide bus recharging, while otherwise the available power can be utilised to supply city e-mobility hubs built around the end stations. High-power off-board chargers with built-in pantograph are considered (150 or 300 kW, see Table 2) [19]. In addition, the slow-to-modestly fast plug-in charging solutions can be considered for a depot, where the charging time can be long in night (Figure 6).

5.3.1. PHEV Fleet Case

Figure 15 shows the PHEV fleet simulation results for different number of end stations equipped with a single fast charger per station (150 kW) and a five work day period. Charging in depot was not considered because the small-capacity PHEV battery (Table 2) can quickly be recharged at the end stations, where the buses rest for a relatively long time (Figures 6 and 7). The results shown in Figure 15 point out that the fuel consumption saving converges to −41% as the number of end station charging spots approaches six. Of course, as the fuel consumption reduces, the electricity consumption grows, but the overall energy cost is reduced by 17% due to cheaper electricity. By conducting PHEV fleet simulations over the five-month period, it has been found that the optimal number of charging stations should be incremented to seven.

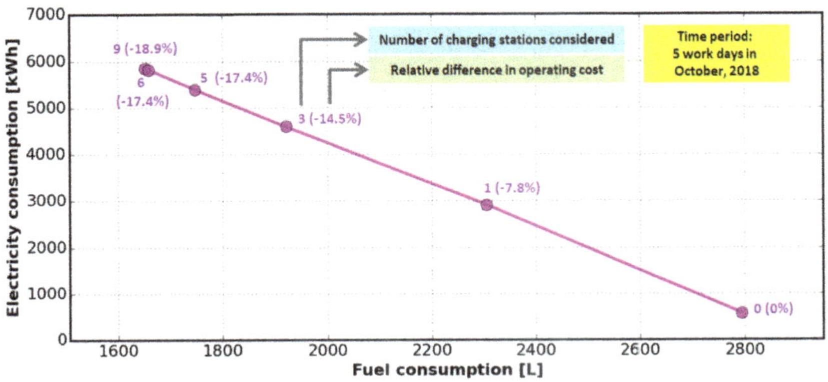

Figure 15. Pareto frontier-like plot showing PHEV bus fleet electricity vs. fuel consumption costs for different number of end stations equipped with a fast charger per station.

5.3.2. Case of BEV Fleet

The BEV fleet simulation results are shown in Table 4. The full five-month period is considered to cover a larger number of "critical" days when reserve buses may be needed. Only scenarios with the number of end-station charging spots being in the vicinity of the optimal one found for the PHEV fleet is examined (around six stations plus depot, each considering a single fast charger). The maximum charging power is set to the levels of 150 kW or 300 kW. Finally, various battery capacities are considered (76, 150 and 250 kWh, as the capacities available for the considered bus [19]).

Table 4. BEV fleet simulation results for different number of charging spots (located at end stations and depot) and reserve buses, and different battery capacities (full five-month period).

Case	Battery Capacity	Charging Power (Number of Charging Stations)	Percentage of Total Electricity Consumed by Reserve Buses	Number of Reserve Buses Required	Number of Bus Swaps Required (Number of Days when Swapping Occurs Out of 152)
BEV 1	76 kWh	300 kW (6)	9.2%	2	558 (106)
BEV 2	76 kWh	300 kW (7)	1.8%	2	94 (54)
BEV 3	76 kWh	300 kW (8)	1.7%	2	90 (52)
BEV 4	150 kWh	150 kW (8)	0.6%	2	4 (3)
BEV 5	250 kWh	150 kW (7)	0.00%	0	0 (0)

The results shown in Table 4 point out that by increasing the number of charging stations, the percentage of total electricity consumed by reserve buses drops from 9.2% (case BEV 1) to 1.8% (case BEV 2). Likewise, the number of bus swaps (concerning reserve bus) drops from 558 in 106 (out of 152) days (BEV 1) to 94 in 54 days (BEV 2). Figure 16 indicates that in the case BEV 2 notable bus swaps occurs only in several days, which are characterised by peak traffic load (typically due to specific needs such as moving tourists from cruising ships to the old city). Similar trends apply to the case of increasing the battery capacity from 76 kWh to 150 kWh and further to 250 kWh (cases BEV 4 and BEV 5, respectively), where the reserve buses are marginally needed in the former case, and not needed in the latter case.

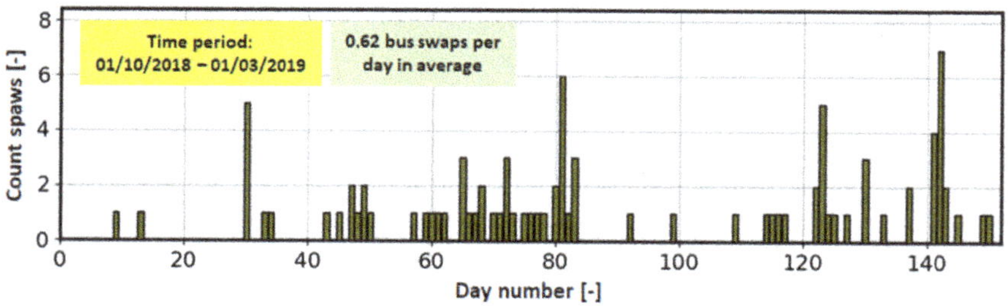

Figure 16. Number of daily bus swaps with reserve buses for BEV 2 case.

Based on the above results, the case BEV 5 might be considered as optimal. However, since increasing of the battery capacity of each bus in a fleet is rather expensive and the need for reserve buses in case BEV 2 is minor (only 1.8%), the case BEV 2 has been adopted as an optimal for final simulations discussed in Section 5.4.

5.4. Comparative Energy Consumption Results

Table 5 gives the energy consumption simulation results for the PHEV and BEV fleets configured in the previous two sections, as well as HEV and CONV fleets, all simulated over the full recording period. The HEV fleet can reduce the fuel consumption by around 50% compared to CONV fleet, owing to regenerative braking, switching the engine off in low speed conditions and placing the engine operating points in the high-efficiency region (Section 4; see also Table 3). Using the PHEV buses provides additional fuel savings, i.e., the fuel consumption reduction compared to CONV case is around 70%. This is due to the use of electricity coming from grid by means of fast charging at end stations. The PHEV fleet electricity consumption equals almost 50% of what is consumed by BEV fleet, which is owing to a relatively low portion of operation in the CS mode (25.5%).

Table 5. Fuel and electricity consumption per fleet type for full five-month period.

	Fleet Type (Total of 10 Buses)			
	CONV	HEV	PHEV	BEV
Fuel Consumption, L	**145,295 (* Ref)**	73,625 (−49.3%)	45,120 (−68.9%)	N/A
Electricity Consumption, kWh	N/A	N/A	145,054 (−45.9%)	**268,035 (* Ref)**

* Ref stands for referent case, and values in brackets represent relative differences with respect to referent case.

The COM also calculates the well-to-wheel CO_2 emissions taking into account the data on CO_2 emissions of power plants, as used in [28] for the cases considering coal, natural gas and renewable energy production. The HEV vs. CONV fleet CO_2 emissions reduction is around 50%, while in the cases of PHEV and BEV fleets, the reduction is from 30% to 65% and from 30% to 93%, respectively, where the lower and higher margins correspond to coal and renewable energy production scenarios.

6. Techno-Economic Analysis Module (TEAM)

6.1. General Description and TCO Model

According to Figure 2, the TEAM uses the simulation data outputted by the COM, as well as the fleet loan payment, insurance, registration, maintenance and similar costs provided by DMM, in order to calculate the fleet TCO (Figure 17). The TCO corresponds to what is in financial terminology called Net Present Value (NPV) of an investment, which is an index that valorises the investment while considering the time value of money. Rates at which the money value decreases or increases over time are in this case modelled by the inflation and discount rates, respectively (where the latter corresponds to the profit that today's money can generate in the future through investments or bank savings). Calculation of future value of money is called compounding, while the opposite approach, in which the NPV of future money is calculated, is referred as discounting. The TCO is calculated by discounting all future expenses, which the investment is expected to generate, to the present time, as shown in Figure 17.

The TCO model components (Figure 17) are divided into three groups depending on the time basis on which the input expenses data are sampled [29], and the corresponding individual costs are given in Table 6 (with no VAT included). The bus service life is considered to be 12 years, the inflation rate 3% and the discount rate 7%. The annually sampled data include registration, maintenance and insurance (RMI) cost, which have been determined for the CONV fleet based on the (past) data provided by the city bus transport operator, and discounted to prices in 2019 according to inflation data [30]. The RMI cost for the HEV, PHEV and BEV fleets are assumed to be 15%, 20% and 40% lower, respectively, when compared to the CONV fleet, because of the significantly reduced CO_2 emissions and simplified maintenance of e-buses [31–33].

The monthly expenses relate to loan payment for purchase of new vehicles and charging infrastructure, including the cost of replacing the e-bus batteries. A general-purpose bank loan is assumed, which is taken over a period of seven years, with a continuous interest rate of 5% and equal monthly annuities. The daily sampled data relate to operating cost, i.e., the fuel and electricity expenses, which are calculated by multiplying fuel and/or electricity consumptions obtained by COM simulations with fuel and/or electricity prices. As in the case of the annually sampled data, the operational cost is adjusted for inflation. Irregular maintenance cost is modelled by a fixed rate occurring every two years. The TEAM also provides the possibility of sensitivity analysis, which allows for the investigation of to what extent variations of a particular parameter affect the TCO. This helps to determine the TCO model reliability, e.g., parameters that cause higher TCO sensitivity should be more reliably estimated. The sensitivity analysis is not considered in this paper.

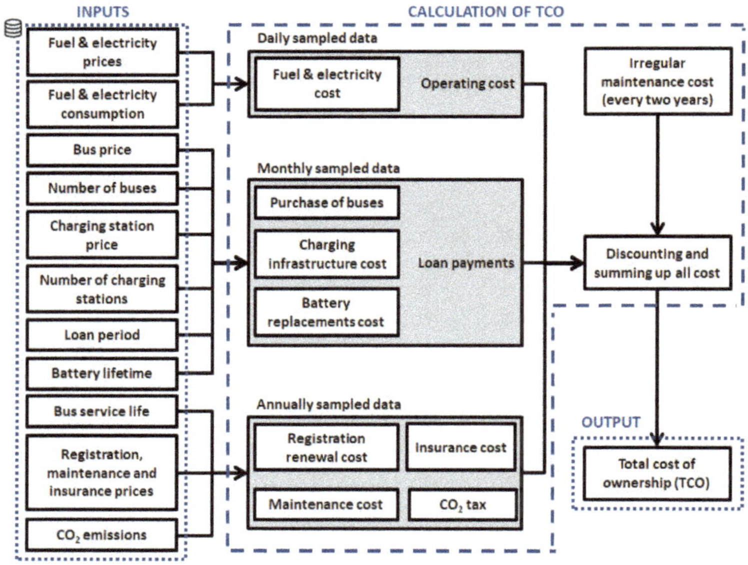

Figure 17. Flowchart of Techno-Economic Analysis Module (TEAM).

Table 6. Input parameters used for TCO calculation for different bus types (no VAT is included).

Purchase Cost [1] of Single Bus (Off-Board Charger)			
Conventional (CONV)	240,000 EUR		
Hybrid Electric (HEV)	400,000 EUR		
Plug-In Hybrid Electric (PHEV)	420,000 EUR		
Battery Electric (BEV)	495,000 EUR		
Infrastructural Cost [2]			
Fast charging station (150 kW; PHEV case)	45,000 EUR (TS) + 80,000 EUR (CS) = 125,000 EUR		
Fast charging station (300 kW; BEV case)	45,000 EUR (TS) + 120,000 EUR (CS) = 165,000 EUR		
Battery Replacement Cost [3]			
Hybrid Electric (HEV), 4.8 kWh	15,000 EUR		
Plug-In Hybrid Electric (PHEV), 19 kWh	25,000 EUR		
Battery Electric (BEV), 76 kWh	80,000 EUR		
Other Parameters			
Bus service life	12 years		
Loan period (buses + charging stations)	7 years		
Battery lifetime	6 years		
Fuel price (mean)	1.0243 EUR/L		
Electricity prices (mean)[4]	High tariff (HT): 0.1215 EUR/kWhLow tariff (LT): 0.1084 EUR/kWh		
Inflation / Discount / Loan rates	3%	7%	5%

[1] Includes incentives (1000 EUR for HEV, 2500 EUR for PHEV and 5000 EUR for BEV) estimated based on [34]. [2] Costs for transformer substation (TS) and charging station (CS) are estimated based on the data provided by local electric utility company and [35], respectively. [3] The battery replacement costs are estimated based on [35] and the replacement is assumed to occur every 6 years because the average bus battery life is 5–12 years [36]. [4] Winter time: 7 a.m. to 9 p.m. (HT), 9 p.m. to 7 a.m. (LT); Daylight saving time: 8 a.m. to 10 p.m. (HT), 10 p.m. to 8 a.m. (LT).

6.2. Simulation Results

The TCO results are given in Figure 18 for different types of city bus fleets and charging configurations selected in Section 5. Different charging scenarios are considered, starting from optimistic Scenario 1 to conservative Scenario 6. In the basic case (Scenario 1), all the e-bus fleets turn out to be competitive with the CONV fleet, which is explained by the influence of high share of fuel cost (see Figure 19) for the particular case of relatively large fleet utilisation (250 km/bus/day in average; see also Figure 5). For the same reason, the e-bus fleets have relatively comparable TCO values. Similar results are obtained if the fuel and electricity prices are randomly sampled (Scenario 2), rather than being constant as given in Table 6. Scenario 3 accounts for the need to use reserve buses in the case of BEV fleet, as found by COM simulations (Section 5). Since in the considered case, BEV 2, the use of two reserve buses results in a marginal increase of electricity consumption (Table 4) and a low number of bus swaps, the use of second reserve bus is very marginal, and is thus excluded from the TCO analysis. Due to the cost of reserve bus, the BEV fleet TCO increases above that of PHEV fleet, but it is still competitive to CONV fleet. When accounting for the e-buses' battery replacement cost (Scenario 4), the TCO of BEV fleet, which has the largest and costliest battery, becomes around 10% higher than that of CONV fleet. If the PHEV- and BEV-type bus electricity consumption is increased by the factor of 40% (Scenario 5) or 100% (Scenario 6) to account for modelling errors (e.g., those related to heating system in winter), the PHEV fleet becomes marginally competitive or uncompetitive, respectively, while the BEV vs. CONV fleet TCO excess tops 23%. This TCO excess in the ultimate BEV case may be compensated for by larger incentives, higher ticket prices (which would reflect better passenger experience), future increase in fuel prices, future decrease of battery prices and similar factors.

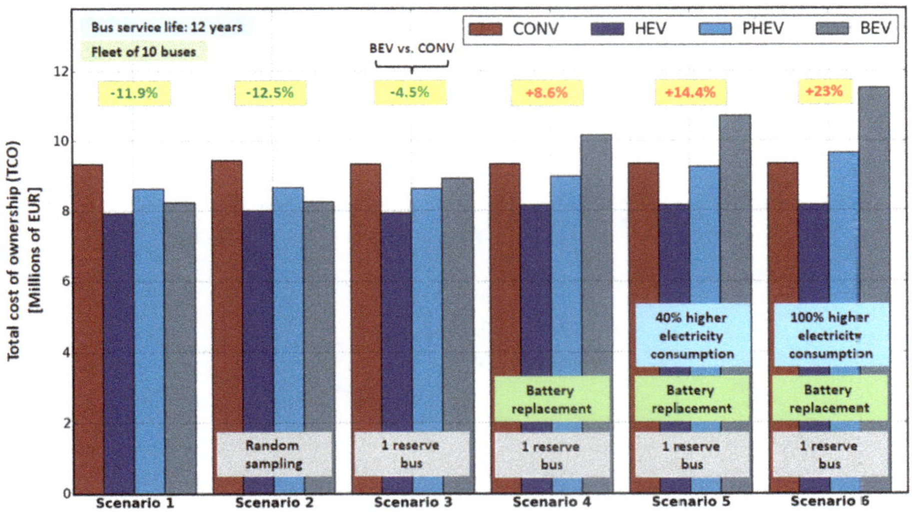

Figure 18. Comparative TCO values for different bus fleet types and electrification scenarios.

The comparative TCO time profiles for different types of bus fleet are shown in Figure 20 for Scenario 4, which is deemed to be most realistic scenario involving the battery replacement and reserve bus cost. The corresponding time profiles of individual TCO costs are shown in Figure 21. For the PHEV, and particularly the BEV fleet, the TCO rapidly rises during the first 7 years due to loan expenses related to the purchase of these expensive buses and corresponding charging infrastructure (Figure 20). Once the loan is paid off, the energy cost becomes dominant, where the efficiency of e-buses and low cost of electricity become beneficial and bring significant savings, as opposed to the CONV case, where the fuel expenses dominate (Figure 21).

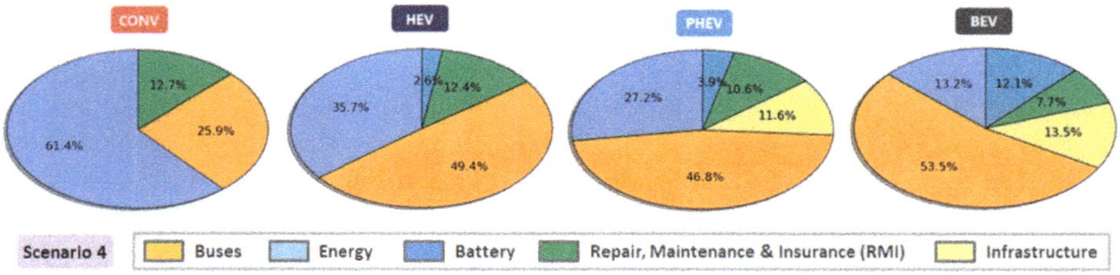

Figure 19. Cost shares for different type of bus fleets and Scenario 4 from Figure 18.

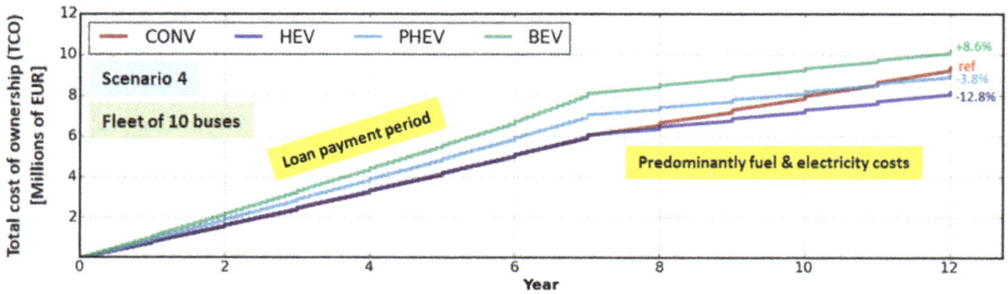

Figure 20. TCO time profile for different bus fleet types and Scenario 4 from Figure 18.

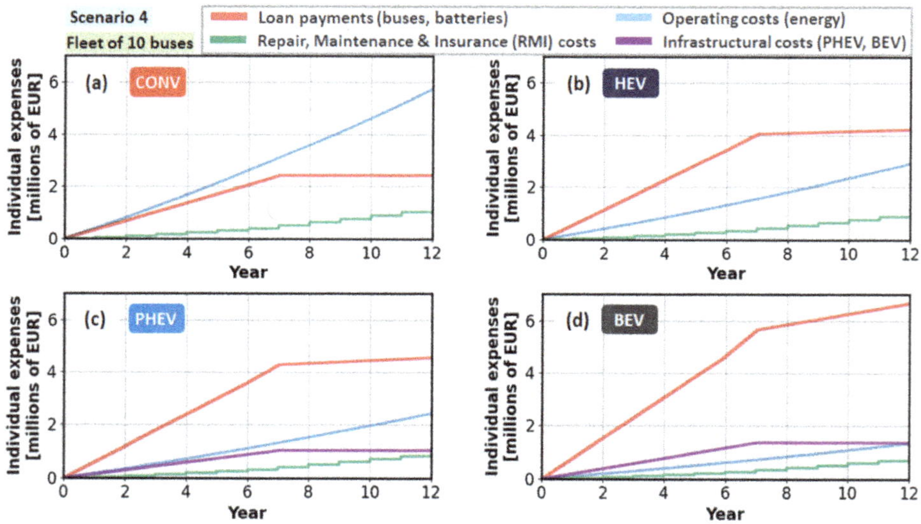

Figure 21. Time profiles of individual TCO costs for different type of bus fleets and Scenario 4 from Figure 18.

Figure 19 shows the percentage shares of individual costs for different types of bus fleets. As the electrification evolves from HEV, via PHEV to BEV buses, the energy (fuel and electricity) cost share monotonically and significantly reduces, but the bus and charging infrastructure cost share increases with similar trends. The PHEV and particularly BEV fleets have lower RMI cost, but this saving is not large enough to compensate for the battery replacement cost.

7. Conclusions

A unique city bus transport electrification tool has been presented, which relies on virtual simulation of conventional and e-bus fleets over recorded driving cycles, including charging management. The tool allows for arbitrary setting of charging station locations, number and parameters of chargers per each station, grid constraints, e-bus battery capacity and similar factors. A special emphasis was on providing a computationally efficient, multiple-maps-based backward looking e-bus model including the vehicle control strategy and heuristic charging management algorithm. In support of electrification planning and decision making, the tool finally provides the total cost of ownership (TCO) of city bus fleets with conventional (CONV) buses and different types of e-buses (HEV, PHEV and BEV). It has been demonstrated through a pilot study related to city bus transport in Dubrovnik, based on which the following main conclusions are drawn.

(1) The considered city bus transport system is such that the city buses are resting in the depot during a relatively short period over the night (typically 3 h), while they are dwelling at end stations for rather significant time (from 10 to 20 min per stay). Therefore, fast charging at end stations (and also in depot for BEV-type buses) relying on stationary chargers equipped with pantograph has been found to be a favourable solution.
(2) The use of a specific, map-based structure of the e-bus model allowed for simulating the bus fleets 20,000 times faster than real time, thus, reducing the full-year 10-bus fleet simulation to a couple of hours on a standard computer workstation.
(3) The comparative virtual simulation results have shown that the use of HEV- and PHEV-type city buses results in reduction of fuel consumption of up to 50% and 70%, respectively, when compared to CONV buses, while BEV buses do not consume fuel at all. The charging system optimisation has shown that the optimal number of end stations equipped with fast chargers is seven (out of 10), where a single reserve bus is marginally needed in the BEV case. The BEV battery capacity can be relatively small (76 kWh) due to the effective opportunity charging and relatively short routes.
(4) The TCO analysis has pointed out that the BEV fleet cannot be competitive to CONV fleet (8.6% higher TCO for BEV vs. CONV), while the HEV fleet is competitive (12.8% lower TCO vs. CONV) and the PHEV fleet is marginally competitive (3.8% lower TCO vs. CONV) in a realistic scenario involving the battery replacement and single reserve bus in the BEV case (Scenario 4). Although the HEV fleet is competitive to the CONV fleet and can reduce the fuel consumption and emissions by up to 50%, it still shares the basic disadvantages of CONV fleet (noisy, no e-drive option in low emission zones, significant emissions).

The future work could be directed to the following tool improvements: (i) optimisation of charging system configuration (e.g., by using genetic algorithm) instead of using the expert knowledge when repeating the virtual simulations for different configurations; (ii) using parallel computing to further increase the numerical efficiency of virtual simulations, particularly when optimisation loop is added; (iii) off-line route rescheduling to avoid or mitigate the use of reserve buses in the BEV case or to redirect the e-buses to designated location for daily recharging; (iv) performing city bus transport electrification analyses using limited set of standard GPS/GPRS tracking data while relying more on the bus schedules; (v) considering other economic models when conducting techno-economic analyses to determine the TCO.

Author Contributions: Conceptualisation, J.T., B.Š. and J.D.; Methodology, J.T., J.S., F.M. and J.D.; Software, J.T., J.S. and F.M.; Validation, J.T., J.S., F.M. and J.D.; Writing—Original Draft Preparation, J.T., J.S., F.M. and J.D.; Writing—Review and Editing, J.D.; Visualisation, J.T., J.S. and F.M.; Supervision, B.Š. and J.D. All authors have read and agreed to the published version of the manuscript.

Funding: It is gratefully acknowledged that this work was supported by the EU European Regional Development Fund trough Interreg CE project SOLEZ ("Smart Solutions supporting Low Emission Zones and other low-carbon mobility policies in EU cities"; https://www.interreg-central.eu/Content.Node/SOLEZ.html).

Acknowledgments: In addition, the research work of the first author has been supported by the European Regional Development Fund under the grant KK.01.1.1.01.0009 (DATACROSS). The authors' appreciation also goes to the city bus transport company Libertas Dubrovnik d.o.o. and the Dubrovnik Development Agency (DURA) for the data support and effective cooperation.

Conflicts of Interest: The authors declare no conflict of interest.

Abbreviations

AMT	Automated Manual Transmission	ECMS	Equivalent Consumption Minimisation Strategy
AT	Automatic Transmission	EV	Electric Vehicle
BEV	Battery Electric Vehicle	GPRS	General Packet Radio Service
CAN	Controller Area Network	GPS	Global Positioning System
COM	Charging Optimisation Module	M/G	Motor/Generator
CONV	Conventional (Diesel) Vehicle	PHEV	Plug-In Hybrid Electric Vehicle
CS	Charge Sustaining (mode)	RB	Rule-Based (controller)
CD	Charge Depleting (mode)	RMI	Registration, Maintenance and Insurance
DMM	Data Management Module	SoC	State of Charge
DPPM	Data Post-Processing Module	TCO	Total Cost of Ownership
EBSM	E-Bus Simulation Module	TEAM	Techno-Economic Analysis Module

References

1. Kumar, R.R.; Alok, K. Adoption of electric vehicle: A literature review and prospects for sustainability. *J. Clean. Prod.* **2020**, *253*. [CrossRef]
2. Das, H.S.; Rahman, M.M.; Li, S.; Tan, C.W. Electric vehicles standards, charging infrastructure, and impact on grid integration: A technological review. *Renew. Sustain. Energy Rev.* **2020**, *120*. [CrossRef]
3. Safoutin, M.J. Predicting the Future Manufacturing Cost of Batteries for Plug-In Vehicles for the U.S. Environmental Protection Agency (EPA) 2017–2025 Light-Duty Greenhouse Gas Standards. *World Electr. Veh. J.* **2018**, *9*, 42. [CrossRef]
4. Tiechert, O.; Chang, F.; Ongel, A.; Lienkamp, M. Joint Optimization of Vehicle Battery Pack Capacity and Charging Infrastructure for Electrified Public Bus Systems. *IEEE Trans. Transp. Electrif.* **2019**, *5*, 672–682. [CrossRef]
5. Zhang, A.; Kang, J.E.; Kwon, C. Multi-day scenario analysis for battery electric vehicle feasibility assessment and charging infrastructure planning. *Transp. Res. Part C* **2020**, *111*, 439–457. [CrossRef]
6. Davidov, S. Optimal charging infrastructure planning based on a charging convenience buffer. *Energy* **2020**, *192*. [CrossRef]
7. Gjelaj, M.; Hashemi, S.; Andersen, P.B.; Traehold, C. Optimal infrastructure planning for EV fast-charging stations based on prediction of user behaviour. *IET Electr. Syst. Transp.* **2020**, *10*, 1–12. [CrossRef]
8. Perera, P.; Hewage, K.; Sadiq, R. Electric vehicle recharging infrastructure planning and management in urban communities. *J. Clean. Prod.* **2020**, *250*. [CrossRef]
9. Cavadas, J.; Homem de Almeida Correia, G.; Gouveia, J. A MIP model for locating slow-charging stations for electric vehicles in urban areas accounting for driver tours. *Transp. Res. Part E* **2015**, *75*, 188–201. [CrossRef]
10. Xiao, D.; An, S.; Cai, H.; Wang, J.; Cai, H. An optimization model for electric vehicle charging infrastructure planning considering queuing behaviour with finite queue length. *J. Energy Storage* **2020**, *29*.
11. Baouche, F.; Billot, R.; Trigui, R.; El Faouzi. Efficient Allocation of Electric Vehicles Charging Stations: Optimization Model and Application to a Dense Urban Network. *IEEE Trans. Intell. Transp. Syst.* **2014**, *6*, 33–43. [CrossRef]
12. Nyman, J.; Olsson, O.; Grauers, A.; Östling, J.; Ohlin, G.; Pettersson, S. A user-friendly method to analyze cost effectiveness of different electric bus systems. In Proceedings of the 30th International Electric Vehicle Symposium & Exhibition, Stuttgart, Germany, 9–11 October 2017.
13. An, K. Battery electric bus infrastructure planning under demand uncertainty. *Transp. Res. Part C* **2020**, *111*, 572–587. [CrossRef]
14. Xylia, M.; Leduc, S.; Patrizio, P.; Kraxner, F.; Silveira, S. Locating charging infrastructure for electric buses in Stockholm. *Transp. Res. Part C* **2017**, *78*, 183–200. [CrossRef]

15. Hoekstra, A.; Vijayashankar, A.; Sundrani, V.L. Modelling the Total Cost of Ownership of Electric Vehicles in the Netherlands. In Proceedings of the 30th International Electric Vehicle Symposium & Exhibition, Stuttgart, Germany, 9–11 October 2017.
16. Škugor, B.; Hrgetić, M.; Deur, J. GPS Measurement-based Road Grade Reconstruction with Application to Electric Vehicle Simulation and Analysis. In Proceedings of the 10th Conference on Sustainable Development of Energy, Water and Environment Systems, Dubrovnik, Croatia, 27 September–2 October 2015.
17. Soldo, J.; Škugor, B.; Deur, J. Optimal Energy Management and Shift Scheduling Control of a Parallel Plug-in Hybrid Electric Vehicle. In Proceedings of the Powertrain Modelling and Control Conference, Loughborough, UK, 10–11 September 2018.
18. Škugor, B.; Cipek, M.; Deur, J. Control Variables Optimization and Feedback Control Strategy Design for the Blended Operating Mode of an Extended Range Electric Vehicle. *SAE Int. J. Altern. Powertrains* **2014**, *3*, 152–162. [CrossRef]
19. Volvo. Available online: https://www.volvobuses.co.uk/ (accessed on 14 May 2020).
20. Guzzella, L.; Sciaretta, A. *Vehicle Propulsion Systems*, 2nd ed.; Springer: Berlin, Germany, 2007.
21. Bin, Y.; Li, Y.; Feng, N. Nonlinear dynamic battery model with boundary and scanning hysteresis. In Proceedings of the ASME 2009 dynamic systems and control conference, Hollywood, CA, USA, 12–14 October 2009; pp. 245–251.
22. Deur, J.; Asgari, J.; Hrovat, D.; Kovač, P. Modeling and Analysis of Automatic Transmission Engagement Dynamics-Linear Case. *J. Dyn. Syst. Meas. Control.* **2006**, *128*, 263–277. [CrossRef]
23. Andersson, C. On Auxiliary Systems in Commercial Vehicles. Doctoral Dissertation, Lund University, Lund, Sweden, November 2004.
24. Škugor, B.; Deur, J.; Cipek, M.; Pavković, D. Design of a Power-split Hybrid Electric Vehicle Control System Utilizing a Rule-based Controller and an Equivalent Consumption Minimization Strategy. *Proc. Inst. Mech. Eng. Part D J. Automob. Eng.* **2014**, *228*, 631–648.
25. ZeEUS. Available online: https://zeeus.eu/publications (accessed on 14 May 2020).
26. Soldo, J.; Škugor, B.; Deur, J. Synthesis of Optimal Battery State-of-Charge Trajectory for Blended Regime of Plug-in Hybrid Electric Vehicles in the Presence of Low-Emission Zones and Varying Road Grades. *Energies* **2019**, *12*, 4296. [CrossRef]
27. Škugor, B.; Soldo, J.; Deur, J. Analysis of Optimal Battery State-of-Charge Trajectory for Blended Regime of Plug-in Hybrid Electric Vehicle. *World Electr. Veh. J.* **2019**, *10*, 75. [CrossRef]
28. Besselink, I.; Oorschot, P.F.; Meinders, E.; Nijmeijer, H. Design of an efficient, low weight battery electric vehicle based on a VW Lupo 3L. In Proceedings of the 25th World Battery, Hybrid and Fuel Cell Electric Vehicle Symposium & Exhibition, Shenzhen, China, 5–9 November 2010; pp. 32–41.
29. Al-Alawi, B.M.; Bradley, T.H. Total cost of ownership, payback, and consumer preference modeling of plug in HEVs. *Appl. Energy* **2013**, *103*, 488–506. [CrossRef]
30. Inflation Calculator. Available online: https://fxtop.com/en/inflation-calculator.php (accessed on 14 May 2020).
31. Aber, J. *Electric Bus Analysis for NYC Transit*; Columbia University: New York, NY, USA, 2016.
32. Potkany, M.; Hlatka, M.; Debnar, M.; Hanzl, J. Comparison of the Lifecycle Cost Structure of Electric and Diesel Buses. *Int. J. Marit. Sci. Technol.* **2018**, *65*, 270–275. [CrossRef]
33. Logtenberg, R.; Pawley, J.; Saxifrage, B. *Comparing Fuel and Maintenance Costs of Electric and Gas Powered Vehicles in Canada*; 2° Institute: Sechelt, BC, Canada, 2018.
34. Elin, K. Charging Infrastructure for Electric City Buses. Master's Thesis, KTH Royal Institute of Technology in Stockholm, Stockholm, Sweden, June 2016.
35. Statista. Available online: https://www.statista.com/statistics/883118/global-lithium-ion-battery-pack-costs (accessed on 14 May 2020).
36. Ranta, M.; Anttila, J.; Pihlatie, M.; Hentunen, A. Optimization of opportunity charged bus operation—A case study. In Proceedings of the 32nd International Electric Vehicle Symposium & Exhibition (EVS32), Lyon, France, 19–22 May 2019.

energies

MDPI

Article

Power Split Supercharging: A Mild Hybrid Approach to Boost Fuel Economy †

Shima Nazari [1,*], Jason Siegel [2], Robert Middleton [2] and Anna Stefanopoulou [2]

[1] Department of Mechanical Engineering, Universiy of California Davis, Davis, CA 95616, USA
[2] Department of Mechanical Engineering, University of Michigan, Ann Arbor, MI 48109, USA; siegeljb@umich.edu (J.S.); rjmidd@umich.edu (R.M.); annastef@umich.edu (A.S.)
* Correspondence: snazari@ucdavis.edu; Tel.: +1-530-752-5801

† This paper is an extended version of our paper published in 2019 WCX World Congress Experience, Detroit, MI, USA, 9–11 April 2019; No. 2019-01-1207.

Received: 17 November 2020; Accepted: 9 December 2020; Published: 14 December 2020

Abstract: This work investigates an innovative low-voltage (<60 V) hybrid device that enables engine boosting and downsizing in addition to mild hybrid functionalities such as regenerative braking, start-stop, and torque assist. A planetary gear set and a brake permit the power split supercharger (PSS) to share a 9 kW motor between supercharging the engine and direct torque supply to the crankshaft. In contrast, most e-boosting schemes use two separate motors for these two functionalities. This single motor structure restricts the PSS operation to only one of the supercharging or parallel hybrid modes; therefore, an optimized decision making strategy is necessary to select both the device mode and its power split ratio. An adaptive equivalent consumption minimization strategy (A-ECMS), which uses the battery state of charge (SoC) history to adjust the equivalence factor, is developed for energy management of the PSS. The A-ECMS effectiveness is compared against a dynamic programming (DP) solution with full drive cycle preview through hardware-in-the-loop experiments on an engine dynamometer testbed. The experiments show that the PSS with A-ECMS reduces vehicle fuel consumption by 18.4% over standard FTP75 cycle, compared to a baseline turbocharged engine, while global optimal DP solution decreases the fuel consumption by 22.8% compared to the baseline.

Keywords: energy management; hybrid electric vehicle; powertrain electrification; equivalent consumption minimization; supercharging; hardware-in-the-loop experiments

1. Introduction

Hybrid electric vehicles (HEVs) are one of the promising solutions for reducing carbon emissions in the transportation sector. During the past two decades, many different architectures for hybridized powertrains have emerged [1]. Unfortunately, despite their relative technology maturity and their proven effectiveness in reducing fuel consumption, the market penetration of HEVs is still poor [2]. The main factor for low sales rates is the higher initial cost of these vehicles compared to traditional vehicles with only internal combustion engines (ICEs). In contrast to expensive full HEVs, which use high-voltage/-power electric machines and electronics, this work investigates an economical low-voltage hybrid system, called a power split supercharger (PSS), as shown in Figure 1.

The PSS, configured with a 9 kW 48 V motor, can drive a supercharger to pressurize the intake air of the engine or it can operate as a regular parallel hybrid system and supply/draw torque directly to/from the crankshaft when the supercharger is locked and bypassed. The inadequate torque of a small naturally aspirated (NA) ICE requires conventional mild hybrid systems to employ larger NA or boosted ICEs for full performance. However, the PSS can provide sufficient boost to a small engine to provide good acceleration while taking advantage of engine downsizing and hybridization to improve efficiency.

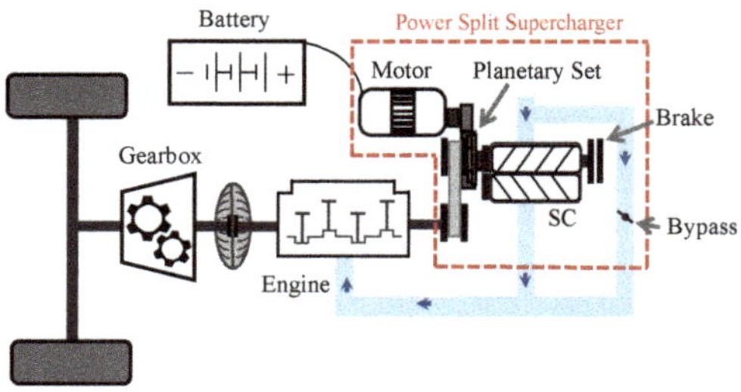

Figure 1. Powertrain schematic with a power split supercharger.

The planetary gear set and the electric motor speed control capability decouple the boost pressure generated by the PSS from the crankshaft speed, resulting in a fast torque response and improved vehicle drive-ability compared to traditional boosting devices, such as turbochargers or mechanical superchargers. Flexible supercharging can also be achieved with an electric supercharger such as the HyBoost system from Valeo [3]. However, powering the supercharger solely with electricity necessitates a larger battery and motor, leading to a higher system cost. Figure 2a shows the required supercharger mechanical power in a 1.6 L gasoline engine studied here, while Figure 2b shows the corresponding motor power in the PSS system for different engine speeds and torques. While for the range of operating points shown the supercharger power is as high as 15 kW, most of this power is supplied by the engine crankshaft. In every operating condition either a small portion of the power comes from the motor or the motor is slightly generating. This characteristic is especially useful for scenarios such as hill climbing, shown in Figure 3, where the supercharger has to provide a continuous boost pressure due to the high requested torque. For the simulated example shown in Figure 3, vehicle cruising at 110 km/h with a road grade of 5° for 20 min, a small SUV with the PSS would slightly charge a 2.5 kW.h battery, while a purely electric supercharger (eSC) would completely deplete the battery, as demonstrated in Figure 3b. The modeled vehicle and engine are explained in further detail in the following sections.

This electric power and energy accessibility problem has pushed vehicle manufacturers to use electric superchargers in combination with a turbocharger, examples of which are Volvo T6 and T8 engines [4]. In these powertrains, the turbocharger can be used during steady state, and the supercharger can make the transients faster. The PSS system, however, can be used as a stand-alone boosting device reducing the system cost in addition to enabling hybrid functionalities such as regenerative braking and start-stop. This work develops an online energy management system for an engine equipped with a PSS and experimentally verifies the fuel economy benefits of the device when it replaces a conventional turbocharger.

While in traditional vehicles the driver's entire requested torque is supplied by an ICE, HEVs need an effective energy management system to determine the power split ratio between the engine and the battery at each time instant. Energy management methods for HEVs are extensively investigated in literature. These methods are often classified as optimization-based methods and rule-based methods [5,6]. Rule-based approaches are usually a set of conditional statements based on simple principles and heuristics; hence, they are easily implementable. As an example, thermostatic control, which is developed for a series HEV [7], turns the engine on or off depending on the battery SoC. Although some rules are derived from optimization results, these methods do not fully exploit the powertrain flexibility and do not guarantee optimal performance. Furthermore, the generated rules are not reusable for a different powertrain configuration or control objective.

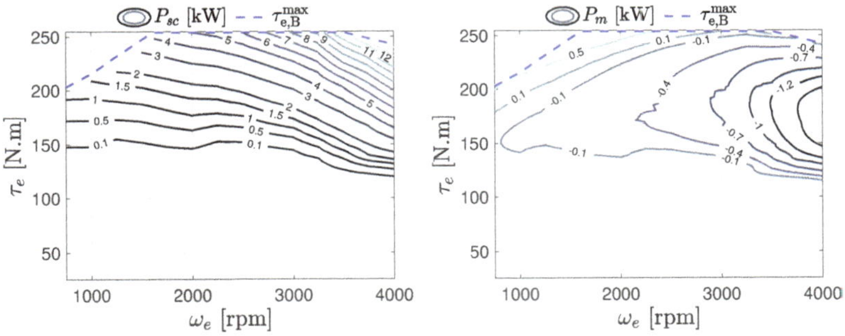

(**a**) supercharger. (**b**) motor.

Figure 2. (**a**) Supercharger mechanical power, (**b**) corresponding motor power in the power split supercharger (PSS) system, both for a 1.6 L gasoline engine.

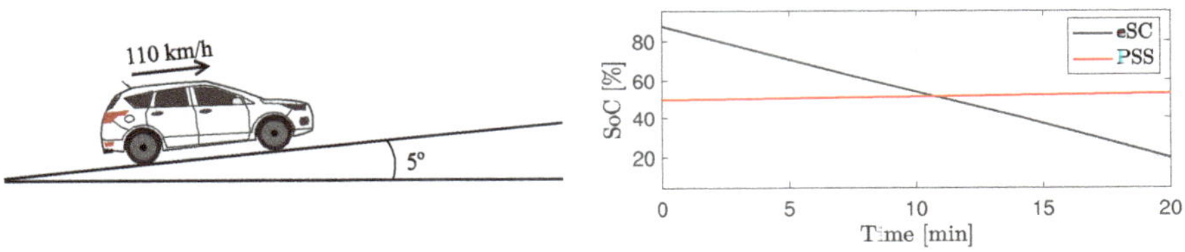

(**a**) (**b**)

Figure 3. (**a**) Vehicle climbing a 5° hill at 110 km/h , (**b**) a 2.5 kWh battery state of charge (SoC) variation for the vehicle with PSS compared to the same vehicle with an electric supercharger (eSC).

Optimization-based approaches can more effectively identify optimum solutions at the price of complexity as they minimize a cost function subject to the system physics and constraints. Various performance metrics such as fuel consumption or emissions can be included in the optimization cost function to achieve different performance goals. The optimization horizon can vary from a single time step, as in equivalent consumption minimization strategy (ECMS) [8], to multiple time steps, such as with model predictive control (MPC) [9], or over the entire drive cycle, as in dynamic programming (DP) [10,11]. Note that only the methods that minimize fuel consumption over the full drive cycle give the global optimum solution; however, these methods are prohibitively computationally expensive while also requiring knowledge of future driver demands. Nevertheless, they provide a criterion for evaluating other energy management algorithms in addition to giving insight into optimal policies.

The charge-sustaining global optimal energy management strategy for a vehicle with the PSS was formulated and solved using DP in a prior work [12], and a simple online energy management system based on ECMS was also previously presented and tested in simulation [13]. This work extends our previous efforts by developing an Adaptive-ECMS (A-ECMS) and documenting the fuel economy benefits of the PSS through advanced hardware-in-the-loop (HIL) experiments. The main contributions of this work are as follows: first, an Adaptive-ECMS energy management system is introduced to select the PSS mode and its power split ratio. Second, the implementation of the hardware-in-the-loop experiments is described in detail, and some practical challenges are explained. Third, the operation of the PSS is demonstrated experimentally, and finally, the effectiveness of the PSS hardware and the developed controllers in fuel consumption reduction of a vehicle is quantified over the standard FTP75 cycle.

After introducing the utilized hardware and models, the global fuel consumption minimization problem is described briefly. An ECMS is formulated for selecting the PSS mode and its power split ratio in Section 3.2 and the Adaptive-ECMS is described in Section 3.3. Section 4 presents the engine dynamometer experimental testbed and the details of HIL implementation. The experimental demonstration of fuel economy results and PSS operation are shown in Section 5, and the paper concludes with the main findings of the work.

2. Experimental Hardware and Model Framework

The baseline engine is a 1.6 L Ford EcoBoost engine, which is a 4 cylinder spark ignition (SI) turbocharged engine. The turbocharger is replaced by the PSS in the alternative powertrain studied in this work. Figure 1 shows a schematic view of the engine with the PSS and other powertrain components. The PSS is configured with a planetary gear set, a roots supercharger, a motor, a bypass valve, and a brake. The sun gear is attached to the supercharger, the ring is connected to the motor, and the carrier is coupled with the engine crankshaft through a set of belt and pulleys. The PSS can enable two distinct operating modes by controlling the motor, bypass, and the brake. In boosting mode, the bypass is closed, the brake is released, and the motor can control the supercharger speed and resulting boost pressure independently of the crankshaft speed. In torque assist mode, the brake locks the sun gear and the supercharger is bypassed. In this mode the planetary gear set acts as a regular gear set, which enables the motor to supply/draw torque to the crankshaft for start-stop, regenerative braking, or assisting the crankshaft. The engine fuel consumption map, shown in Figure 4, is produced using a high-fidelity GT-Power model, which is described in detail and validated against engine dynamometer experiments elsewhere [14]. In Figure 4, $\tau_{e,\mathrm{NA}}^{\max}$ is the maximum torque that the NA engine can produce, $\tau_{e,B}^{\max}$ is the maximum engine torque when the PSS is in boosting mode, and $(\tau_{e,\mathrm{NA}} + \tau_{\mathrm{TA}})^{\max}$ is the powertrain maximum torque in torque assist mode (maximum motor torque added to the crankshaft).

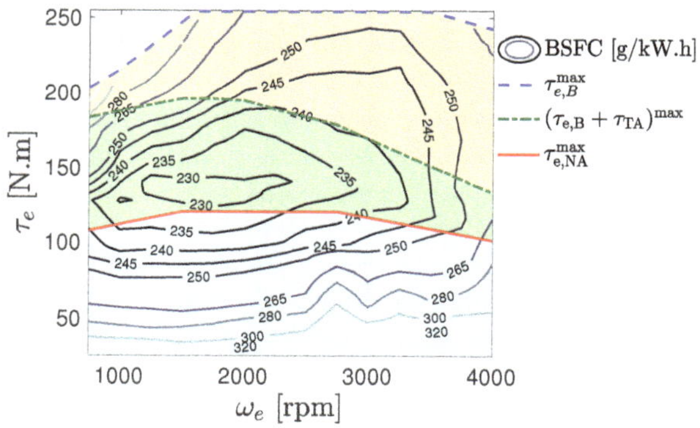

Figure 4. Brake specific fuel consumption (BSFC) map for the engine with the power split supercharger (PSS). The naturally aspirated engine maximum torque, $\tau_{e,\mathrm{NA}}^{\max}$, the powertrain maximum torque in torque assist mode, $(\tau_{e,\mathrm{NA}} + \tau_{\mathrm{TA}})^{\max}$, and the maximum engine torque during boosting mode, $\tau_{e,\mathrm{B}}^{\max}$, are also represented.

The modeled vehicle is a MY2015 Ford Escape crossover SUV. The drivetrain model includes the crankshaft dynamics, a friction clutch, a torque converter, and a 6-speed automatic transmission. The model details and control strategy are described in a prior work [14]. The driver model is a gain-scheduled proportional + integral (PI) controller and uses a 1 s preview of the tracking error and vehicle acceleration. A 1.2 kWh lithium-ion battery is assumed for the rest of this study. An open circuit voltage with a resistance (OCV-R) is used to model the battery and compute its state of charge dynamics, detailed in [12].

3. Energy Management System

3.1. Global Fuel Consumption Minimization

The optimal energy management problem for a vehicle with a PSS and with the full driving profile preview was formulated and solved using DP elsewhere [12,15]. In this work the DP solution was used as a benchmark to evaluate the effectiveness of the online energy management algorithm; thus, only a summary of the DP formulation is presented. The cost function for the global fuel consumption minimization problem is given in Equation (1). Different terms from left penalize the fuel flow rate, the gear shifts, the engine cranking (for start-stop), and the PSS mode, respectively,

$$\min\left\{\sum_{k=1}^{N}\left(\dot{m}_f(k)T_s+\alpha|n_g(k)-n_g(k-1)|+\beta(\max(x_e(k)-x_e(k-1),0))+\lambda(1-u_{\text{br}}(k))\right)\right\} \quad (1)$$

where k refers to the kth step time, N is the problem horizon, which is the full drive cycle, T_s is the sampling time equal to 1 s, $\dot{m}_f$ is the fuel flow rate, n_g represents the gear number, x_e stands for the engine on/off state, and u_{br} is the PSS brake position used to indicate the PSS mode, where $u_{\text{br}}=0$ is the boosting mode and $u_{\text{br}}=1$ the torque assist mode. The coefficient α controls the gear shift frequency, β is the engine cranking fuel penalty, and λ has a very small value to enforce the brake locked as the default mode. The full detail of the problem constraints are presented in the original work [15]. The battery state of charge, the engine on/off state, and the gear number are the modeled states. The latter two had to be modeled as states to be penalized in the objective function. The control inputs for this problem are the PSS mode, the commanded torque assist from the electric motor, the engine on/off command, and the gear shift command. A MATLAB-based dynamic programming function [16] was used to solve this problem.

In the prior work the manufacturer map for the electric motor was used to estimate the fuel economy, and no loss was assumed for the planetary gear set and pulleys. However, the experiments showed that both the motor efficiency and its torque limits are different from the manufacturer map. Hence, new experimentally validated maps were produced to update the results in this work. Figure 5 shows the measured efficiency from/to the electric power, measured by an AVL battery emulator, to/from the engine-dynamometer crankshaft, measured using a torque meter. The maximum and minimum motor torque limits are also presented. Compared to the prior maps, the losses were up to 15% more, especially at low speed and negative torques. The minimum motor torque was also slightly higher at lower engine speeds. Both of these reduced the recuperated power from regenerative braking during a cycle. The DP results presented later in Table 1 are updated with the new map.

Table 1. Drive cycle fuel consumption results for Ford Escape MY2015.

Powertrain	Result Type	Energy Management	FC	ΔFC	ΔSoC
			[L/100 km]	[%]	[%]
Turbocharged	Simulation	DP	6.76	-	
Engine + PSS	Simulation	DP	5.22	22.8	0.0
Turbocharged	Experiment	-	7.29	-	-
Engine + PSS	Experiment	A-ECMS	5.95	18.4	1.1

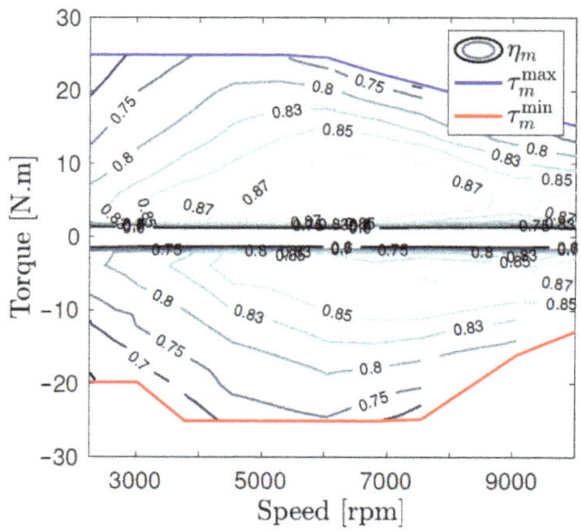

Figure 5. Experimentally generated motor and gear set map.

3.2. Equivalent Consumption Minimization Strategy

The equivalent consumption minimization strategy assigns an equivalent fuel flow rate to the electric energy consumption by using an equivalence factor and minimizes the sum of the engine and motor fuel flow rate at only the current time step [8,17]. As this strategy does not need any preview information it can be implemented online. The motor torque, τ_m, is calculated as

$$\tau_m = \underset{\tau_m}{\text{argmin}}(\dot{m}_f(\tau_e, \omega_e) + \alpha_{eq} P_m(\tau_m, \omega_m)) \tag{2}$$

in which τ_e and ω_e represent the engine torque and speed respectively, ω_m is the motor speed, and α_{eq} is the equivalence factor. The energy management system (EMS) of a vehicle with a PSS has to select the PSS mode first. Only if the torque assist mode is selected will the optimum motor torque need to be determined in the next step to minimize the powertrain fuel consumption. When the boosting mode is selected the motor torque is not an optimization parameter but is instead used to control the boost pressure and, hence, the engine torque.

Boosting is only justified when the requested crankshaft torque, τ_{crk}^d, is larger than the torque limit that the naturally aspirated engine can produce, $\tau_{e,NA}^{max}$ (above the blue area in Figure 4), because simultaneous boosting and throttling is not an efficient policy [18]. Therefore, it is fuel efficient to lock and bypass the supercharger when $\tau_{crk}^d < \tau_{e,NA}^{max}$. On the other hand, due to the small motor size, the NA engine with direct torque assist from the motor cannot produce a torque higher than $(\tau_{e,NA} + \tau_{TA})^{max}$, shown in Figure 4; thus, when a high torque in the yellow area of Figure 4 is requested, the PSS has to work in boosting mode. Finally, when the requested torque is smaller than the maximum powertrain torque in torque assist mode and larger than the NA engine torque limit (green area in Figure 4), the requested torque can be achieved through either mode. A consumption minimization rule is introduced to determine the PSS mode that produces the minimum equivalent fuel consumption as follows:

$$u_{br} = \begin{cases} 0 & \text{if} \quad \tau_{crk}^d > (\tau_{TA} + \tau_{e,NA})^{max} \\ 1 & \text{if} \quad \tau_{crk}^d \leq \tau_{e,NA}^{max} \\ \underset{u_{br}=0,1}{\text{argmin}}(\dot{m}_{f,eq}) & \text{otherwise} \end{cases} \tag{3}$$

in which u_{br} is the PSS brake position used to represent the PSS mode, and $\dot{m}_{f,eq}$ is the equivalent fuel flow rate of the engine and motor, computed for each mode as

- Torque assist mode ($u_{br} = 1$):

$$\dot{m}_{f,\text{eq}} = \min_{\tau_\text{m}^d} \left(\dot{m}_f(\tau_e^d, \omega_e) + \alpha_\text{eq} P_m(\tau_\text{m}^d, \omega_m) \right) \quad (4)$$

$$\tau_e^d = \tau_\text{crk}^d - \tau_\text{TA}^d \quad (5)$$

$$\tau_m^d = \frac{g_R}{n_\text{im} n_\text{ri}(g_S + g_R)} \tau_\text{TA}^d \quad (6)$$

where the superscript d refers to the desired or commanded values, g_R is the ring gear teeth number, g_S is the sun gear teeth number, n_{ri} is the ring-to-idler gear ratio, and n_im is the idler-to-motor gear ratio. The variable τ_TA represents the torque assist from the motor on the crankshaft and is related to the motor torque through (6).

- Boosting mode ($u_\text{br} = 0$):

$$\dot{m}_{f,\text{eq}} = \dot{m}_f(\tau_e^d, \omega_e) + \alpha_\text{eq} P_m(\tau_e^d, \omega_e) \quad (7)$$

$$\tau_e^d = \tau_\text{crk}^d \quad (8)$$

Equation (8) indicates that during boosting mode the entire crankshaft requested torque has to be supplied by the engine. In this mode the supercharger pressure ratio is decoupled from the engine operating speed. Keeping the throttle valve open in the boosted condition reduces the engine losses and increases the efficiency. Adopting this strategy, the steady-state motor power for driving the supercharger can be mapped into engine operating points shown in Figure 2b. This map is used to calculate the equivalent fuel flow rate of the motor in boosting mode.

The solution to Equations (2) and (3) is computed for various values of the equivalence factor α_eq. Figure 6a–c shows the solution to Equation (3) for equivalence factors of 0.13, 0.18, and 0.23 kg/kWh, respectively. The green color in these plots indicates boosting mode, while the red color shows torque assist mode. Figure 7a–c presents the optimum motor torque during torque assist mode generated from (2) for the same equivalence factors.

The equivalence factor represents the relative value of the electric power. A smaller equivalence factor uses the torque assist mode more often and uses the motor to assist the crankshaft over a larger operating region (more red color in Figures 6a and 7a). A higher equivalence factor increases the penalty for electric power, which causes the ECMS controller to use the boosting mode more often in Figure 6b,c while also using the motor to generate more energy, often at low loads (more green color in Figure 7b,c).

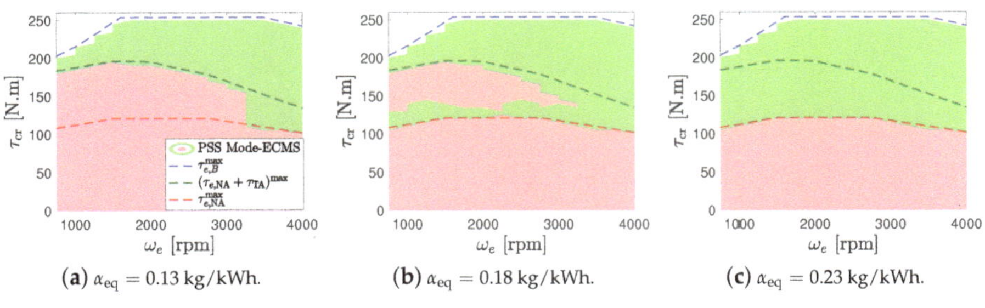

Figure 6. Equivalent consumption minimization strategy (ECMS)-generated PSS mode for different equivalence factors. The green color represents boosting mode, and the red color indicates torque assist mode. (**a**) $\alpha_\text{eq} = 0.13$ kg/kWh, (**b**) $\alpha_\text{eq} = 0.18$ kg/kWh, and (**c**) $\alpha_\text{eq} = 0.23$ kg/kWh. Increasing the equivalence factor shifts the optimal strategy from torque assist to favor boosting mode.

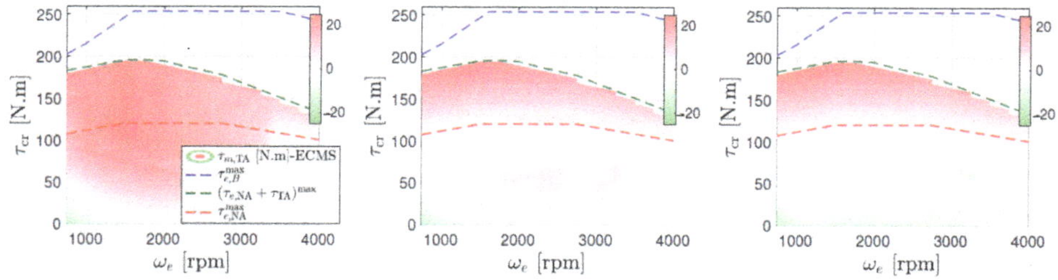

(a) $\alpha_{\text{eq}} = 0.13$ kg/kWh. (b) $\alpha_{\text{eq}} = 0.18$ kg/kWh. (c) $\alpha_{\text{eq}} = 0.23$ kg/kWh.

Figure 7. ECMS generated motor torque during torque assist mode for different values of equivalence factor (**a**) $\alpha_{\text{eq}} = 0.13$ kg/kWh, (**b**) $\alpha_{\text{eq}} = 0.18$ kg/kWh, and (**c**) $\alpha_{\text{eq}} = 0.23$ kg/kWh.

3.3. Adaptive-ECMS

The traditional ECMS method requires tuning the equivalence factor offline for every driving profile to ensure that the battery state of charge remains within the desired/operational limit. However, in real-world applications the future velocity profile is not known. Therefore, it is necessary to tune the ECMS factor in real time to ensure acceptable operation of the energy management system, especially when starting from an unfavorable initial condition. A-ECMS adjusts the equivalence factor based on drive cycle prediction, driving pattern recognition, or feedback from the battery state of charge [19]. In this work a modification of the approach that uses the SoC feedback [20] is adopted. The equivalence factor is adjusted as

$$\alpha_{\text{eq}}(k) = \begin{cases} 0.25 & \text{if} \quad \text{SoC} < 40\% \\ 0.10 & \text{if} \quad \text{SoC} > 60\% \\ \max\Big(\min\Big(\alpha_{\text{eq}}(k-1) + q\big(SoC(k) - SoC(k-1)\big), 0.25\Big), 0.10\Big) & \text{otherwise} \end{cases} \tag{9}$$

where q is a constant coefficient. The suitable sampling time for Equation (9) depends on the relative size of the energy storage and energy consumption. Simulations showed that a sampling time of less than 1 min can keep the SoC in 40–60% range for the tested drive cycle. However, a sampling time of 15 s was selected for the experiments for a tighter control over SoC, without adding a significant computational effort. Further investigations into sampling time dependence on drive cycle and vehicle parameters are left for future development. Figure 8 shows the A-ECMS implementation for hardware-in-the-loop experiments presented in the next section. P_b represents the battery power.

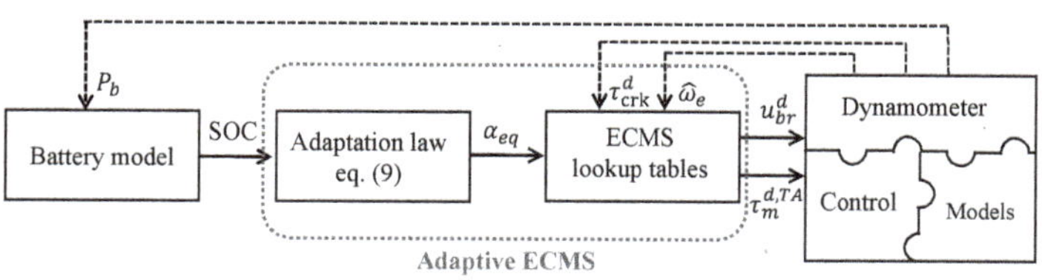

Figure 8. A-ECMS implementation for hardware-in-the-loop experiments.

4. Experimental Setup

4.1. Testbed

Figure 9 shows a picture of the engine-dynamometer experimental testbed, which includes an AVL AC transient dynamometer, an instrumented 1.6 L Ford EcoBoost engine, and the power split supercharger by EATON. An AVL SESAM-FTIR emission measurement bench is used to measure exhaust gas species and calculate fuel flow rate along with a hot wire air flow rate meter. An AVL battery simulator is used to power the PSS and measure the motor current and voltage. A rapid prototyping electronic control system (RPECS^{TM}) from Southwest Research Institute (SwRI) was used to integrate all engine controllers, the online energy management system, and implement real-time vehicle and driver models.

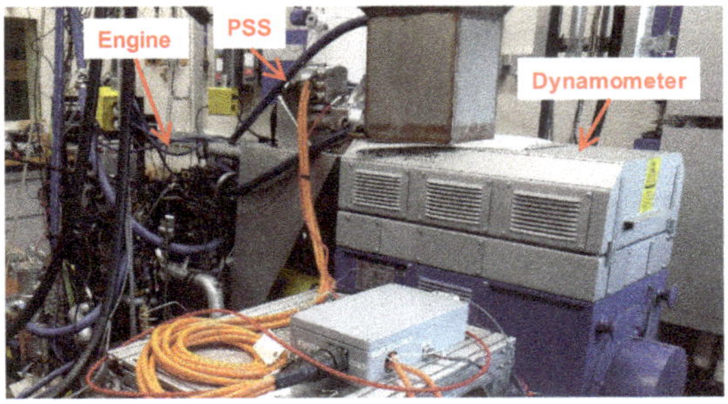

Figure 9. Engine-dynamometer experimental testbed.

4.2. *Implementing A-ECMS and Hardware-in-the-Loop Experiments*

In the hardware-in-the-loop experiments the engine and the PSS are the physical hardware, and the vehicle, the driver, and the battery are models coded in RPECS^{TM}. The AC dynamometer is programmed to play the role of the vehicle body and follow a drive cycle speed profile. In these experiments the driver model issues an accelerator or brake pedal based on the vehicle velocity tracking error. From this command the energy management system and the low-level controllers compute the actuator positions for the engine, dynamometer, and the PSS, including the throttle position (u_{θ}^{d}), motor torque (τ_{m}^{d}), the PSS brake command (u_{br}^{d}), the supercharger bypass command (u_{bp}^{d}), and the engine speed (ω_{e}^{d}). The produced crankshaft torque ($\hat{\tau}_{\text{crk}}$) is measured and fed back into the vehicle longitudinal dynamics to calculate the next vehicle speed. This feedback system, presented also in Figure 10, permits velocity tracking and an accurate drive cycle fuel economy measurement. Note that the AC dynamometer can track either a desired speed or a desired torque profile. The torque tracking mode is not always safe because, in this case, the crankshaft speed is determined by the torque balance between the engine and the dynamometer. Operating in this mode in case of a subsystem failure, communication delay, or software bug can result in over speeding the engine, damage, or complete destruction. Therefore, in this work the dynamometer was always used in the speed control mode, and as shown later in the experimental results, the dynamometer controller did an impeccable job in tracking the engine speed set point. The following sections present HIL implementation for the two propulsion modes, locked and unlocked torque converter. All models and the energy management system are implemented in a 5 ms loop in RPECS^{TM}.

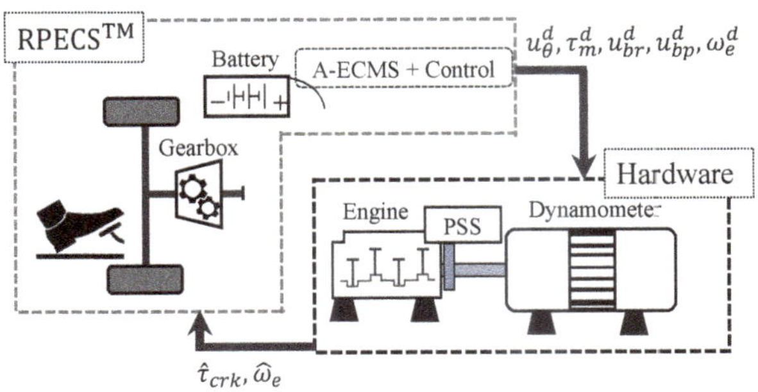

Figure 10. Hardware-in-the-loop (HIL) implementation.

4.2.1. Hardware-in-the-Loop Implementation for Locked Torque Converter

When the torque converter (TC) is locked, the vehicle speed is computed from the longitudinal vehicle dynamics and the measured crankshaft torque as follows:

$$M\frac{d}{dt}v = F_t - F_b - F_l \tag{10}$$

$$F_t = \frac{1}{R_w}(i_g\hat{\tau}_{\text{crk}} - \tau_{\text{loss}}) \tag{11}$$

$$\omega_e^d = \frac{v}{i_g R_w} \tag{12}$$

where M is the vehicle mass, v is the vehicle velocity, F_t is the traction force, F_b is the braking force, F_l is the road load, i_g is the gear ratio, τ_{loss} is torque loss in the transmission, R_w is the wheel radius, and ω_e^d is the engine speed computed from the vehicle speed and fed back into the dynamometer. The details of the vehicle and transmission models are presented elsewhere [14].

The requested tractive torque, τ_{trc}^d, when the accelerator pedal is active is linearly mapped to the pedal position, u_{acc}:

$$\tau_{\text{trc}}^d = u_{\text{acc}}(\tau_{e,\text{B}}^{\max} - \tau_e^{\min}) + \tau_e^{\min} \tag{13}$$

where $\tau_e^{\min}$ is the minimum engine torque, and $\tau_{e,\text{B}}^{\max}$ is from Figure 4. The requested braking torque on the gearbox inlet shaft, τ_{brk}^d, is computed from the brake pedal position, u_{brk} produced by the driver model:

$$\tau_{\text{brk}}^d = \frac{u_{\text{brk}}\tau_{\text{brk}}^{\max}}{i_g} \tag{14}$$

where $\tau_{\text{brk}}^{\max}$ is the maximum braking torque on the wheels.

The PSS optimal mode, u_{br}^d, during traction (which is $\tau_{\text{crk}}^d > 0$) and the optimum motor torque in torque assist mode, $\tau_m^{d,\text{TA}}$, are computed offline and stored in look up tables based on the requested crank torque, the measured engine speed, and the equivalence factor:

$$u_{\text{br}}^d = \Gamma(\tau_{\text{crk}}^d, \hat{\omega}_e, \alpha_{\text{eq}}) \tag{15}$$

$$\tau_m^{d,\text{TA}} = \Lambda(\tau_{\text{crk}}^d, \hat{\omega}_e, \alpha_{\text{eq}}) \tag{16}$$

The desired engine torque is

$$\tau_e^d = \begin{cases} \tau_{\text{crk}}^d & \text{if} \quad u_{\text{br}}^d = 0 \\ \tau_{\text{crk}}^d - \tau_m^{d,\text{TA}}\frac{(g_R+g_S)n_{\text{im}}n_{\text{ri}}}{g_R} & \text{if} \quad u_{\text{br}}^d = 1. \end{cases} \tag{17}$$

Finally, the desired intake manifold pressure is computed from the engine speed and the desired engine torque:

$$p_{\text{im}}^d = \Xi(\tau_e^d, \omega_e). \tag{18}$$

When the desired manifold pressure is less than the ambient pressure the supercharger is bypassed ($u_{\text{bp}}^d = 1$) and the intake throttle is used to control the intake manifold pressure, while when the desired intake manifold pressure is higher than the ambient pressure the throttle is wide open and the supercharger speed is controlled by the motor to achieve the desired intake manifold pressure. Both the throttle controller and the supercharger speed controller are feedforward PI controllers. The supercharger speed controller has an inner PI controller to manipulate the motor torque to achieve the desired supercharger speed. The details of the low-level controllers are presented elsewhere [14].

The motor torque during regenerative braking, $\tau_m^{d,\mathrm{Reg}}$, is computed by

$$\tau_m^{d,\mathrm{Reg}} = \max(\tau_{crk}^d \frac{g_R}{(g_R + g_S) n_{\mathrm{im}} n_{\mathrm{ri}}}, \tau_m^{\min}) \tag{19}$$

where $\tau_m^{\min}$ is the minimum motor torque shown in Figure 5. The PSS mode is set to torque assist mode during braking. Note that Equation (19) is the solution to Optimization (2) since simultaneous generation from the engine and regenerative braking is prohibited in here ($\tau_e = \tau_e^{\min}$ during braking). Finally, the commanded motor torque, τ_m^d, comes from either the PI controller ($\tau_m^{d,\mathrm{PI}}$), regenerative braking, or the torque assist (from A-ECMS) depending the PSS mode and the requested tractive torque sign:

$$\tau_m^d = \begin{cases} \tau_m^{d,\mathrm{PI}} & \text{if} \quad u_{\mathrm{br}}^d = 0 \\ \tau_m^{d,\mathrm{TA}} & \text{if} \quad u_{\mathrm{br}}^d = 1, \tau_{\mathrm{trc}}^d \geq 0 \\ \tau_m^{d,\mathrm{Gen}} & \text{if} \quad u_{\mathrm{br}}^d = 1, \tau_{\mathrm{trc}}^d < 0 \end{cases} \tag{20}$$

4.2.2. Hardware-in-the-Loop Implementation for Unlocked Torque Converter

When the torque converter is unlocked, there is no mechanical coupling between the engine and the wheels. However, in order to use the dynamometer in the speed control mode with an unlocked torque converter, it is assumed that the engine speed is equal to its idling speed when the TC unlocks ($\omega_e^d = \omega_{e,\mathrm{idle}}$). The minimum engine torque to hold the idle speed is calculated from the torque converter K-factor (K) and torque ratio (TR), which are functions of turbine to pump speed ratio (SR):

$$\mathrm{SR} = \frac{\omega_{\mathrm{tct}}}{\omega_{\mathrm{tcp}}} \tag{21}$$

$$\tau_{\mathrm{tcp}} = (\frac{\omega_{\mathrm{tcp}}}{K})^2 \tag{22}$$

$$\tau_{\mathrm{tct}} = \tau_{\mathrm{tcp}} \times \mathrm{TR} \tag{23}$$

where ω_{tct} is the torque converter turbine speed, τ_{tct} is the turbine torque, τ_{tcp} is the pump torque, and ω_{tcp} is the pump speed. Given that ω_{tcp} is equal to the engine idling speed when the torque converter unlocks, the minimum torque on the crankshaft when the turbine speed drops to less than engine idling speed can be computed as a function of turbine speed, $\tau_{\mathrm{tcp}}^*(\omega_{\mathrm{tct}})$. Accordingly, the minimum torque on the crankshaft, $\tau_{\mathrm{crk}}^{\min}$, is computed by

$$\tau_{\mathrm{crk}}^{\min} = \begin{cases} \tau_{\mathrm{tcp}}^*(\omega_{\mathrm{tct}}) & \text{if} \quad \omega_{\mathrm{tct}} \leq \omega_{e,\mathrm{idle}} \\ -\infty & \quad \text{otherwise.} \end{cases} \tag{24}$$

The requested torque on the crankshaft is

$$\tau_{\mathrm{crk}}^d = \max(\tau_{\mathrm{trc}}^d, \tau_{\mathrm{brk}}^d, \tau_{\mathrm{crk}}^{\min}). \tag{25}$$

Equation (25) imposes some positive torque demand on the crankshaft at low vehicle speed to maintain the engine idling speed, and it disables regenerative braking under these conditions. Similar to the locked torque converter case, Equations (15)–(18) and (20) are used to determine u_θ^d, τ_m^d, u_{bp}^d, and u_{br}^d in this mode. The engine speed is set equal to the idling speed, $\omega_e^d = \omega_{e,\mathrm{idle}}$, and (11) has to be corrected to include the torque converter torque ratio,

$$F_t = \frac{1}{R_w}(i_g \hat{\tau}_{\mathrm{crk}} \times \mathrm{TR} - \tau_{\mathrm{loss}}). \tag{26}$$

4.2.3. Engine Start-Stop

In a vehicle the engine is connected to the transmission, and the transmission clutch is open during engine starts. On the engine dynamometer the engine is permanently connected to the dynamometer with a large inertia. Therefore, it is not possible to mimic the engine start-stops with a dynamometer. To emulate the start-stop behavior, the stopped portions of the drive cycle, where the engine is turned off, are removed from the velocity profile for the vehicle with the PSS, and a fuel penalty is added for each start-stop event. The next section shows the resulted velocity profiles in addition to other experimental results.

5. Experimental Results

Figure 11 shows the velocity tracking for FTP75 drive cycle from the HIL experiments for the baseline turbocharged engine and the engine when the PSS replaces the turbocharger. In addition to the vehicle speed and reference speed, v^{ref}, the standard minimum velocity threshold , v^{min}, is also plotted, showing that both engines successfully follow the cycle profile. The following sections document the fuel consumption and PSS operation details during the HIL experiments.

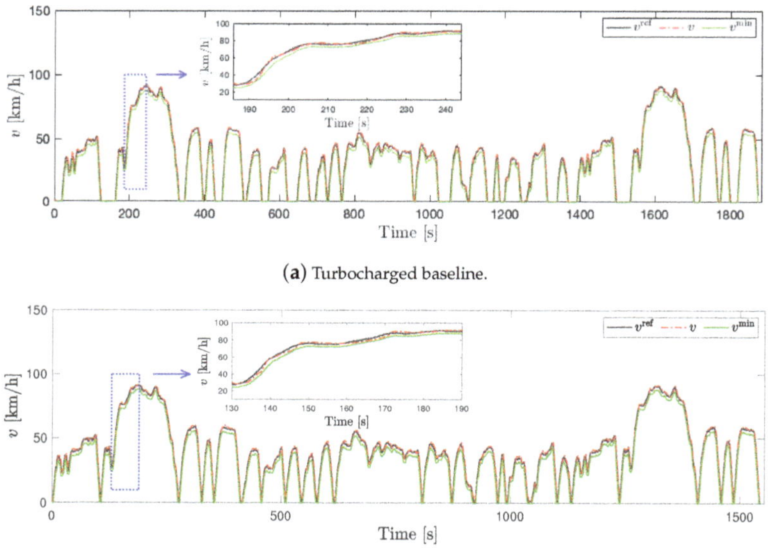

(**a**) Turbocharged baseline.

(**b**) Engine with PSS.

Figure 11. Velocity tracking during hardware-in-the-loop experiments, (**a**) vehicle with turbocharged engine, (**b**) vehicle with the PSS, stopped portions of the cycle removed to emulate start-stop.

5.1. *Fuel Consumption Reduction with PSS*

Table 1 summarizes the experimental fuel consumption (FC) results along with the predicted global minimum fuel consumption, produced with DP and a simplified vehicle model. The gearshifts of the baseline turbocharged engine are also optimized by DP in results shown in the first row of Table 1. The same gearshift strategy is used for the turbocharged engine and the engine with PSS during experiments. DP predicts that the engine with the PSS consumes 22.8% less fuel compared to the turbocharged engine. The HIL experiments were repeated three times for the PSS and two times for the baseline turbocharged engine, and the mean FC values are reported in the table. The FC values varied from 5.92 to 5.99 l/100 km for the engine with PSS and from 7.20 to 7.37 l/100 km for the turbocharged engine. The HIL experiments show that the engine with the PSS consumed 18.4% less fuel compared to the baseline turbocharged engine on average, which is only 4.4% higher than the global minimum FC from DP. This result substantiates the effectiveness of the implemented energy management system considering that A-ECMS does not use any preview information and only minimizes its cost function at a current time step. There is some offset between the absolute values of FC in simulations versus experiments. The reason is that the simulations use a fuel consumption map produced by GT-Power simulations, which is shown to accurately predict the fuel consumption variation with load and speed and between different engine configurations, but has a constant offset compared to the experimentally measured fuel consumption [14].

Figure 12a shows the battery state of charge (SoC) variation during the HIL experiment. Starting from 50% SoC, the battery SoC maintained between 44% to 51% during the experiment, showing the possibility of further battery size and system cost reduction. Figure 12b shows the equivalence factor. The adaptation rule (9) can keep the SoC between 40% and 60%, but still the initial value of $\alpha_{eq}(0)$ was tuned to get a final SoC value close to 50%. Finally, the fuel mass was corrected as follows to account for the small ΔSoC between the start and end of the cycle,

$$W_f^{\text{cor}} = W_f + UC_n\bar{\alpha}_{\text{eq}}\Delta SoC \tag{27}$$

where W_f is the fuel mass, W_f^{cor} is the corrected fuel mass, U is the battery open circuit voltage, C_n is the battery capacity, and $\bar{\alpha}_{\text{eff}}$ is the average equivalence factor during the experiment from Figure 12b.

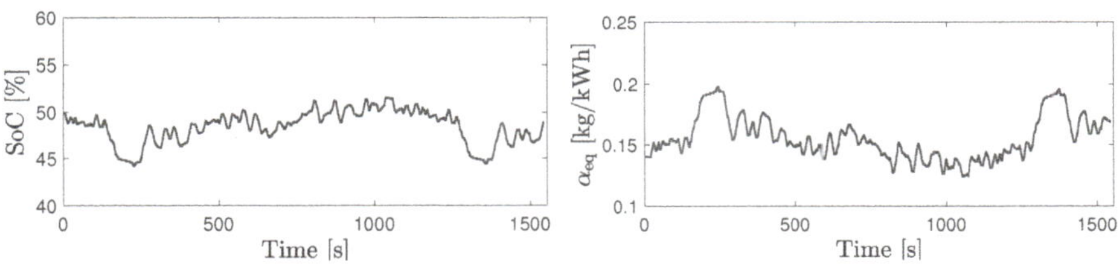

(**a**) Battery state of charge. (**b**) Equivalence factor.

Figure 12. The battery state of charge and A-ECMS equivalence factor variation during experiments, (**a**) state of charge, (**b**) equivalence factor.

5.2. Power Split Supercharger Operation during a Transient Drive Cycle

This section presents the details of the hardware and energy management system operation during the HIL experiments. Figure 13 shows the intake manifold pressure, p_{im}. Figure 13a shows p_{im} over the entire cycle along with the PSS brake position. The intake manifold pressure increased to more than the ambient pressure (around 100 kpa) only during few instances, which correspond to vehicle accelerations where the PSS switches to boosting mode ($u_{\text{br}} = 0$). Figure 13b shows p_{im} variation during a portion of the cycle in more detail (marked with blue square in Figures 11b and 13a) on top of the desired signal value, $p_{\text{im}}^{\text{des}}$, and the supercharger bypass valve position. As mentioned before, when $p_{\text{im}}^{\text{des}}$ is less than the ambient pressure, the supercharger is bypassed and the throttle controls p_{im}. When $p_{\text{im}}^{\text{des}}$ increases to more than the ambient pressure, the throttle opens wide, the bypass valve closes, and the supercharger controls p_{im}. With the current controller gains, the 0→90% response time to achieve full boost is around 1 s, but it can be reduced to around 0.7 s with more rigorous calibration and gain scheduling.

Figure 14 shows the engine speed, the motor speed, and the supercharger brake position for the same portion of drive cycle. The motor speed is multiplied by $\frac{R}{n_{\text{im}}n_{\text{ri}}(g_R+g_S)}$, which corresponds to the gear ratio between the motor and crankshaft when the supercharger is locked. During the boosting mode the supercharger speed, ω_{sc}, is related to the motor and crankshaft speed by

$$\omega_e = \frac{g_S}{(g_S+g_R)}\omega_{\text{sc}} + \frac{R}{n_{\text{im}}n_{\text{ri}}(g_R+g_S)}\omega_m \tag{28}$$

when the PSS switches to boosting mode the brake opens and the motor has to decrease its own speed by applying some negative torque to increase the supercharger speed. The motor torque and power during the selected interval are shown in Figure 15. The motor torque is the reported value by the motor control unit, and the motor power, P_m, is measured by the AVL battery simulator. The time gap corresponds to a vehicle acceleration; hence, the motor is assisting the crankshaft when PSS is in torque assist mode (positive motor torque). When switching to boosting mode the motor initially applies some negative torque to speed up the supercharger, and then the motor torque is controlled to track the desired intake manifold pressure. Finally, when boosting is not required, the motor speed increases by applying a positive torque, and when the motor speed is high enough (supercharger speed close to zero) the supercharger is locked again. The brake position is also shown on these plots to distinguish between boosting and torque assist modes. Note that the motor power sign depends on both its torque and speed signs because the PSS motor can rotate in both directions.

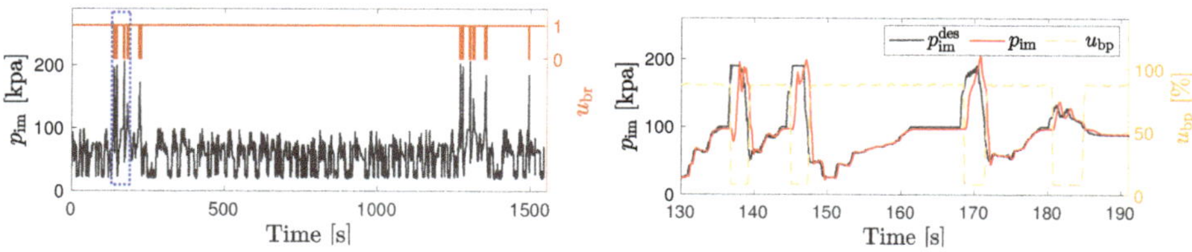

(**a**) Intake manifold pressure during the entire cycle. (**b**) Intake manifold pressure during selected acceleration.

Figure 13. Intake manifold pressure during FTP75 cycle, (**a**) intake manifold pressure and PSS mode over the entire cycle, (**b**) intake manifold pressure, desired intake manifold pressure, and supercharger bypass during t = 130 s to t = 190 s.

The final piece of the HIL experiments is controlling the desired engine speed. Figure 16 shows the commanded engine speed and its actual value controlled by the dynamometer. As seen, the dynamometer can perfectly track the desired engine speed.

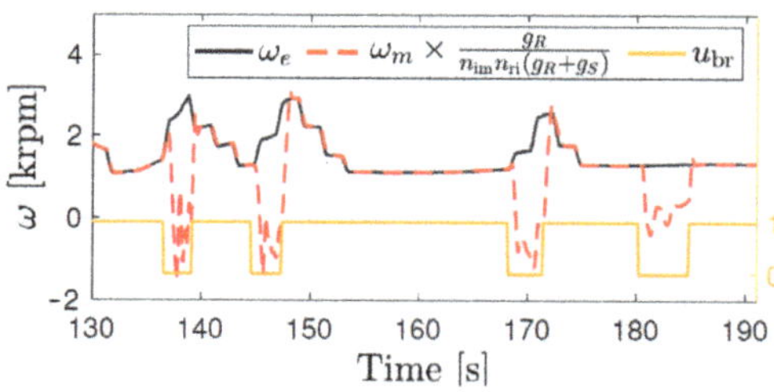

Figure 14. Engine and motor speed for the hardware-in-the-loop experiments during t = 130 s to t = 190 s.

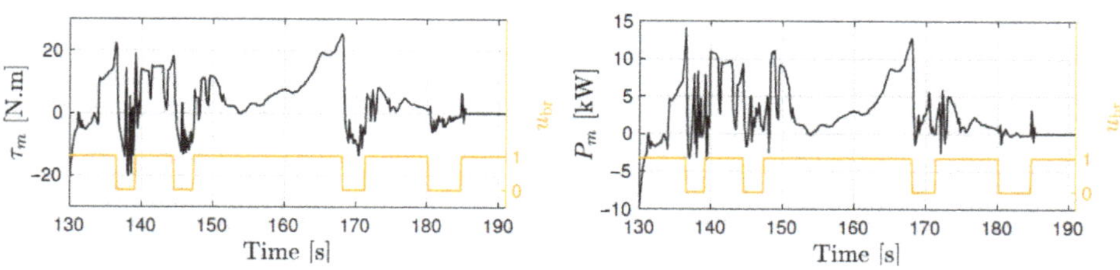

(**a**) Motor torque

Figure 15. Motor operation for the hardware-in-the-loop experiments during t = 130 s to t = 190 s (**a**) motor torque, (**b**) motor power.

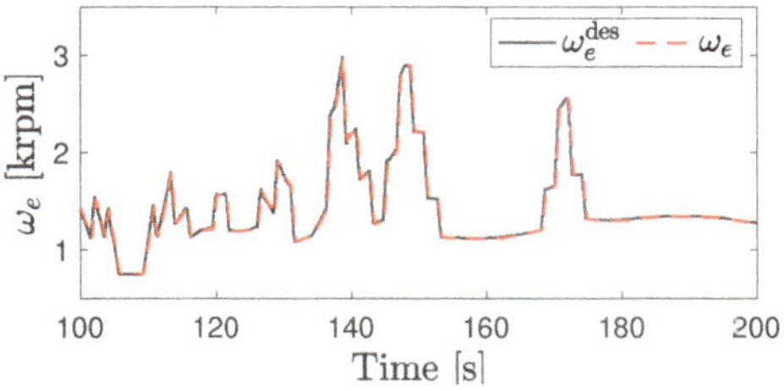

Figure 16. Engine speed and its desired value for the hardware-in-the-loop experiments during $t = 130$ s to $t = 190$ s.

6. Conclusions

This work presented optimal energy management and hardware-in-the loop experiments for a novel low-voltage hybrid system that can be used either as a flexible supercharger or as a parallel hybrid system, enabling start-stop, regenerative braking, and torque assist. An adaptive equivalent consumption minimization strategy from the literature was modified and customized to the PSS system for selecting both the device mode and the power split ratio in hybrid mode. It was shown that when the relative cost of the electric power is higher, the algorithm chooses to use the supercharger across a wider range of operating points, while when the electric power is relatively cheaper the energy management system supplies the motor torque directly to the engine crankshaft. The implementation of drive cycle hardware-in-the loop experiments on an engine dynamometer testbed was discussed in detail, and some of practical aspects were explained. It was shown that the new device with the developed energy management system decreased Ford Escape fuel consumption by 18.4% compared to a baseline turbocharged engine over the FTP75 standard cycle, which is only 4.4% less than the global optimal solution from dynamic programming. Finally, the details of the PSS operation and mode transitions during experiments were shown and discussed in detail. Possible future research directions would be further analysis of the A-ECMS adaptation law and testing the PSS and the developed A-ECMS algorithm in a vehicle.

Author Contributions: Conceptualization, S.N., J.S., R.M. and A.S.; methodology, S.N.; software, S.N. and R.M.; validation, S.N., J.S. and R.M.; formal analysis, S.N.; writing—original draft preparation, S.N.; writing—review and editing, R.M. and A.S.; supervision, A.S.; project administration, A.S.; funding acquisition, A.S. and J.S. All authors have read and agreed to the published version of the manuscript.

Funding: The information, data, or work presented herein was funded in part by the Advanced Research Projects Agency-Energy (ARPA-E), U.S. Department of Energy, under Award Number DE-AR0000659. The views and opinions of authors expressed herein do not necessarily state or reflect those of the United States Government or any agency thereof.

Acknowledgments: The authors would like to thank the EATON corporation and Southwest Research Institute for providing technical support instrumental to the success of this work.

Conflicts of Interest: The authors declare no conflict of interest. The funders had no role in the design of the study; in the collection, analyses, or interpretation of data; in the writing of the manuscript, or in the decision to publish the results.

Abbreviations

The following abbreviations are used in this manuscript:

A-ECMS	adaptive equivalent consumption minimization strategy
DP	dynamic programming
ECMS	equivalent consumption minimization strategy
EMS	energy management system
eSC	electric supercharger
FC	fuel consumption
HEV	hybrid electric vehicle

HIL	hardware-in-the-loop
ICE	internal combustion engine
MPC	model predictive control
NA	naturally aspirated
PI	proportional+integral
PSS	power split supercharger
RPECS	rapid prototyping electronic control system
SC	supercharger
SI	spark ignition
SoC	state of charge
TC	torque converter

References

1. Ehsani, M.; Gao, Y.; Miller, J.M. Hybrid electric vehicles: Architecture and motor drives. *Proc. IEEE* **2007**, *95*, 719–728. [CrossRef]
2. Bureau of Transportation Statistics. Available online: https://www.bts.gov/content/gasoline-hybrid-and-electric-vehicle-sales (accessed on 31 March 2020).
3. King, J.; Heaney, M.; Saward, J.; Fraser, A.; Criddle, M.; Cheng, T.; Morris, G.; Bloore, P. HyBoost: An Intelligently Electrified Optimised Downsized Gasoline Engine Concept. In *Proceedings of the FISITA 2012 World Automotive Congress*; Springer: Berlin/Heidelberg, Germany, 2013; pp. 483–496.
4. Volvo. Available online: https://www.volvocars.com (accessed on 1 March 2020).
5. Panday, A.; Bansal, H.O. A review of optimal energy management strategies for hybrid electric vehicle. *Int. J. Veh. Technol.* **2014**. [CrossRef]
6. Enang, W.; Bannister, C. Modelling and control of hybrid electric vehicles (A comprehensive review). *Renew. Sustain. Energy Rev.* **2017**, *74*, 1210–1239. [CrossRef]
7. Gao, J.; Sun, F.; He, H.; Zhu, G.G.; Strangas, E.G. A comparative study of supervisory control strategies for a series hybrid electric vehicle. In Proceedings of the 2009 Asia-Pacific Power and Energy Engineering Conference, Wuhan, China, 27–31 March 2009; pp. 1–7.
8. Paganelli, G.; Delprat, S.; Guerra, T.M.; Rimaux, J.; Santin, J.J. Equivalent consumption minimization strategy for parallel hybrid powertrains. In Proceedings of the Vehicular Technology Conference, IEEE 55th Vehicular Technology Conference, VTC Spring 2002, Birmingham, AL, USA, 6–9 May 2002; Volume 4, pp. 2076–2081.
9. Borhan, H.; Vahidi, A.; Phillips, A.M.; Kuang, M.L.; Kolmanovsky, I.V.; Di Cairano, S. MPC-based energy management of a power-split hybrid electric vehicle. *IEEE Trans. Control. Syst. Technol.* **2012**, *20*, 593–603. [CrossRef]
10. Lin, C.C.; Peng, H.; Grizzle, J. A stochastic control strategy for hybrid electric vehicles. In Proceedings of the American Control Conference, Boston, MA, USA, 30 June–2 July 2004; Volume 5, pp. 4710–4715.
11. Lin, C.C.; Peng, H.; Grizzle, J.W.; Kang, J.M. Power management strategy for a parallel hybrid electric truck. *IEEE Trans. Control. Syst. Technol.* **2003**, *11*, 839–849.
12. Nazari, S.; Siegel, J.B.; Stefanopoulou, A. Optimal Energy Management for a Mild Hybrid Vehicle with Electric and Hybrid Engine Boosting Systems. *IEEE Trans. Veh. Technol.* **2019**, *68*, 3386–3399 [CrossRef]
13. Nazari, S.; Middleton, R.; Siegel, J.; Stefanopoulou, A. *Equivalent Consumption Minimization Strategy for a Power Split Supercharger*; No. 2019-01-1207; SAE Technical: Detroit, MI, USA, 2019.
14. Nazari, S.; Middleton, R.; Sugimori, K.; Siegel, J.; Stefanopoulou, A. Assessing a hybrid supercharged engine for diluted combustion using a dynamic drive cycle simulation. *SAE Int. J. Altern. Powertrains* **2018**, *7*, 351–368. [CrossRef]
15. Nazari, S.; Siegel, J.; Stefanopoulou, A. Optimal energy management for a hybrid electric vehicle with a power split supercharger. In Proceedings of the 2018 IEEE Vehicle Power and Propulsion Conference (VPPC), Chicago, IL, USA, 27–30 August 2018; pp. 1–6.
16. Sundstrom, O.; Guzzella, L. A generic dynamic programming Matlab function. In Proceedings of the Control Applications, (CCA) & Intelligent Control, (ISIC), St. Petersburg, Russia, 8–10 July 2009; pp. 1625–1630.

17. Paganelli, G.; Tateno, M.; Brahma, A.; Rizzoni, G.; Guezennec, Y. Control development for a hybrid-electric sport-utility vehicle: strategy, implementation and field test results. In Proceedings of the American Control Conference, Arlington, VA, USA, 25–27 June 2001; Volume 6, pp. 5064–5069.
18. Nazari, S.; Stefanopoulou, A.; Martz, J. Minimum backpressure wastegate control for a boosted gasoline engine with low pressure external EGR. In Proceedings of the ASME 2016 Dynamic Systems and Control Conference, American Society of Mechanical Engineers, Ann Arbor, MI, USA, 12–16 October 2016; p. V002T20A002.
19. Onori, S.; Serrao, L. On Adaptive-ECMS strategies for hybrid electric vehicles. In Proceedings of the international scientific conference on hybrid and electric vehicles, Malmaison, France, 6–7 December 2011; Volume 67.
20. Sun, C.; Sun, F.; He, H. Investigating adaptive-ECMS with velocity forecast ability for hybrid electric vehicles. *Appl. Energy* **2017**, *185*, 1644–1653. [CrossRef]

energies

MDPI

Article

Energy Optimization of Electric Vehicles by Distributing Driving Power Considering System State Changes

In-Gyu Jang [1], Chung-Seong Lee [2] and Sung-Ho Hwang [1,*]

1 School of Mechanical Engineering, Sungkyunkwan University, Suwon 16419, Korea; neoera78@gmail.com
2 Central R&D Center, Mando Corporation, Seongnam 13486, Korea; chungseong.lee@halla.com
* Correspondence: hsh0818@skku.edu; Tel.: +82-31-290-7464; Fax: +82-31-290-5889

Abstract: In a battery-electric vehicle, a representative electric vehicle, there is a growing demand for performance and one-charge mileage improvement. As an alternative to such improvements, the capacity of the battery has been increased; however, due to the corresponding increase in the weight of the battery and the limited space in the vehicle, increasing the capacity of the battery also has limitations. Therefore, researches are being actively conducted to improve system operation efficiency to overcome such limitations. This paper proposes a distributing method of the driving forces to a battery-powered electric shuttle bus for last-mile mobility equipped with the decentralized driving system while taking into account voltage changes of the input terminals due to changes in the battery charge. The system operation efficiency changes were compared and evaluated by performing energy consumption analysis using 'Manhattan Bus Driving Cycle' at low voltage condition (SOC 20%). Various analyzes were performed and compared, such as the uniform distribution method of driving forces of the front and rear wheels (*Uniform*), the optimization method without considering the input terminal voltage change (V_{norm} = 90 V), and the optimization method considering the input terminal voltage change (V_{dclink}). As a result, it shows that the proposed algorithm can improve 6.0% compared to the conventional uniform driving force distribution method (*Uniform*). Moreover, it shows that the real-time optimization method without considering the input voltage change (V_{norm} = 90 V) can improve 5.3% compared to the uniform distribution method. The proposed method can obtain an additional 0.7% increase in total cost compared to the existing optimization method, which shows that the vehicle system has cost-effectiveness by reducing the battery capacity required to achieve the same mileage.

Keywords: driving force distribution; decentralized traction system; 4WD electric vehicle; energy efficiency; traction control; efficiency optimization

Citation: Jang, I.-G.; Lee, C.-S.; Hwang, S.-H. Energy Optimization of Electric Vehicles by Distributing Driving Power Considering System State Changes. *Energies* **2021**, *14*, 594. https://doi.org/10.3390/en14030594

Academic Editor: Islam Safak Bayram

Received: 21 December 2020
Accepted: 22 January 2021
Published: 25 January 2021

Publisher's Note: MDPI stays neutral with regard to jurisdictional claims in published maps and institutional affiliations.

1. Introduction

Due to their high energy density and convenience, fossil fuels have been exploited for vehicles for a long time. However, this exploitation has been accompanied by global problems of air pollution in the form of fine particulates resulting from vehicle exhausts, including NO_x, as well as global warming, which is attributable to carbon dioxide emissions.

Therefore, countries worldwide have strengthened regulations on the fuel efficiency and emissions of vehicles. Additionally, some countries and cities have even prohibited the driving and sale of vehicles equipped with internal combustion engines [1]. To cope with these regulations, the global automobile industry has developed and released various kinds of eco-friendly vehicles, denoted as 'xEV'. Among them, electric vehicles have been spotlighted as alternatives that can help reducing air pollution due to their zero emission. There has also been a rapid increase in the number of electric vehicle models that have been released into the market, as well as the sales [2]. The market outlooks estimate that the share of electric vehicles could reach approximately 28% of all motor vehicles in the motor vehicle industry by 2030 [3].

The charging time and driving distance per battery charge still remain as issues that can be further improved. In case of the charging time issue, the charging time for 400 km driving range of electric vehicle equipped with 400 V system is 29 min, but, the charging time can be reduced to less than 15 min for the same driving range by changing the voltage level from 400 V to 800 V [4]. In case of driving distance, the driving distance between 80 km and 200 km for a battery-powered electric shuttle bus equipped with 30.7 kWh and 33 kWh, respectively, has been presented [5,6]. To extend the driving distance provided by a single charge of an electric vehicle battery, various methods have been introduced, including one that developed a high-power-density battery (solid state battery) to increase the energy capacity [7], as well as another study that installed two different driving systems on the front and rear wheels and optimized the power distribution of the front and rear wheels [8]. Furthermore, Xibo et al. proposed the traction force distribution method to minimize power losses for permanent magnet (PM)-type traction motors for a front and rear wheel driven electric vehicle [9]. Studies on independent four-wheel driving systems intending to increase the performance efficiency of electric vehicles have also shown promising results. For example, Park et al. conducted a study on the optimization of the driving energy and systematic stability of an electric vehicle equipped with a four-wheel drive system by employing fuzzy logic [10]. The efficiency of the driving system of an electric vehicle depends substantially on the temperature and input voltage of the driving system [11], and the input voltage is also dependent on the voltage level at the terminal of the battery because of the wire connection between the terminal of the battery and the input terminal of the driving system. The voltage level at the terminal and capacity of the battery vary according to the elapsed driving hours, and changes in the input voltage affect the output of the driving system. However, the studies mentioned above [8–10] did not account for changes in the system state.

This paper presents methods to increase the energy efficiency of a battery-powered electric shuttle bus equipped with a decentralized four-wheel drive system. First, the specification of a battery-powered electric shuttle bus will be derived. Then, the analysis of the input voltage effects on the driving system efficiency will be conducted, and an algorithm to obtain the optimal distribution of driving torque to the front and rear wheels by accounting for the varying input voltages will be proposed. The effects and gain of the algorithm on the driving efficiency of a battery-powered electric shuttle bus will be verified by Matlab/Simulink simulation.

This paper is organized as follows: The specifications for a driving system and requirements for a battery-powered electric shuttle bus are defined in Section 2, in Section 3, design results for the drive system are described. The efficiency changes are also examined with the input voltage changes in Section 3. The algorithm used to allocate the driving torque to the front and rear wheels while securing optimal system efficiency is explained in Section 4. In Section 5, the simulation model for a battery-powered electric shuttle bus is built and the effects on the system efficiency in low voltage condition is analyzed by comparing with uniform distribution method and also with a fixed voltage optimization method through simulation according to 'Manhattan Bus cycle'. Finally, an effectiveness of proposed algorithm is analyzed for a battery-powered electric shuttle bus.

2. Electric Vehicle System Design

In this section, the requirement of the target vehicle is defined and then, the requirement for propulsion system is specified by using longitudinal dynamics.

2.1. Vehicle Requirements

The target vehicle selected for the simulation is a battery-powered electric shuttle bus. This bus drives according to a predetermined interval to connect key places to respective final destinations as a means of last-mile transportation. The driving route for the target vehicle has a one-way interval of 3.7 km, wherein the simulation design included an average driving speed of 15 km/h, a driving time of approximately 15 min, a daily operation of 8 h, and up to 15 passengers. The specifications of the target vehicle are summarized in Table 1.

Table 1. Specifications of the target vehicle.

Attribute	Unit	Value	Attribute	Unit	Value
Passengers	people	12–15	Dimension	m	L 4.94, W 2.10, H 2.65
Gross/Curve Vehicle Weight	kg	3500/2400	Frontal Area	m^2	4.73
Air Drag Coefficient	–	0.4	Tire	–	225 50 R18
Battery	–	15 kWh (73–108 V)	No. of Motors	EA	4

Table 2 presents the performance requirements of the target vehicle. By considering the fact that the vehicle drives at a low speed along the last-mile interval in the downtown regions of cities, the following performance requirements were set for the vehicle: a maximum driving speed of 45 km/h [5,6], an autonomous driving speed of 25 km/h [5,6], a maximum climbing capacity enabling propulsion on a road with a maximum gradient of 28%, continuous driving at 5 km/h on a road with an identical gradient, and a maximum acceleration or deceleration capability of 3 m/s^2.

Table 2. Performance requirements of the target vehicle.

Attribute	Unit	Value	Attribute		Unit	Value
Maximum Speed	km/h	45	Gradeability	Take-off	%	28
Operating Speed	km/h	25		5 km/h		25
Acceleration	m/s^2	3		25 km/h		12
Deceleration	m/s^2	3	-	-	-	-

The electric shuttle bus has an independent four-wheel drive system with the following power transmission architecture. Figure 1 shows the configuration of the power system of the target autonomous electric vehicle.

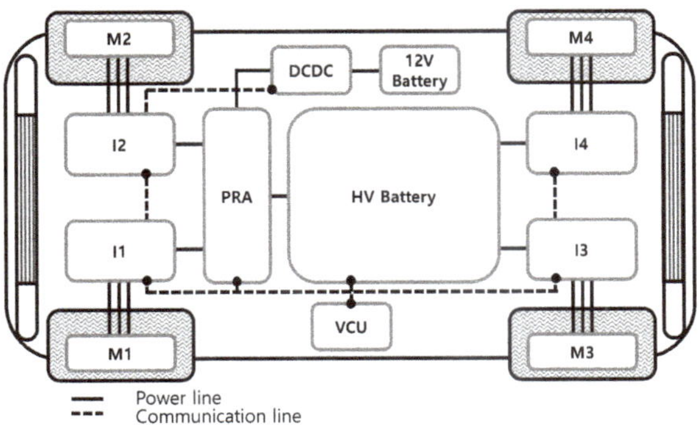

Figure 1. System architecture of the target electric shuttle bus. VCU: Vehicle Control Unit; M1 to M4: In-wheel Motor; I1 to I4: Inverter; PRA: Power Relay Assembly; DCDC: DCDC Converter; HV: High Voltage.

2.2. Propulsion System Sizing

To fulfill the power requirements for the desired performance of the target vehicle, it is necessary to set the design objectives of the power system. Thus, according to the power performance requirements of the target vehicle, the correlation thereof with the attributes of the power system needs to be identified. As shown in Figure 2, the correlations between the indicators of the power performance of the target vehicle and the attributes of the power system are defined.

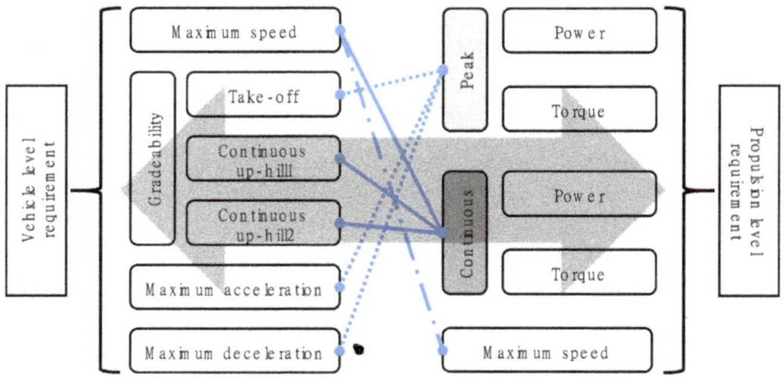

Figure 2. Assignment of vehicle performance requirements to propulsion system requirements.

The maximum speed signifies the maximum continuous driving speed of the target vehicle on a flat road (gradient 0%) corresponding to the continuous output of its power system. As an attribute of a vehicle that corresponds to the continuous output of its power system, the performance of that vehicle continuously driving along a gradient road (gradeability) could be taken into account. In addition, the performance in terms of the maximum acceleration and deceleration, along with the maximum gradeability (gradient), correspond to the instantaneous output of the power system.

The forces (F_x) acting on the driving vehicle can be differentiated as the driving force of the vehicle (F_t) and the forces of air resistance (F_{aero}), rolling resistance ($F_{rolling}$), and climbing resistance (F_{grade}). These are illustrated in Equations (1)–(5). The balance relationship between forces, corresponding to the power performance objectives of the vehicle in question, can be represented in terms of longitudinal vehicle dynamics (illustrated in Equation (1)), from which the forces required for each wheel can be determined [12]:

$$\sum F_x = M\ddot{x} \tag{1}$$

$$F_x = F_t - F_{aero} - F_{rolling} - F_{grade} \tag{2}$$

$$F_{aero} = \frac{1}{2}\rho C_d A_f \dot{x}^2 \tag{3}$$

$$F_{rolling} = C_r M g \cos \varnothing \tag{4}$$

$$F_{grade} = M g \sin \varnothing \tag{5}$$

Here, ρ represents the air density of 1.293 kg/m^3; C_d is a coefficient of air resistance; A_f and $\dot{x}$ represent the front area and velocity of the vehicle, respectively; C_r denotes the coefficient of rolling resistance, which is 0.01; and M, g, and $\varnothing$ represent the mass of a vehicle, gravitational acceleration, and longitudinal gradient of the road, respectively. The other specifications were borrowed from the vehicle specifications presented in Table 1.

Table 3 and Figure 3 present the results of the power output and torque required for the vehicle, which were obtained using Equations (1)–(5) and Equations (6) and (7), respectively. By accounting for changes in the loading condition, air pressure, and tire diameter of the vehicle, a speed of 2 km/h and a 1% margin for the gradient angle were added:

$$T_t = F_t R_t \tag{6}$$

$$P = T_t \omega_t = F_t \dot{x} \tag{7}$$

Here, T_t and R_t represent the wheel torque and tire radius, respectively.

Table 3. Propulsion system requirements corresponding to vehicle performance.

Vehicle Performance		Unit	Value	Allocation	Propulsion Requirement	Unit	Value
Maximum Speed		km/h	45	C [1]	Power	kW	7.2
					Torque	Nm	183
					Speed	rpm	377
Gradeability	Take-off	%	28	P [2]	Torque	Nm	3275
	5 km/h		25	C	Power	kW	12.5
					Torque	Nm	2971
	25 km/h		12	C	Power	kW	33.5
					Torque	Nm	1598
Peak Acceleration		m/s²	3	P	Torque	Nm	2533
Peak Deceleration		m/s²	3	P	Torque	Nm	2533

[1] C: Continuous; [2] P: P signifies the 'Peak'.

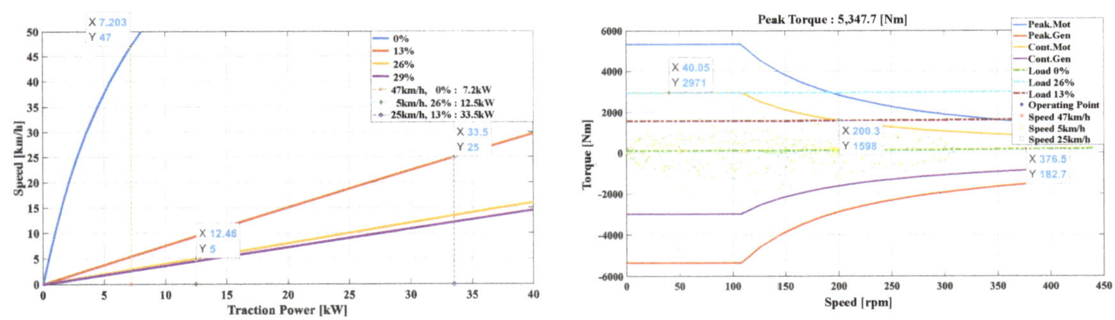

(a) (b)

Figure 3. Sizing results of (**a**) power-gradient-speed and (**b**) torque-speed chart with the requirement point.

An operating speed of 25 km/h was selected to satisfy the requirement that the maximum speed remain below 45 km. Based on the above results, the largest value among the values corresponding to the continuous output was selected by using Equations (8) and (9), by which the continuous maximum torque of 2971 Nm and the continuous maximum output of 33.5 kW were derived. Additionally, a base speed of 108 rpm was determined by Equation (7), which is the TN characteristic expression of the driving system. Here, the instantaneous maximum torque and instantaneous maximum output are dependent upon the cooling mechanism of the motor. For example, in the case of water cooling, these are approximately 1.8 to 2 times the continuous peak torque and continuous output, respectively. In this paper, the value of 1.8 was used, resulting in a maximum torque of 5348 N and an instantaneous maximum output of 60.4 kW:

$$T_t = F_t R_t \tag{8}$$

$$T_{max} = \max(T_{\max\_speed}, T_{grade\_25\%}, T_{grade\_12\%}) \quad (9)$$

As shown in Figure 1, four driving motors are mounted on the vehicle. Furthermore, the required performance of each driving motor in the single driving system was derived by dividing the entire torque and output of the vehicle into four shares, as presented in Table 4.

Table 4. Unit propulsion system requirements.

Propulsion Requirement		Unit	Value
Peak	Torque	Nm	1337
	Power	kW	15.1
Continuous	Torque	Nm	743
	Power	kW	8.4
Speed	Maximum	rpm	377
	Base	rpm	108

2.3. Propulsion System Design

This section describes the specifications of the driving motor that is capable of satisfying the requirements of the output characteristics in the drive system derived from Section 2.2. The correlation between the V_{dclink} value and the efficiency map of the motor is analyzed.

Table 5 lists the specifications of the motor used in this study. Figure 4 illustrates the dimensions and shapes for the cross-sectional view of the motor. Figure 5 shows the efficiency map of the motor corresponding to V_{dclink} values of 105 V and 75 V, respectively. To develop the efficiency map of the motor, the finite element method (FEM) was used to obtain the inductance L_d and L_q on the dq-axis, as well as the magnetic flux ψ_a for a permanent magnet.

Table 5. Design specifications of the in-wheel motor.

Classification	Unit	Value	Classification	Unit	Value
Type	-	IPMSM	Rotor diameter	mm	200
Phase/Pole/Slot	-	3/12/18	Rotor length	mm	34
Stator outer diameter	mm	260	Magnet thickness	mm	5
Stator inner diameter	mm	201.6	Magnet B_r	T	1.37
Slot opening width	mm	3	Magnet μ_r	-	1.05
Slot opening depth	mm	2	Rated torque	Nm	125

Based on the FEM results, the information needed for the efficiency map of the motor is derived through the following Equation of torque (10) and Equations of voltage (11)–(13) [13], by using the equivalent circuit (Figure 6) on the dq-axis of the motor [14]:

$$T = \tfrac{3}{4}N_p\left\{\psi_a\, i_{oq} + (L_d - L_q) i_{od}\, i_{oq}\right\} = \tfrac{3}{4}N_p\left\{\psi_a\, i_a \cos\beta + \tfrac{1}{2}(L_q - L_d)\, i_a^2 \sin 2\beta\right\} \quad (10)$$

$$v_{od} = -\omega_e L_q I_{oq}, \quad v_{oq} = \omega_e L_d I_{od} + \omega_e \psi_a, \quad v_o = \sqrt{v_{od}^2 + v_{oq}^2} \quad (11)$$

$$v_d = R_a I_{od} + \left(1 + \frac{R_a}{R_c}\right) v_{od}, \quad v_q = R_a I_{oq} + \left(1 + \frac{R_a}{R_c}\right) V_{oq} \quad (12)$$

$$V_{limit} = V_{dclink} \times \eta_{inv} \geq \sqrt{v_d^2 + v_q^2} \quad (13)$$

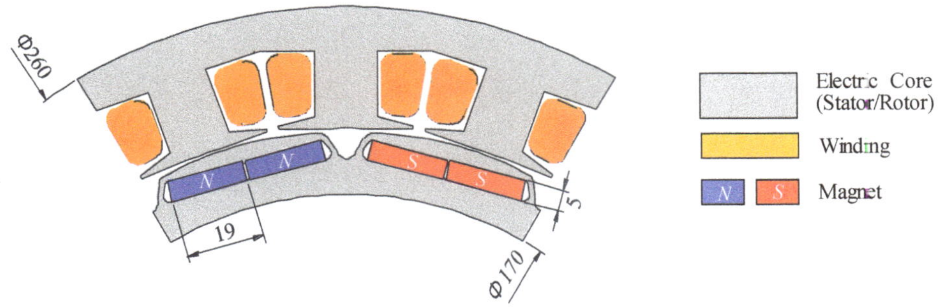

Figure 4. In-wheel motor model.

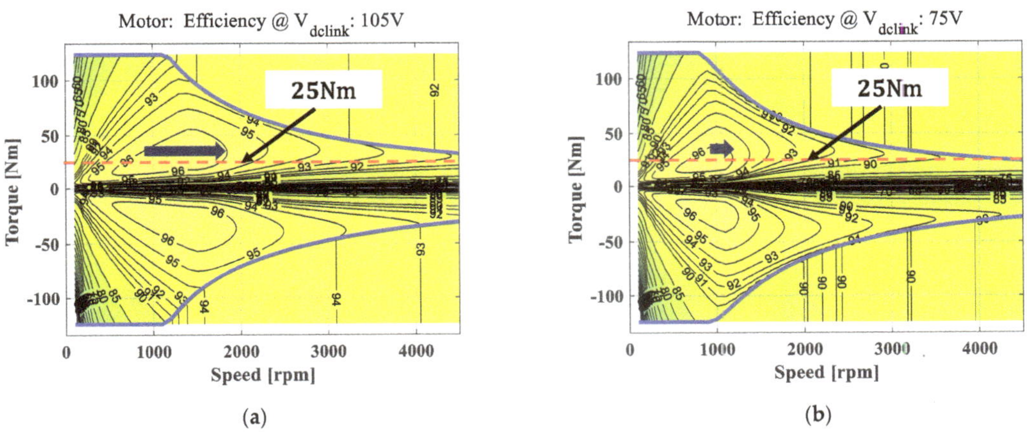

Figure 5. Efficiency map according to V_{dclink}: (**a**) V_{dclink} = 105 V and (**b**) V_{dclink} = 75 V.

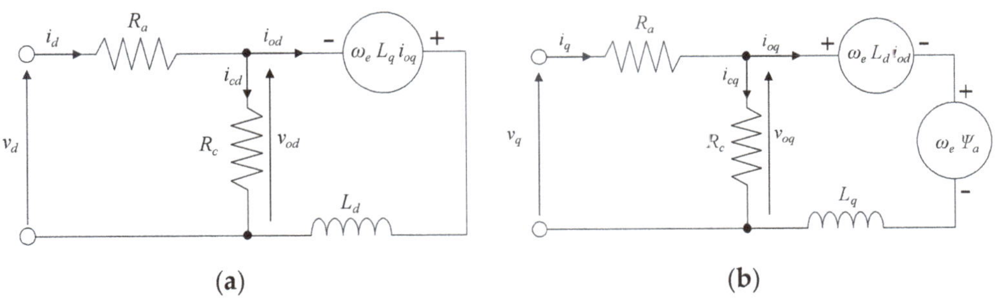

Figure 6. Equivalent circuit: (**a**) *d-axis* and (**b**) *q-axis* [14].

The meanings of the symbols expressed in Equations (10)–(13) are as follows:

N_p: Number of poles; ψ_a: Magnetic flux of a permanent magnet; L_d and L_q: Inductance on the dq-axis;

i_{od} and i_{oq}: Current on the dq-axis; i_{cd} and i_{cq}: Current for the core resistance R_c on the dq-axis;

i_d and i_q: Input current on the dq-axis; i_a: Input current; β: Current angle;

ω_e: Electric angular velocity; ω_m: Mechanical angular velocity; v_{od} and v_{oq}: Voltage on the dq-axis;

R_a: Phase resistance for the winding; R_c: Core loss resistance; and η_{inv}: Inverter efficiency.

Equation (10) for torque T can be derived using Faraday's law [15]. In order to meet the voltage limit by considering the maximum speed of the vehicle, the given voltage limit V_{limit} is required to satisfy equation (13). In Equation (13), V_{dclink} denotes the peak voltage in the terminal of the inverter input for the battery. The inverter efficiency is assumed as η_{inv} = 0.95 to derive V_{limit} [13]

The motor efficiency η_m is derived by the maximum torque per ampere (MTPA) control methodology to generate the maximized efficiency. The efficiency of the motor is reduced by the losses generated from the motor. These losses can be distinguished into the copper loss, iron loss, and mechanical loss. The copper loss P_{cu} is due to the input current of the stator coil. The iron loss for the electrical core of the stator and rotor P_{iron} results from the eddy current of the core, which is proportional to the rotating speed of the motor. P_{cu} is calculated using Equation (14), and P_{iron} is calculated using Equation (15) by deriving the iron loss resistance R_c after conducting finite element analysis by using the loss information of the electrical core (15). The mechanical loss is excluded in this study due to the fact that the measurements are indispensable. Therefore, the efficiency of the motor can be expressed as shown in Equation (16) [13]:

$$P_{cu} = i_a^2 R_a \tag{14}$$

$$P_{iron} = \frac{v_o^2}{R_c} \tag{15}$$

$$\eta_m = \frac{P}{P + P_{loss}} = \frac{T\omega_m}{T\omega_m + P_{cu} + P_{iron}} \tag{16}$$

3. Analysis of the Propulsion System Efficiency

As illustrated in Figure 5, an analysis of the efficiency and output for a torque of 25 Nm over the entire speed interval for the whole efficiency map was conducted to identify changes in the efficiency for the two V_{dclink} models. By using Equations (10)–(16), the causes of the changes in efficiency according to V_{dclink} were analyzed. Figure 7 shows the analyzed results. The efficiency of the V_{dclink} = 75 V model decreased compared to that of the V_{dclink} = 105 V model, in accordance with the increase of speed after exceeding the base speed. In the drive region for the torque of 25 Nm as shown in Figure 5, the mechanical output at each speed was identical.

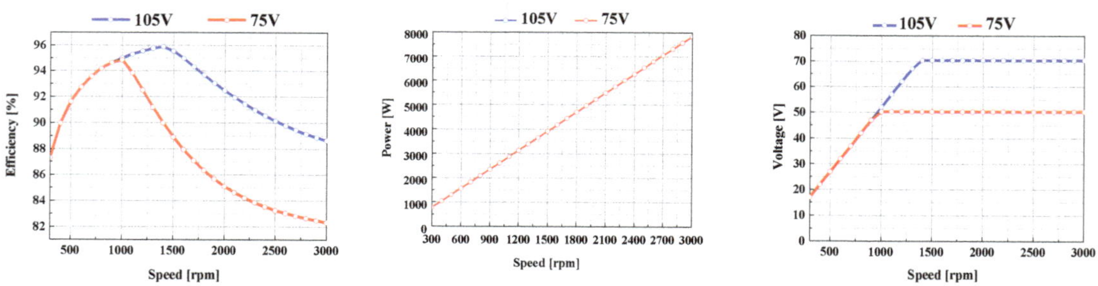

(a) (b) (c)

Figure 7. Efficiency and power according to V_{dclink}: (**a**) efficiency, (**b**) power, and (**c**) voltage.

The difference in the loss characteristics, which caused the difference in the efficiency of the two V_{dclink} models, is described by using Figure 8. The phase resistance R_a of the two V_{dclink} models is identical, since the two models use the same motor. However, the phase currents used to generate the torque of 25 Nm are different for different V_{dclink} values, as shown in Figure 8a. The requirement of the two models is to meet the identical power condition, as shown in Figure 7b. In the case of V_{dclink} = 105 V, the voltage limit is reached at the base speed of 1450 rpm. However, in the case of V_{dclink} = 75 V, the voltage limit is reached at the base speed of 1000 rpm. Considering the voltage limit condition after reaching the base speed of the two models, the V_{dclink} = 75 V model requires more current I_a than the V_{dclink} = 105 V model in order to satisfy the identical power output condition $P = V_{dclink}\, I_a$.

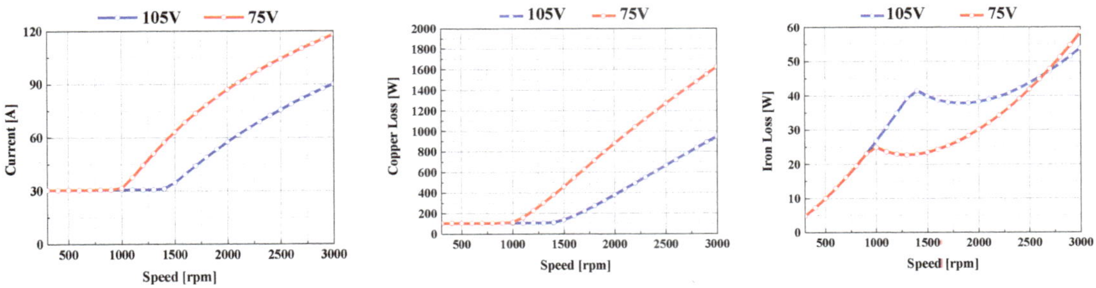

(a) (b) (c)

Figure 8. Current and loss according to V_{dclink}: (**a**) current, (**b**) copper loss, and (**c**) iron loss.

In this paper, the power factor was assumed to be 1. Therefore, in the case of V_{dclink} = 75 V, a higher input current is required than in V_{dclink} = 105 V, as shown in Figure 8a. For this reason, the copper loss increased significantly compared to that of V_{dclink} = 105 V, as shown Figure 8b. In terms of the iron loss, the magnitude is significantly smaller than that of the copper loss, as shown in Figure 8c. Therefore, in the case of V_{dclink} = 75 V, the main cause of the reduced efficiency is that the increased copper loss due to the phase current is dominant.

4. Control Strategy

As illustrated in Figure 9, the real-time optimization algorithm for the power distribution to the front and rear wheels, which showed the lowest energy consumption to create the required driving torque under a given speed, is presented. The 'Virtual Driver' creates the required driving torque (T_d) in order to follow the given speed (V_r), and the 'Energy Optimization' creates the 'set-point' of the driving torque of each system divided into four wheels with minimum energy consumption while satisfying the required driving torque. Here, V_r, T_d, T_{m1}–T_{m4}, T_{tot}, γ_{opt}, P_M, P_{dcdc}, P_{batt}, V_m, V_{dclink}, and $\dot{SOC}$ signify the required speed, required torque by the driver, target torque of the driving system, total driving torque, ratio of the optimal energy distribution for the front and rear driving torques, required electric driving torque of the driving system, electric power consumption of the 12 V power system, total power consumption of the battery, vehicle speed, input voltage of the driving system, and rate of change in the electric energy of the battery, respectively.

The real-time optimization method presented in this paper is explained concretely below, and the effects of changing V_{dclink} were verified by comparing two cases of power consumption, where the changes in V_{dclink} were taken into account or not taken into account for the specified point (speed, required torque).

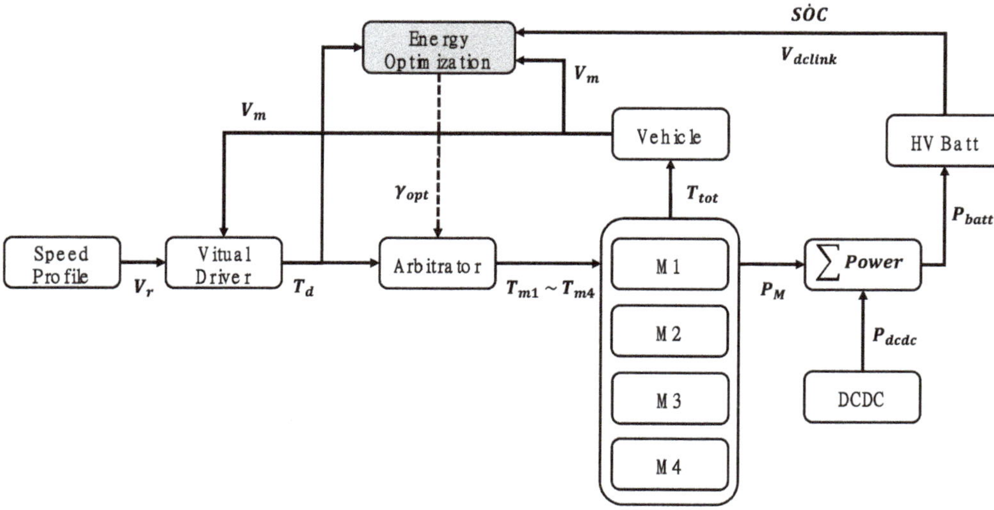

Figure 9. Energy optimization concept diagram.

4.1. Optimal Front/Rear Torque Distribution Concept

The real-time optimization of the driving energy can be modeled using Equations (17)–(23), shown below. The system state variable SOC and the controlled input of the ratio γ of the driving torques to the front and rear wheels are used for modeling. Here, the required torque ($T_{d,f}$) for the front and rear wheel drive of the system varies according to the distribution ratio to the front and rear wheels.

Thus, the operating point varies accordingly, resulting in changes in the efficiencies (η_f, η_r) of the front and rear wheel drive systems. Additionally, the electric power (P_M) used by the drive system varies in accordance with the changing distribution ratio of the driving torque. Here, as examined in Section 3, the efficiency (η) of drive system as a map can be modeled with speed and torque. In this study, the efficiency (η) map of the drive system is modeled with Equation (22) by taking the change of the input voltage (V_{dclink}) into account. The input voltage (V_{dclink}) modeled by internal resistance circuit of the battery is as a function of SOC and the rate of SOC (23) [16]; in this way, the input voltage (V_{dclink}) of the given state is considered. In addition, the braking stability based on the ideal braking torque distribution was designed. Then, the regenerative braking torques distributed to the front and rear axles were limited by the maximum generated torque of the corresponding traction system.

By employing Equation (18), which represents the cost function, the total consumption of energy can be minimized by minimizing the consumption of electric power ($\dot{E}$) required at each moment by the driving system in every time interval 10 ms, as expressed in Equation (19):

$$\dot{SOC} = f(SOC(t),\ P_d(t), \gamma(t)) \tag{17}$$

$$J^* = \min_{\gamma(t),\ t_0 \le t \le t_f} \int_{t_0}^{t_f} \dot{E}(P_d(t),\ \gamma(t), t)dt \tag{18}$$

$$\dot{E}(P_d(t),\ \gamma(t), t) = P_M(t) + P_{dcdc}(t) \tag{19}$$

$$P_M(t) = P_{M,f}(t) + P_{M,r}(t) \tag{20}$$

$$P_{M,f}(t) = \frac{P_d(t)\gamma(t)}{\eta_f(t)},\quad P_{M,r}(t) = \frac{P_d(t)(1-\gamma(t))}{\eta_r(t)} \tag{21}$$

$$\eta_f(t) = g\left(T_{d,f}(t), \omega_f(t), V_{dclink}(t)\right), \quad \eta_r(t) = g(T_{d,r}(t), \omega_r(t), V_{dclink}(t)) \tag{22}$$

$$V_{dclink}(t) = h\left(SOC(t),\ \dot{SOC}\right) \tag{23}$$

subject to:

$$SOC(t) \in \{SOC_{min},\ SOC_{max}\}$$
$$\gamma(t) \in \{\gamma_{min},\ \gamma_{max}\}$$
$$P_{M,f} \in \{0,\ P_{max}\},\ P_{M,r} \in \{0,\ P_{max}\}$$
$$P_{dcdc}(t) = Constant$$

Figure 10 illustrates the method of distribution of the driving torque to optimize the real-time energy consumption. When the torque (T_d) required by a driver is given at a certain speed (V_m), then the combination of the available driving torques of $T_{d,f}$ and $T_{d,r}$ are created by the array (γ) of the driving torque distribution ratio. By exploiting the efficiency map reflecting the voltage of the input terminal, the efficiency vectors of the front and rear wheel corresponding to each created combination of the driving torque can be generated. Consequently, the energy consumption P_M at each element in the array of the distribution ratio of the driving torque can be calculated. Additionally, as expressed in Equation (24), the array of the minimum consumption of energy ($\dot{E}$) can be extracted to derive the distribution ratio (γ_{opt}) of the distribution driving torque to the front and rear wheels with minimized energy consumption:

$$\gamma_{opt}(t) = \text{argmin}\left\{\dot{E}(P_{batt}(t),\ \gamma(t), t)\right\} \tag{24}$$

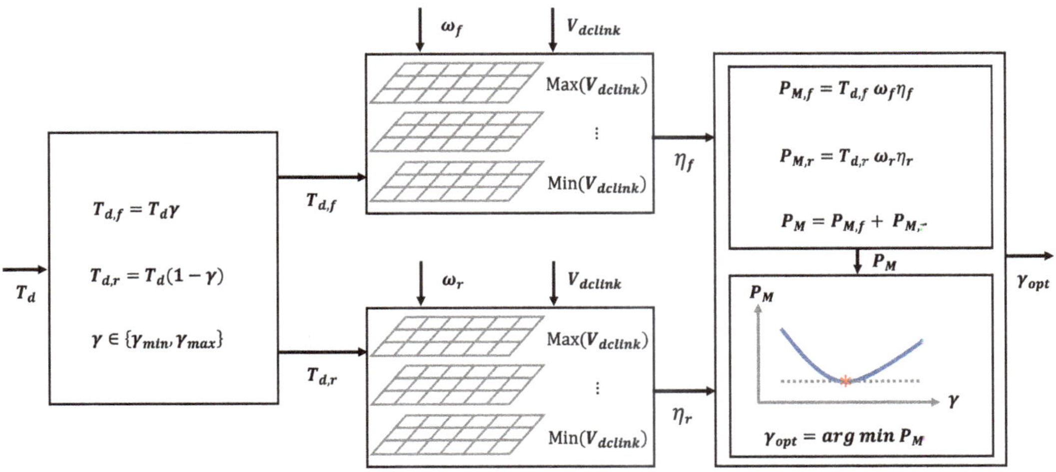

Figure 10. Concept of energy optimization.

4.2. Analysis of Optimal Front/Rear Torque Distribution

In this section, the effectiveness of the algorithm presented in this study will be analyzed by comparing two cases: one in which the input voltage changes are taken into account and one in which they are not. Figure 11 illustrates the results of the optimal distribution ratios of driving torques to the front and rear wheels corresponding to each voltage at the input terminal according to the respective motor speed and driving torque required by the driver.

As illustrated in Figure 5, when a driver requires a driving torque that is below that of peak efficiency at the rotating speed of driving, the optimal distribution ratio (γ_{opt}) of the driving torque to the front and rear wheels becomes zero. In contrast, when the required driving torque exceeds the peak efficiency, the ratio becomes $\gamma_{opt} = 0.5$. As a result, the driving torque is distributed uniformly to the front and rear wheels. However, since the driving torque required by a driver is limited to four times the maximum torque of the unit driving system, the optimal distribution ratio of the driving torque to the front and rear wheels will be limited to $\gamma_{opt} = 0.5$ when the driving torque required by a driver exceeds 50% of the available maximum torque, due to the limit set for the system. As illustrated in the map showing the efficiency characteristics of the unit driving system (Figure 5), this phenomenon is attributed to the rapid decrease in the efficiency characteristics in the low-torque domain, resulting in minimized driving in that domain.

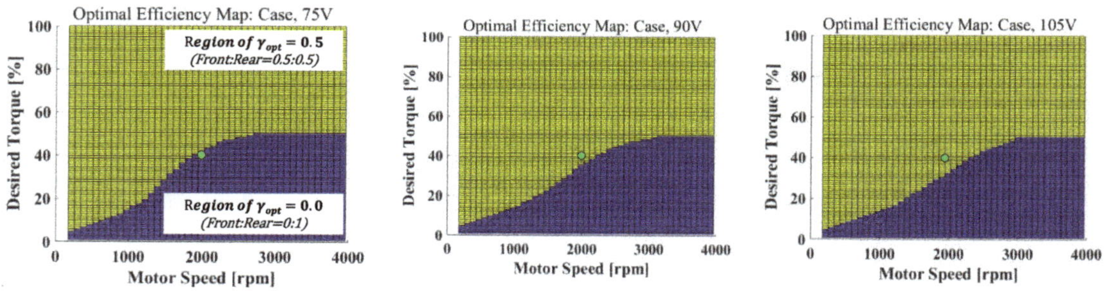

(a) (b) (c)

Figure 11. Optimal front/rear distribution ratios with regard to the speed and desired torque: (**a**) at V_{dclink} = 75 V, (**b**) at V_{dclink} = 90 V (V_{norm}), and (**c**) at V_{dclink} = 105 V.

The effects of the voltage variation at an input terminal on the optimal distribution of the driving torque to the front and rear wheels were verified at a system input terminal voltage of V_{dclink} = 75 V. The driving situation is assumed that motor speed is 2000 rpm (approx. 23 km/h) and the torque required by the driver is 40% of the entire driving torque. The operating point of the front and rear wheels and the power consumption for the optimal distribution ratio of the driving torque to the front and rear wheels derived at V_{dclink} = 90 V (V_{norm}), with a system input terminal voltage of V_{dclink} = 75 V, are illustrated in Figure 12.

When applying the optimal driving torque distribution to both the front and rear wheels, derived at the input terminal voltage of V_{dclink} = 90 V (V_{norm}), approximately 76 [W] more power was used, as compared to what was when applying the optimal driving torque distribution derived from the case when the input terminal voltage was V_{dclink} = 75 V. This result shows that consideration of the input terminal voltage is necessary to optimize system efficiency.

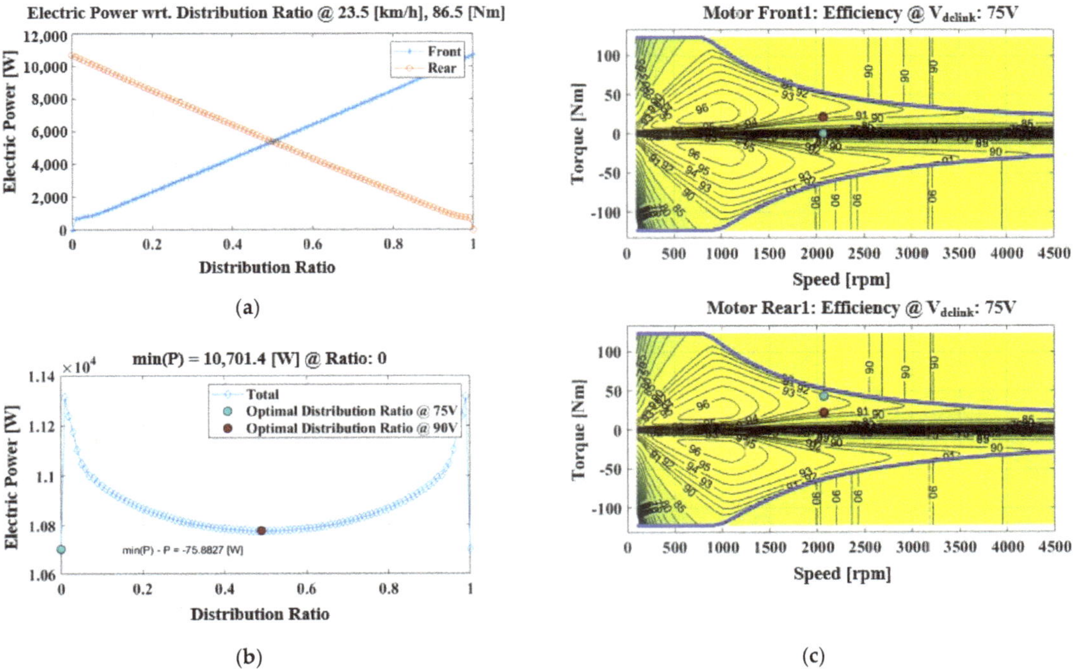

Figure 12. The difference of power consumption on optimal distribution with V_{dclink}; (**a**) The desired electric power of front and rear traction motor with respect to the distribution ratio, (**b**) Total electric power with respect to the distribution ratio, (**c**) Operating points of front and rear traction motor.

5. Simulation Results

To verify the algorithm presented in this paper, simulation model for a battery-powered electric shuttle bus in the longitudinal direction and driving simulation of the vehicle were carried out according to the Manhattan Bus Cycle as shown in Figure 13. The Manhattan Bus Cycle uses a maximum driving speed of 40.9 km/h, peak acceleration/deceleration of 0.2 g, and driving time of 1089 s with an average driving speed of 11 km/h; these are similar to the operating conditions and specifications of the target vehicle [17]. The simulation model built with Matlab/Simulnk is shown in Figure 14. All parameters used in the simulation are described as Tables 1 and 2 in Section 2. The battery model based on the internal resistance of the battery, Equation (25), is used to calculate the V_{dclink} of the driving system. The V_{dclink} of the driving system can be derived by Kirchhoff's current law [16]. Here, no voltage drop and losses between the battery output terminal and the input terminal of the driving system is assumed. And the state of charge (SOC) can be calculated by Equation (26) which is the ratio of the charged current over the full charged capacity.

$$V_{dclink} = V_{oc} - I_b R_i, \quad P_{batt} = V_{dclink} I_b \tag{25}$$

$$SOC_k = SOC_{k-1} + \frac{\Delta I_{b_k}}{Q_0} \tag{26}$$

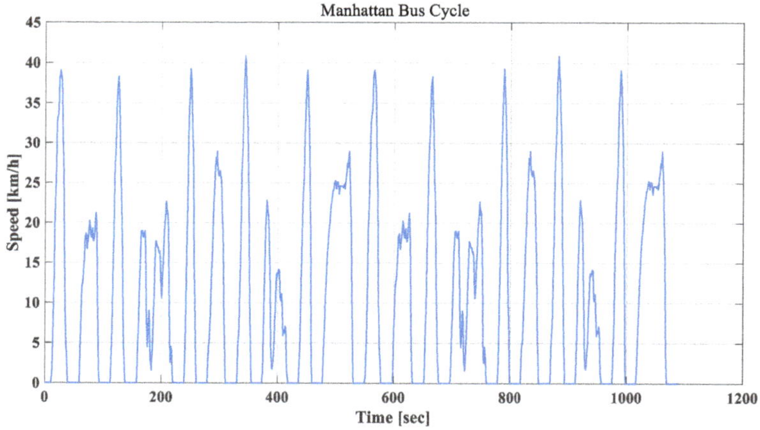

Figure 13. Manhattan Bus Cycle [17].

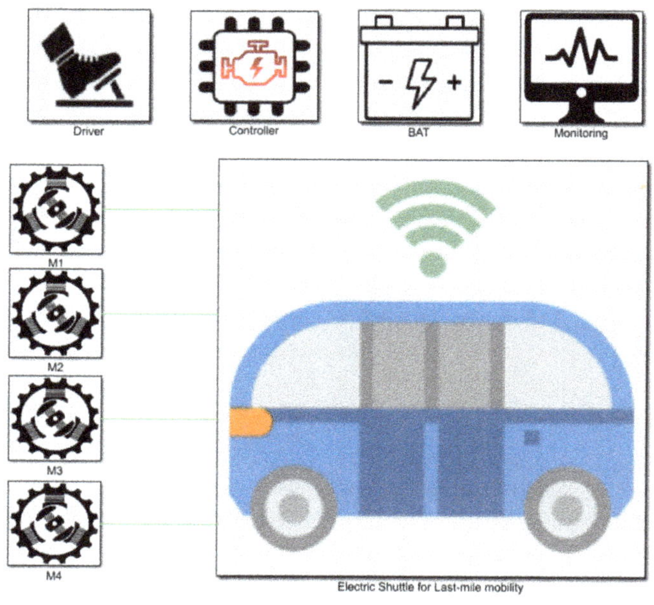

Figure 14. Simulation model: M1 to M4: In-wheel Motor.

Here, V_{oc}, I_b, R_i, SOC_k, SOC_{k-1}, ΔI_{b_k} and Q_0 represent open circuit voltage of the battery, current of the circuit, internal resistance of the battery, state of charge at time interval k, state of charge at time interval $k-1$, variation of the current at time interval $k-1$ and nominal battery capacity of the battery, respectively.

In total, three simulations corresponding to each case were carried out. All simulation was conducted with time interval of 10 ms. To compare the performance of the presented algorithm for varying input terminal voltages, a low–voltage condition (instead of the nominal voltage of 90 V) and an initial condition where the battery SOC was 20% were used. As described in Section 2, the target vehicle is a battery-powered electric vehicle for last-mile mobility. Once the target vehicle is charged, the mission of the target vehicle is to drive the predefined for root without additional charge of the battery. So, the simulation condition assumes that the battery of the vehicle is depleted to 20% of SOC. Here, the initial value of the input terminal voltage of the driving system was 82 V. The conditions employed for the simulation are as follows: a case with an uniform distribution, a case with an optimal distribution of the driving torque based on a nominal voltage of 90 V, and a case with the optimal distribution of the driving torque reflecting the real-time voltage of the input terminal. The following additional conditions were assumed for the simulation: inverter and decelerator efficiencies of 95% and an average electric component load of 400 W.

Table 6 presents the analysis results. The designed vehicle was considered in the driving simulation carried out based on the Manhattan Bus Cycle. In the case of uniform distribution of the required driving torque by the driver to both the front and rear wheels, the energy efficiency was 4.58 km/kWh, whereas the energy efficiency for the case where the optimal driving torque distribution was applied to both the front and rear wheels without considering the effect of the input terminal voltage was 4.83 km/kWh. This indicates an increase in efficiency of approximately 5.3%, compared to the uniform distribution. Furthermore, for the final case of applying the optimal distribution of the driving torque to both the front and rear wheels while considering the input terminal voltage (as presented in this study), the energy efficiency was found to be 4.86 km/kWh, indicating an improvement of approximately 6.0% compared to the case with uniform distribution. This shows that an additional 0.7% improvement in the energy efficiency was achieved by taking the input terminal voltage into account.

Table 6. Simulation results.

Test Case	SOC_{start} [%]	$V_{dclink,start}$ [V]	J [kJ]	Energy Efficiency [km/kWh]	Benefit [%]
Uniform distribution	20	82	2609.60	4.58	-
Optimization@V_{norm}	20	82	2477.60	4.83	+5.3
Optimization@V_{dclink}	20	82	2380.60	4.86	+6.0

Figure 15 shows the operating points of the front and rear wheels under the conditions of the three simulations listed in Table 6. Due to the uniform distribution of the driving torque to the left and right single-axle torque, the left and right sides thereof represent the operating points of the front-right (FR) and rear-right (RR) wheels. In addition, the operating points were marked on the efficiency map of the 90 V condition to generalize the marking of operating points on the efficiency map, despite the fact that the efficiency map varied with changes in the input terminal voltage. Figure 15a illustrates the uniform distribution of the driving torque, wherein the operating points of FR and RR are identical. The operating point in (b) represents the results of allocating the driving torque to the front and rear wheels according to the driving speed of the vehicle and the driving torque required by the driver by following (b) presented in Figure 11. Meanwhile, (c) represents the results of applying the driving torque derived by applying the method presented in Figure 10, according to varying the input terminal voltage between (a) and (b) in Figure 11, to both the front and rear wheels. Altogether, the operating points of (b) and (c) in Figure 15 appear to be similar to each other. As illustrated in Figure 15c, the operating point varied with the application of the optimal distribution ratio of the driving torque as the input terminal voltage varied. Due to the accumulation of changes in the operating points, an additional 0.7% improvement in efficiency was secured.

Figure 16 represents the results of the simulations corresponding to (a)–(c) in Figure 15. In terms of the distribution ratio, when the input voltage is taken into account, it is operated at a better efficiency point than the nominal voltage of V_{dclink} = 90 V (V_{norm}) by delaying transition to the uniform distribution. As a result, the accumulated consumption of energy decreased, ultimately resulting in an increased final energy efficiency.

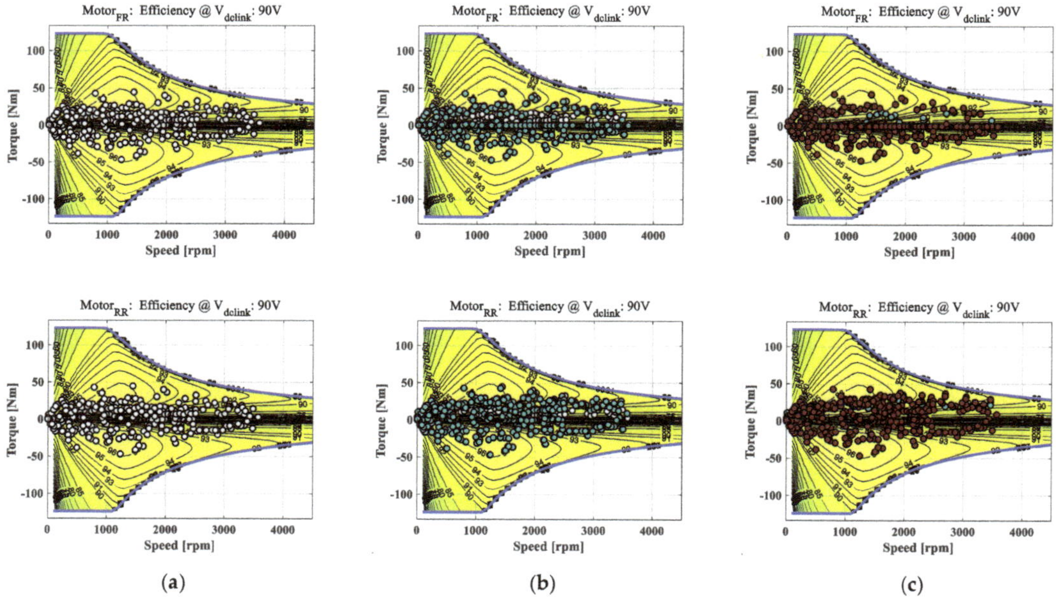

Figure 15. Operating points: (**a**) Uniform distribution, (**b**) Optimization@V_{norm} (90 V), and (**c**) Optimization@V_{dclink}.

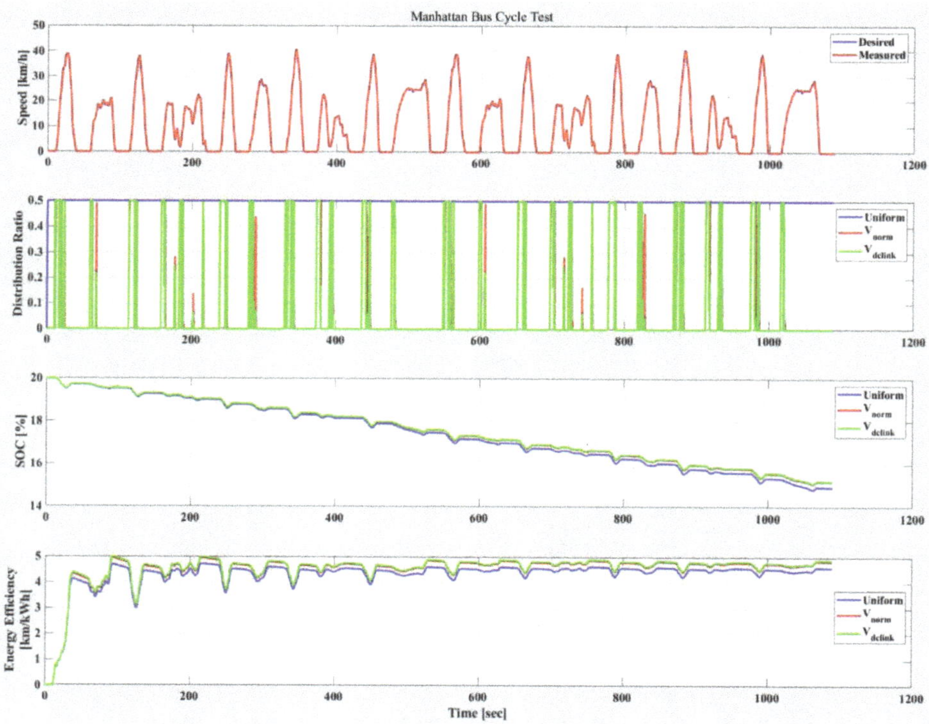

Figure 16. Simulation results.

6. Conclusions

In this paper, a method for improving the energy efficiency was presented by considering the input terminal voltage of the driving system of a battery-powered electric shuttle bus equipped with a decentralized driving system according to the battery's SOC. The proposed algorithm was verified by conducting simulations of the vehicle driving efficiency according to the Manhattan Bus Cycle. The conclusions are as follows:

- After presenting the correlation between the requirements of the power performance and the driving system's attributes of a battery-powered electric shuttle bus designed for last-mile mobility, the performance requirements of the driving system were specified.
- Furthermore, the design results of the driving motor with the aim of satisfying the specification of the driving system was presented, and the changes in the efficiency characteristics according to the varying input voltages were examined by analyzing the loss characteristics according to the input terminal voltage for the designed driving motor. The analyses showed that more electricity consumption was needed to generate an identical output power with a reduced input voltage, indicating the simultaneous increase in copper loss and decrease in the efficiency of the driving motor.
- To improve the energy efficiency of a battery-powered electric shuttle bus equipped with a decentralized driving system, an algorithm an algorithm that can allocate the driving torque in real-time considering the input terminal voltage was proposed. By applying the proposed algorithm, the effect on efficiency was verified by performing a driving simulation of the vehicle along the Manhattan Bus Cycle at low voltage condition (SOC 20%). When the proposed algorithm is applied, the fuel economy is 4.86 km/kWh, and the efficiency of +6.0% is improved compared to the value of 4.58 km/kWh obtained by uniformly distributing the driving torque to the front and rear wheels (*Uniform*). Furthermore, an additional efficiency gain of +0.7% was shown compared to the value of 4.83 km/h obtained with the algorithm that allocates the driving torque based on the nominal voltage level (V_{norm} = 90 V) without taking the input terminal voltage into account. This indicates the opportunity for an additional improvement in efficiency by exploiting the software approach for a battery-powered electric shuttle bus equipped with an identical driving system on each wheel. It makes sense because our approach of extending mileage using software, taking into account the actual voltage conditions of a battery-powered electric shuttle, can be widely implemented. This concept will be implemented to the real target electronic control unit based on the 32-bit micro-processor and verified with vehicle test in the future study.

Author Contributions: Conceptualization, I.-G.J.; methodology, I.-G.J.; software, I.-G.J.; validation, I.-G.J., C.-S.L.; data curation, I.-G.J., C.-S.L.; writing—original draft preparation, I.-G.J.; writing—review and editing, S.-H.H.; visualization, I.-G.J., C.-S.L.; supervision, S.-H.H. All authors have read and agreed to the published version of the manuscript.

Funding: This research received no external funding.

Acknowledgments: This research was supported by the MSIT (Ministry of Science and ICT), Korea, under the ITRC (Information Technology Research Center) support program (IITP-2020-2018-0-01426) supervised by the IITP (Institute for Information & Communications Technology Planning & Evaluation).

Conflicts of Interest: The authors declare no conflict of interest.

References

1. The End of the Road? An Overview of Combustion-Engine Car Phase-Out Announcements across Europe. Available online: https://theicct.org/sites/default/files/publications/Combustion-engine-phase-out-briefing-may11.2020.pdf (accessed on 14 October 2020).
2. IEA. *Global Electric Car Sales by Key Markets, 2010–2020*; IEA: Paris, France; Available online: https://www.iea.org/data-and-statistics/charts/global-electric-car-sales-by-key-markets-2015-2020 (accessed on 8 January 2021).
3. Electric Vehicle Outlook 2020. Available online: https://bnef.turtl.co/story/evo-2020 (accessed on 19 September 2020).

4. Jung, C. Power Up with 800-V Systems: The benefits of upgrading voltage power for battery-electric passenger vehicles. *IEEE Electrif. Mag. Mar.* **2017**, *5*, 53–58. [CrossRef]
5. Navya. Available online: https://navya.tech/en/solutions/moving-people/self-driving-shuttle-for-passenger-transportation/#autonomous (accessed on 8 January 2021).
6. EASY Mile. Available online: https://easymile.com/vehicle-solutions/ez10-passenger-shuttle (accessed on 8 January 2021).
7. Jürgen, J.; Wolfgang, J.; Zeier, G. A solid future for battery development. *Nat. Energy* **2016**, *1*, 1–4.
8. Kim, J. Optimal Power Distribution of Front and Rear Motors for Minimizing Energy Consumption of 4-Wheel-Drive Electric Vehicles. *Int. J. Automot. Technol.* **2016**, *17*, 319–326. [CrossRef]
9. Yuan, X.; Jiabin, W.; Kalhana, C. Torque distribution strategy for a front and rear wheel driven electric vehicle. *IEEE Trans. Veh. Technol.* **2012**, *61*, 3365–3374. [CrossRef]
10. Jinhyun, P.; In Gyu, J.; Sung-Ho, H. Torque Distribution Algorithm for an Independently Driven Electric Vehicle Using a Fuzzy Control Method: Driving Stability and Efficiency. *Energies* **2018**, *11*, 3479.
11. Nan, Z.; Nigel, S.; Rong, Y.; Ran, G. Investigation of DC-Link Voltage and Temperature Variations on EV Traction System Design. *IEEE Trans. Ind. Appl.* **2017**, *53*, 3707–3718.
12. Rajesh, R. *Vehicle Dynamics and Control*, 2nd ed.; Springer Science & Business Media: Berlin, Germany, 2012; pp. 87–101.
13. Takeda, Y.; Morimoto, S.; Honda, Y. *Design and Control of Internal Permanent Magnet Synchronous Motor*; Ohmsha, Ltd.: Tokyo, Japan, 2001; pp. 120–140.
14. Hu, J.; Jia, M.; Xiao, F.; Fu, C.; Zheng, L. Motor Vector Control Based on Speed-Torque-Current Map. *Appl. Sci.* **2020**, *10*, 78. [CrossRef]
15. Hendershot, J.R.; John, T.; Miller, J.E. *Design of Brushless Permanent-Magnet Machines*; Motor Design Books: Venice, FL, USA, 2012; pp. 78–85.
16. Hongwen, H.; Rui, X.; Jinxin, F. Evaluation of Lithium-Ion Battery Equivalent Circuit Models for State of Charge Estimation by an Experimental Approach. *Energies* **2011**, *4*, 582–598.
17. Emission Test Cycles: Manhattan Bus Cycle. Available online: https://www.nrel.gov/transportation/drive-cycle-too (accessed on 8 January 2021).

energies

Article

Stochastic Drift Counteraction Optimal Control of a Fuel Cell-Powered Small Unmanned Aerial Vehicle

Jiadi Zhang [1], Ilya Kolmanovsky [1] and Mohammad Reza Amini [2,*]

[1] Department of Aerospace Engineering, University of Michigan, Ann Arbor, MI 48109, USA; jiadi@umich.edu (J.Z.); ilya@umich.edu (I.K.)
[2] Department of Naval Architecture and Marine Engineering, University of Michigan, Ann Arbor, MI 48109, USA
* Correspondence: mamini@umich.edu

Abstract: This paper investigates optimal power management of a fuel cell hybrid small unmanned aerial vehicle (sUAV) from the perspective of endurance (time of flight) maximization in a stochastic environment. Stochastic drift counteraction optimal control is exploited to obtain an optimal policy for power management that coordinates the operation of the fuel cell and battery to maximize the expected flight time while accounting for the limits on the rate of change of fuel cell power output and the orientation dependence of fuel cell efficiency. The proposed power management strategy accounts for known statistics in transitions of propeller power and climb angle during the mission, but does not require the exact preview of their time histories. The optimal control policy is generated offline using value iterations implemented in `Cython`, demonstrating an order of magnitude speedup as compared to `MATLAB`. It is also shown that the value iterations can be further sped up using a discount factor, but at the cost of decreased performance. Simulation results for a 1.5 kg sUAV are reported that illustrate the optimal coordination between the fuel cell and the battery during aircraft maneuvers, including a turnpike in the battery state of charge (*SOC*) trajectory. As the fuel cell is not able to support fast changes in power output, the optimal policy is shown to charge the battery to the turnpike value if starting from a low initial *SOC* value. If starting from a high *SOC* value, the battery energy is used till a turnpike value of the *SOC* is reached with further discharge delayed to later in the flight. For the specific scenarios and simulated sUAV parameters considered, the results indicate the capability of up to 2.7 h of flight time.

Keywords: air mobility; fuel cell hybrid aircraft; stochastic optimal control; energy management; drift counteraction optimal control

Citation: Zhang, J.; Kolmanovsky, I.; Amini, M.R. Stochastic Drift Counteraction Optimal Control of a Fuel Cell-Powered Small Unmanned Aerial Vehicle. *Energies* **2021**, *14*, 1304. https://doi.org/10.3390/en14051304

Academic Editor: Aldo Sorniotti

Received: 31 January 2021
Accepted: 22 February 2021
Published: 27 February 2021

1. Introduction

With the growing market for unmanned aerial vehicles (UAVs), a wide range of industries and organizations, including military, government, industrial, and recreational users, deploy this technology across the globe [1–3]. Among different types of UAVs, small unmanned aerial vehicles (sUAVs) [4] are attractive for military, aerial photography, and environmental monitoring applications due to their small size and flexible operation [5]. Considering the (i) hardware and weight constraints, (ii) limited onboard energy storage, and (iii) performance requirements for sUAVs, improving their endurance (maximizing their flight time) is of great importance for extending the duration of their missions, which could involve surveillance, search and rescue, disaster relief, traffic control, and precision agriculture, thereby motivating the development of novel propulsion systems and the implementation of optimal control policies for power and energy management. Among different propulsion systems for such sUAVs, a hybrid propulsion system consisting of a polymer electrolyte membrane fuel cell (PEMFC) and a battery has been proposed for long duration missions, e.g., in [6–9]. Other propulsion systems may incorporate energy harvesters, such as in [10]. In this paper we focus on novel approaches to the energy management of sUAVs through optimal coordination between the PEMFC and battery for the previously proposed fuel cell hybrid propulsion system.

Rule-based (e.g., thermostat-like on-off control [11]), dynamic programming-based [12], and model predictive control (MPC) [13] methods have been considered for the energy management of hybrid aircraft. As in automotive energy management applications [14], the use of simple rule-based strategies may not provide optimal performance, while the conventional formulations of MPC and dynamic programming do not directly address the flight time maximization objective. Furthermore, deterministic variants of MPC and dynamic programming may require an accurate preview of the propeller power and climb angle over a long horizon and are computationally demanding if optimization has to be performed online. Similarly, the Pontryagin maximum principle (PMP)-based guidance solutions [15] need accurate characterization of the flight environment.

In this paper, we consider a different approach to the problem of endurance maximization for a hybrid UAV with a polymer electrolyte membrane fuel cell (PEMFC) based on an application of stochastic drift counteraction optimal control (SDCOC) [16], which directly addresses the problem of maximizing the time to constraint violation in a stochastic environment. In our case, the objective is to maintain the vehicle flying for a maximum amount of time by coordinating the fuel cell and the battery to provide the requested propeller power subject to the limited amount of fuel and battery state of charge (*SOC*) onboard the vehicle. The transitions in aircraft climb angle and propeller power are modeled stochastically by a Markov chain with the transition probabilities determined from historical data representing typical missions of an sUAV. Then, a control policy that minimizes a cost functional reflective of expected time-to-violate constraints is determined offline through value iterations; this control policy is then deployed onboard for the online coordination of the fuel cell and the battery in the sUAV.

In a preliminary conference paper [17] by the second author of this paper, the application of SDCOC for power management of a hybrid sUAV with a direct methanol fuel cell (DMFC) was considered. While the DMFC is often considered as a suitable power source for ground vehicles [18] and has certain advantages, PEMFCs are more appealing for air mobility applications [6,7] due to their relatively lower operating temperature, allowing for a quick start-up [19], higher efficiency (up to 60% [18,20]) and power density, and higher safety due to the use of the solid electrolyte [18].

Different from [17], in this paper, we consider the application of SDCOC to the power management of a hybrid sUAV with a PEMFC rather than a DMFC. To accommodate a different fuel cell and an sUAV, the fuel cell model is changed and improvements are made to the models used to compute the propeller power and thrust, as well as the evolution of the *SOC*.

More importantly, the lack of the ability of the PEMFC to rapidly change its power output imposes a stringent operating constraint (rate limit on PEMFC power output), which was not treated in [17], but is treated in this paper. This rate limit increases the complexity of the problem as an extra state needs to be introduced in the model and handled in SDCOC, and it also changes the optimal policies and the optimal response of the system. For instance, as the fuel cell is not able to support fast changes in power output, the optimal policy is shown to charge the battery to a turnpike value if starting from a low initial state of charge value. If starting from a high *SOC*, the battery energy is used till a turnpike value of the state of charge is reached with further discharge delayed to a later phase of the flight. In either case, the high frequency chattering of fuel cell load demand power in [17], which cannot be supported by the PEMFC, is eliminated.

Additionally, in this paper, the value iterations are implemented in `Cython` rather than `MATLAB`, with an order of magnitude speedup as compared to the `MATLAB` implementation. As value iterations are frequently used to solve dynamic programming problems in different applications and `Python` is becoming increasingly popular, our results on the ten-fold speedup with `Cython` without a substantive increase of the code complexity are of reference value to other researchers considering the computational implementation of dynamic programming.

Furthermore, a discount factor is introduced into the cost function of SDCOC, and its impact on the convergence speed of the value iterations is illustrated. It is shown that this discount factor results in the faster convergence of value iterations, but the performance of the control policy (in terms of exit time) is decreased.

While the SDCOC theory was developed in [16], that reference did not address the fuel cell or sUAV application studied in this paper. Our approach to representing motor power demand and climb angle by a Markov chain with a finite number of states follows [21], which is the first (to the authors' knowledge) paper proposing the use of stochastic dynamic programming for automotive powertrain control applications; that paper also did not address the fuel cell or sUAV application studied in the present paper, nor the drift counteraction problem formulation.

The remainder of this paper is organized as follows: Section 2 describes the sUAV sub-systems and their models. Section 3 presents an integrated model of the hybrid system and defines the problem in a form suitable for SDCOC. Section 4 summarizes SDCOC, and Section 5 reports the results. Finally, Section 6 presents concluding remarks.

2. Physical Description of the Systems and Model

An sUAV with a series hybrid propulsion system, shown in Figure 1, was chosen in which the power supplied by the battery and the power supplied by the PEMFC are combined to meet the propeller motor power demand. The PEMFC uses hydrogen as the fuel, which is stored in the tank, and air from the atmosphere. A fraction of the energy generated by the PEMFC can be used to charge the battery. The fuel cell pack and battery pack are sized large enough so that they are able to meet the sUAV's mean power demand individually, should either one not be operating properly.

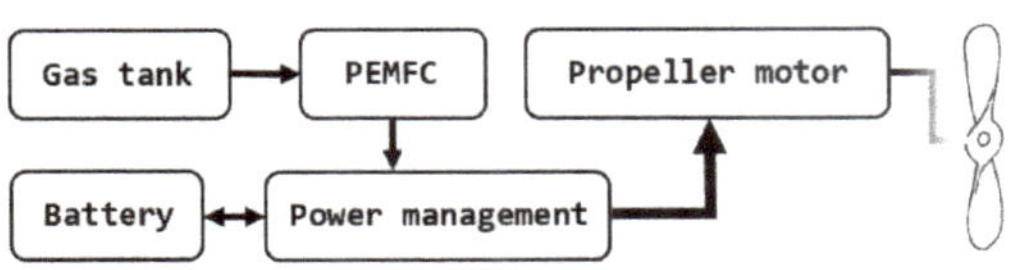

Figure 1. A diagram of a fuel cell-powered series hybrid small UAV (sUAV).

The model used in this paper for generating the SDCOC policies captures the battery's SOC dynamics, the fuel cell's hydrogen rate dynamics, and the fuel cell load power dynamics. Thus, the states of this model are the SOC, the mass of hydrogen remaining in the gas tank, and the fuel cell load demand power. The motor power of the sUAV and climb angle are treated as operating variables, and the SDCOC controller determines changes in the fuel cell load demand power. This system level model has been implemented by combining component submodels and characterizations available from the literature; our methodology is generic and can accommodate changes in these component models.

2.1. sUAV Dynamics

A control-oriented dynamic model of the sUAV is used for SDCOC law development. The sUAV is constrained to a longitudinal flight path in a vertical plane [22]. Table 1 defines the notations for the variables used in the model. Table A1 in Appendix A lists the model parameter values, partly based on [23–25]. The development of lightweight electric components (batteries, fuel cells, motors) for sUAVs is an active area of research; see, e.g., [26,27]. In our model, we assumed that such lightweight components are available to be consistent with the assumed sUAV's weight.

Table 1. List of variables used in the sUAV model.

Variable	Description	Unit
v	Velocity of the sUAV	m/s
γ	Climb angle	deg
T	Thrust force	N
α	Angle of attack	deg
L	Lift force	N
D	Drag force	N
C_L	sUAV coefficient of lift	\
C_D	sUAV coefficient of drag	\
ρ_{air}	Air density	kg/m^3
P_{sUAV}	Power required by the sUAV	W
N	Angular speed of the electric motor	RPM
P_P	Power generated by the propeller	W
$P_{P,ideal}$	Ideal propeller power	W
η_P	Propulsive efficiency	\
U_M	Electric motor driver's input voltage	V
I_M	Electric motor driver's input current	A
P_M	Elector motor driver's input power	W
η_M	Motor efficiency	\
$P_{FC,total}$	Total power of the fuel cell	W
$P_{FC,load}$	Load demand power of the fuel cell	W
P_{aux}	Power required by the auxiliaries	W
U_{FC}	Single cell voltage	V
I_{FC}	Single cell current	A
i_{FC}	Single cell current density	A/cm^2
U_{act}	Activation polarization	V
U_{ohm}	Ohmic losses	V
U_{conc}	Concentration polarization	V
U_{OC}	Equivalent open circuit voltage of a single fuel cell	V
R'_{FC}	Modified single fuel cell resistance	Ω
$\tilde{R}'_{FC}$	Variable defined in Equation (18)	$\Omega \cdot cm^2$
$U_{B,OC}$	Open circuit voltage of the battery	V
SOC	Battery's state of charge	\
P_B	Power of the battery	W
SOC_0	Initial SOC	\
S_f	Split fraction	\
u	Control input	\
ΔP_{FC}	Defined in Equation (25)	W
m_{FR}	Mass of fuel remaining	kg

Using a flat Earth coordinate system, the longitudinal equations of motion of the sUAV are given by:

$$\dot{v} = \frac{Tcos(\alpha) - D}{m} - gsin(\gamma), \tag{1}$$

$$\dot{\gamma} = \frac{Tsin(\alpha) + L}{mv} - \frac{gcos(\gamma)}{v}, \tag{2}$$

where v is the velocity of the sUAV and γ is the climb angle. The lift L and drag force D are characterized as:

$$L = \frac{1}{2}\rho_{air}v^2 S_{ref} C_L,\ D = \frac{1}{2}\rho_{air}v^2 S_{ref} C_D, \tag{3}$$

where $C_L = C_{L0} + C_{L,\alpha}\alpha$, $C_D = C_{D0} + KC_L{}^2$. Neglecting vertical acceleration (i.e., with $L = mg$), solving Equations (1) and (2) yields the thrust required by the sUAV,

$$T = \left[\left(m\dot{v} + mgsin(\gamma) + \frac{1}{2}\rho_{air}v^2 S_{ref} C_{D0} + \frac{2Km^2g^2}{\rho_{air}v^2 S_{ref}}\right)^2 + (mv\dot{\gamma} + mgcos(\gamma) - mg)^2\right]^{\frac{1}{2}}. \tag{4}$$

Here, ρ_{air} is a function of altitude. The power required by the sUAV is then given by:

$$P_{sUAV} = Tv. \quad (5)$$

2.2. Propeller Model

The propeller model is used to relate the torque and angular velocity generated by the electric motor to the power required by the sUAV and the velocity of the sUAV, respectively [22]. With the propulsive efficiency given by η_P, the power required to drive the propeller is:

$$P_P = \frac{P_{sUAV}}{\eta_P}. \quad (6)$$

According to the disk actuator theory, the ideal propeller power is:

$$P_{P,ideal} = \frac{1}{2}Tv\left(1 + \sqrt{1 + \frac{8T}{\pi\rho_{air}v^2{d_P}^2}}\right).$$

In general, the actual power required would be about 15% greater than this [28], which means $P_P = 1.15P_{P,ideal}$. Thus, η_P can be calculated as:

$$\eta_P = \frac{P_{sUAV}}{1.15P_{P,ideal}} = \frac{2}{1.15 + 1.15\sqrt{1 + \frac{8T}{\pi\rho_{air}v^2{d_P}^2}}}. \quad (7)$$

Combing Equation (7) with (5) and (6) yields:

$$P_P = \frac{1.15P_{sUAV}}{2} + \frac{1.15P_{sUAV}}{2}\sqrt{1 + \frac{8P_{sUAV}}{\pi\rho_{air}v^3{d_P}^2}}.$$

2.3. Electric Motor Model

Electric motors used in sUAV applications exhibit high speed and high torque, as well as high power-to-weight ratios [29]. Assuming the power factor is equal to unity and the magnetic losses can be neglected, the output power of the motor is given by:

$$P_P = (U_M - I_M R_M)(I_M - I_{M,0}). \quad (8)$$

The angular velocity of the motor in revolutions per minute (RPM) can be expressed as:

$$N = (U_M - R_M I_M)K_V, \quad (9)$$

which should be equal to the RPM of propeller $N = \frac{v}{Jd_P}$. From Equations (8) and (9), the motor current, I_M, is:

$$I_M = \frac{P_P K_V}{N} + I_{M,0}.$$

The motor power and motor efficiency are given by, respectively,

$$P_M = U_M I_M, \quad \eta_M = \frac{P_P}{P_M}.$$

2.4. Fuel Cell Model

A PEMFC system is the primary power source for the sUAV. The total power generated by the fuel cell stack is calculated as:

$$P_{FC,total} = n_{FC}U_{FC}I_{FC}. \quad (10)$$

This power must cover the load demand $P_{FC,load}$ and the power required for auxiliaries [18], P_{aux},

$$P_{FC,total} = P_{FC,load} + P_{aux}, \tag{11}$$

where P_{aux} is the total power required for the compressor motor, the hydrogen circulation pump, the humidifier water circulation pump, the coolant pump, the cooling fan motor, and the bias power, P_0. After simplifications, P_{aux} could be written as a function of the fuel cell current [30],

$$P_{aux} = P_0 + n_{FC}\kappa_{FC}I_{FC}. \tag{12}$$

The fuel cell current is a function of the current density and the fuel cell area,

$$I_{FC} = i_{FC}A_{FC},$$

where i_{FC} could be obtained by solving the equation,

$$U_{FC} = U_{rev} - U_{act} - U_{ohm} - U_{conc}. \tag{13}$$

The reversible cell potential U_{rev} is related to the molar specific Gibbs free energy Δg_f and the number of ions passed in the reaction n_e [24],

$$U_{rev} = \frac{\Delta g_f}{n_e F}.$$

The activation polarization U_{act} is a result of the energy required to initiate the reaction, which can be described by the semi-empirical Tafel equation [31–33],

$$U_{act} = c_0 + c_1 ln(i_{FC}),$$

where c_0 and c_1 depend on temperature. When the current density is small, this equation can be modified [34] as:

$$U_{act} = c_0(1 - e^{-c_1 i_{FC}}), \tag{14}$$

where $c_0 = -5.8 \times 10^{-4}\hat{T} + 0.5736$ and $c_1 = \frac{R\hat{T}}{n_e \alpha_{FC} F}$.

The ohmic losses U_{ohm} are due to the resistance to the flow of (i) ions in the membrane and in the catalyst layers and (ii) electrons through the electrodes [18],

$$U_{ohm} = i_{FC}\tilde{R}_{FC}, \tag{15}$$

where $\tilde{R}_{FC} = R_{FC}A_{FC}$.

The concentration polarization U_{conc}, is given by

$$U_{conc} = d_0 e^{d_1 i_{FC}}. \tag{16}$$

With the parameters given in Appendix A, the polarization curve of a single cell is plotted in Figure 2. In reality, the current density could be controlled within a certain range. After excluding the very low current densities ($i_{FC} < 0.1$ A/cm^2), (13) could be linearized [34,35] as:

$$U_{FC} = U_{OC} - \tilde{R}_{FC}i_{FC}, \tag{17}$$

where U_{OC} is the voltage at which the linearized curve crosses the y-axis, which should not be confused with U_{rev}.

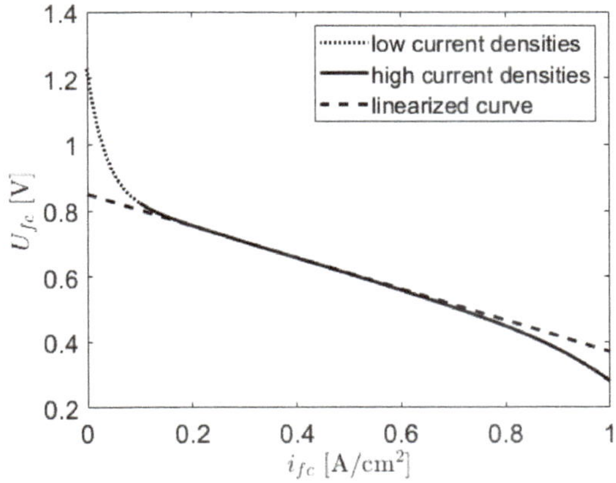

Figure 2. Polarization curve for a given PEMFC.

Unlike ground vehicles, the sUAV changes its orientation during the flight, which would change the inner resistance of the fuel cell by about five times [36] from horizontal to vertical. To this end, (17) is modified to account for this effect as:

$$U_{FC} = U_{OC} - \tilde{R}'_{FC} i_{FC}, \tag{18}$$

where $\tilde{R}'_{FC} = \tilde{R}_{FC}(1 + k_0 sin(k_1|\gamma|))$. Combining Equation (18) with Equations (10)–(12) yields:

$$n_{FC} R'_{FC} I^2_{FC} - (n_{FC} U_{OC} - n_{FC} \kappa_{aux}) I_{FC} + P_{FC,load} + P_0 = 0, \tag{19}$$

where $R'_{FC} = \tilde{R}'_{FC} / A_{FC}$. Overall, I_{FC} can be expressed as:

$$I_{FC} = \frac{n_{FC}(U_{OC} - \kappa_{aux}) - \left[n^2_{FC}(U_{OC} - \kappa_{aux})^2 - 4 n_{FC} R'_{FC}(P_{FC,load} + P_0)\right]^{\frac{1}{2}}}{2 n_{FC} R'_{FC}}. \tag{20}$$

2.5. Battery Model

The battery model represents a pack of Model 21700 lithium polymer battery cells. The battery pack is assembled in such a way that the cells are connected in series. According to [37], the open-circuit voltage of the battery can be estimated as:

$$U_{B,OC} = SOC(U_{B,max} - U_{B,min}) + U_{B,min}. \tag{21}$$

The battery power and the fuel cell load demand power sum up to provide the electrical power to the motor such that:

$$P_M = P_B + P_{FC,load}. \tag{22}$$

Further, the current drawn from the battery set is obtained by solving:

$$P_B = n_B(U_{B,OC} I_B - I^2_B R_{B,int}), \tag{23}$$

which should not exceed its maximum discharge current $I_{B,max}$.

The battery Coulombic efficiency in the battery model is assumed to be 100%. Thus, the SOC is satisfied as:

$$SOC(t) = SOC_0 - \int_{t_0}^{t} \frac{I_B(t)}{C_B} dt. \tag{24}$$

where t, t_0, and SOC_0 are the current time, initial time, and initial SOC, respectively.

3. Hybrid System Model and Problem Formulation

3.1. Hybrid System

The fuel cell load demand power, which will be indicated as P_{FC} in the following section, is the only variable under control. Due to the output characteristic of the PEMFC, the change of P_{FC} is chosen to be 5% of its maximum power, which depends on γ according to (18). The fuel cell load demand power dynamics are then:

$$P_{FC}(t_{n+1}) - P_{FC}(t_n) = u \cdot \Delta P_{FC}, \tag{25}$$

where $u \in \{-1, 0, 1\}$, $\Delta P_{FC} = 5\% P_{FC,max}$, and $P_{FC}(t_n)$ is the fuel cell load demand power at $t = t_n$. Here, three different values of u correspond to decreasing, sustaining, or increasing P_{FC}. According to Equation (19), the maximum load cell power can be calculated as:

$$P_{FC,max} = \frac{4n_{FC}R'_{FC}P_0 - (n_{FC}U_{OC} - n_{FC}\kappa_{aux})^2}{-4n_{FC}R'_{FC}}. \tag{26}$$

Using Equations (25) and (26), the final expression representing the fuel cell load demand power dynamics is given by:

$$\frac{dP_{FC}}{dt} = u \cdot 5\% \frac{4n_{FC}R'_{FC}P_0 - (n_{FC}U_{OC} - n_{FC}\kappa_{aux})^2}{-4n_{FC}R'_{FC}}. \tag{27}$$

The SOC dynamics are obtained by differentiating both sides of (24) with respect to time,

$$\frac{dSOC}{dt} = -\frac{I_B}{C_B}. \tag{28}$$

Combing (28) with (21) and (23) yields,

$$\frac{dSOC}{dt} = \frac{-n_B U_{B,OC} + \sqrt{(n_B U_{B,OC})^2 - 4n_B R_{B,int} P_B}}{2n_B R_{B,int} C_B}, \tag{29}$$

where $U_{B,OC} = SOC(U_{B,max} - U_{B,min}) + U_{B,min}$.

The motor power and battery power are related by:

$$P_B = S_f P_M, \tag{30}$$

where S_f is referred to as the split fraction, which could be calculated from (22) as:

$$S_f = \frac{P_M - P_{FC}}{P_M}.$$

Using Equations (29) and (30), the final expression representing the SOC dynamics is given by:

$$\frac{dSOC}{dt} = \frac{-U_{B,OC} + \sqrt{(U_{B,OC})^2 - \frac{4R_{B,int}S_f P_M}{n_B}}}{2R_{B,int}C_B}, \tag{31}$$

where the internal resistance $R_{B,int}$ and the battery capacity C_B are assumed to be constant [38].

The mass of remaining fuel dynamics is obtained from Faraday's law as:

$$\frac{dm_{FR}}{dt} = -\frac{n_{FC}I_{FC}}{n_e F}M_h, \tag{32}$$

where I_{FC} is calculated from P_{FC} as shown in (20).

Equations (25), (29), and (32) are the final form of the state equations used in this study, where the states of the system are $x = [SOC, M_{FR}, P_{FC}]$, the control is $u \in \{-1, 0, 1\}$, the outputs of the system are $y = [S_f, P_B]$, and the operating variables are $w = [P_M, \gamma]$. These operating variables are treated as measured disturbances in the model.

Based on the above modeling assumptions and parameters in Table A1, the maximum fuel cell output power is 795 W at $\gamma = 0$ deg, 496.14 W at $\gamma = \pm 10$ deg, and 335.71 W at $\gamma = \pm 20$ deg. The theoretical maximum power for the battery series (of eight batteries) is 2940 W, due to the limitation of the discharge current (35 A); the maximum power of the battery is 1176 W at any climb angle.

3.2. Problem Formulation

The forward Euler method is used in this paper to approximate the time derivatives. During each time segment Δt, the motor power of the sUAV is w_1 and the climb angle is w_2.

The following updated equations approximately model the sUAV hybrid propulsion system:

$$SOC(t_{n+1}) = SOC(t_n) + \frac{dSOC}{dt}(t_n)\Delta t,$$
$$M_{FR}(t_{n+1}) = M_{FR}(t_n) + \frac{dM_{FR}}{dt}(t_n)\Delta t,$$

where $SOC(t_n)$ and $M_{FR}(t_n)$ are the state of charge and the mass of hydrogen remaining at $t = t_n$.

The system is controlled by the change of the fuel cell load demand power ΔP_{FC} at each discrete time instant. Thus, the fuel cell power is modeled as:

$$P_{FC}(t_{n+1}) = P_{FC}(t_n) + u\Delta P_{FC}(t_n).$$

The motor power and climb angle are typically unknown a priori. In this paper, a Markov chain model is used to describe the evolution of w_1 and w_2 with the transition probabilities identified from the historical data. Once particular w_1 and w_2 values are encountered, a prediction of their probability distribution over the next time segment will be made using the Markov chain model.

The objective of the stochastic endurance maximization problem is to determine a control law that maximizes the time the sUAV can travel before the system states exit a prescribed set,

$$G = \Big\{(SOC, M_{FR}, P_{FC}) : SOC_{min} \leq SOC \leq SOC_{max}, \quad M_{FR,min} \leq M_{FR} \leq M_{FR,max}, \;\; 0 \leq P_{FC} \leq P_{FC,max}\Big\}. \tag{33}$$

The constraints on the SOC and M_{FR} in Equation (33) reflect the minimum and maximum values of the battery state of charge and the mass of fuel, respectively. The constraints on P_{FC} are reflective of the fact that the fuel cell load demand power cannot (i) exceed the maximum power of the fuel cell and (ii) be negative.

The optimal control policy developed in this paper through the application of DCOC specifies the change in fuel cell load power over one step, $\Delta P_{FC}(t) = P_{FC}(t+1) - P_{FC}(t)$, as a function of $SOC(t)$, the mass of hydrogen fuel left, $M_{FR}(t)$, and the current fuel cell load power, $P_{FC}(t)$. The battery power complements fuel cell power in matching propeller requested power.

3.3. Markov Chain Modeling

A Markov chain model [39] is used to represent the evolution of $\boldsymbol{w}$ (in our case, $\boldsymbol{w} = [w_1, w_2]$). The transition probabilities of the Markov chain are defined as:

$$p_{ij} = prob\{\boldsymbol{w}(t_{n+1}) \in W_j \mid \boldsymbol{w}(t_n) \in W_i\}, \tag{34}$$

where W_i and W_j $(i, j = 1, \cdots, N)$ are cells partitioning the feasible range of the operating conditions. The state dependence of the transition probabilities adds flexibility in reflecting the typical motor power and climb angle profiles of an sUAV.

The p_{ij}'s can be obtained from the statistical analysis of the historical flight data,

$$p_{ij} = \frac{M_{ij}}{M_i}, \tag{35}$$

where M_{ij} is the total number of transitions from the cell W_i to the cell W_j (i.e., $\boldsymbol{w}(t_n) \in W_i, \boldsymbol{w}(t_{n+1}) \in W_j$), while M_i is the total number of transitions from W_i to any other cell, including W_i [21].

4. Control Law Construction

Here, we adopt the SDCOC framework from [16], which is applied to a discrete-time model with the following form,

$$\boldsymbol{x}(t_{n+1}) = f(\boldsymbol{x}(t_n), \boldsymbol{u}(t_n), \boldsymbol{w}(t_n)), \tag{36}$$

where $\boldsymbol{x}(t_n)$ is the state vector, $\boldsymbol{u}(t_n)$ is the control vector, and $\boldsymbol{w}(t_n)$ is the vector of operating variables, which is not known until the time instant t_n. The system has both control constraints and state constraints imposed as $\boldsymbol{u}(t_n) \in U$ and $\{\boldsymbol{x}(t_n), \boldsymbol{w}(t_n)\} \in G$, respectively, where U and G are specified sets. A Markov chain with a finite number of states is used to represent transitions in $\boldsymbol{w}(t_n) \in W = \{\boldsymbol{w}^p : p \in P\}$. Here, P is the size of the grid for $\boldsymbol{w}$. The transition probability from $\boldsymbol{w}(t_n) = \boldsymbol{w}^i \in W$ to $\boldsymbol{w}(t_{n+1}) = \boldsymbol{w}^j \in W$ is denoted by p_{ij}, expressed in (34). In a discounted variant of SDCOC, the objective is to determine a control function $\boldsymbol{u}(\boldsymbol{x}, \boldsymbol{w})$ such that, with $\boldsymbol{u}(t_n) = \boldsymbol{u}(\boldsymbol{x}(t_n), \boldsymbol{w}(t_n))$, a cost functional of the form,

$$J^{\boldsymbol{x}_0, \boldsymbol{w}_0, \boldsymbol{u}} = \mathbb{E}_{\boldsymbol{x}_0, \boldsymbol{w}_0}\left[\sum_{t=0}^{\tau^{\boldsymbol{x}_0, \boldsymbol{w}_0, \boldsymbol{u}}(G)-1} \delta^t \cdot 1\right], \tag{37}$$

is maximized. Here, $\tau^{\boldsymbol{x}_0, \boldsymbol{w}_0, \boldsymbol{u}}(G) \in \mathbb{Z}^+$ represents the first time instant when the trajectories of $\boldsymbol{x}(t_n)$ and $\boldsymbol{w}(t_n)$, which are denoted by $\{\boldsymbol{x}^u, \boldsymbol{w}^u\}$ and result from applying the control $\boldsymbol{u}(t_n) = \boldsymbol{u}(\boldsymbol{x}(t_n), \boldsymbol{w}(t_n))$ with values in the set U, exit the prescribed compact set G. δ is a discount factor [40]. For $\delta = 1$, (37) maximizes the exit time, i.e., the time till the prescribed constraints become violated. The use of the discount factor $0 < \delta < 1$ facilitates faster convergence of the value iterations. Note that $\{\boldsymbol{x}^u, \boldsymbol{w}^u\}$ is a random process, $\tau^{\boldsymbol{x}_0, \boldsymbol{w}_0, \boldsymbol{u}}(G)$ is a random variable, and $\mathbb{E}_{\boldsymbol{x}_0, \boldsymbol{w}_0}[\cdot]$ denotes the conditional expectation given the initial values of $\boldsymbol{x}$ and $\boldsymbol{w}$.

To solve (37), the value iterations approach is used, which produces a sequence of value function approximations, V_n, at specified grid-points $\boldsymbol{x} \in \{\boldsymbol{x}^k : k \in K\}$,

$$V_0(\boldsymbol{x}^k, \boldsymbol{w}^i) \equiv 0,$$

$$V_n(\boldsymbol{x}^k, \boldsymbol{w}^i) = \max_{\boldsymbol{u}^m, m \in M}\left\{\sum_{j \in J} F_{n-1}(f(\boldsymbol{x}^k, \boldsymbol{u}^m, \boldsymbol{w}^i), \boldsymbol{w}^j) \cdot p_{ij} \cdot \delta^t + 1\right\},$$

where $\boldsymbol{u} \in \{\boldsymbol{u}^m : m \in M\}$ is a specified grid for $\boldsymbol{u}$. Here, K and M are the size of the grid for $\boldsymbol{x}$ and $\boldsymbol{u}$, respectively. In each iteration, once the values of V_{n-1} at the grid-points have been determined, linear or cubic interpolation is employed to approximate $V_{n-1}(f(x^k, u^m, w^i), w^j)$ as $F_{n-1}(x, w^i) = \text{Interpolate}[V_{n-1}](x, w^i)$, if $(x, w^i) \in G$, and $F_{n-1}(x, w^i) = 0$, if $(x, w^i) \notin G$. A termination criterion of the form $|V_n(x, w^i) - V_{n-1}(x, w^i)| \leq \epsilon$ for all $x \in \{x^k : k \in K\}$ and $i \in P$, where $\epsilon > 0$ is sufficiently small, is used.

Once an approximation of the value function, V_*, is available, an optimal control law is determined as:

$$u_*(x, w^i) \in \left\{ u:\ V_*(x, w^i) - \sum_{j \in J} V_*(f(x, u, w^i), w^j) \cdot p_{ij} \cdot \delta - 1 \leq \epsilon \right\}.$$

5. Control Law Computations and Results

5.1. sUAV Configuration and Model Parameters

The model was parameterized for a 1.5 kg sUAV [23] that can be used for aerial photography and environmental monitoring applications. The minimum and maximum SOC values were set to $SOC_{min} = 0.2$ and $SOC_{max} = 0.8$. The minimum and maximum values of M_{FR} were set as $M_{FR_{min}} = 2$ g and $M_{FR_{max}} = 9$ g. For the value iterations, the SOC grid was chosen with a step size of 0.05, and the M_{FR} grid was chosen with a step size of 0.5 g. The grid for the control variable u was set as $\{-1, 0, 1\}$.

The transition probabilities for the operating variables (motor power and climb angle) were obtained from the time histories of the sUAV motor power and climb angle using (35) and assuming a time step $\Delta t = 1$ s. These time histories were based on a scenario in which an sUAV follows a moving ground vehicle that sUAV operators are interested in monitoring. In this scenario, the ground vehicle, and consequently the sUAV, is assumed to be traveling with the velocity profile defined by concatenating the EPA Highway Cycle [41] nine times. For the sUAV, the speed profile is modified to remain above the stall speed while avoiding extreme acceleration values.

The climb angle time history, shown in Figure 3, was obtained from the Google Earth elevation profile for a path from Monroe, West Virginia, to Princeton, West Virginia, with the help of GPS visualizing software [42]. See [43] for the assessment of the accuracy of such extracted profiles.

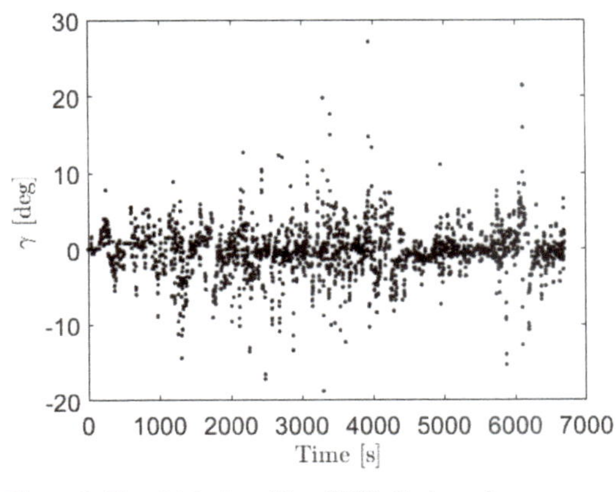

Figure 3. Time histories of the sUAV climb angle.

Figure 4 provides the time histories of the sUAV motor power calculated based on the equations in Section 2.3. The trajectories in Figures 3 and 4 were used to compute the transition probabilities.

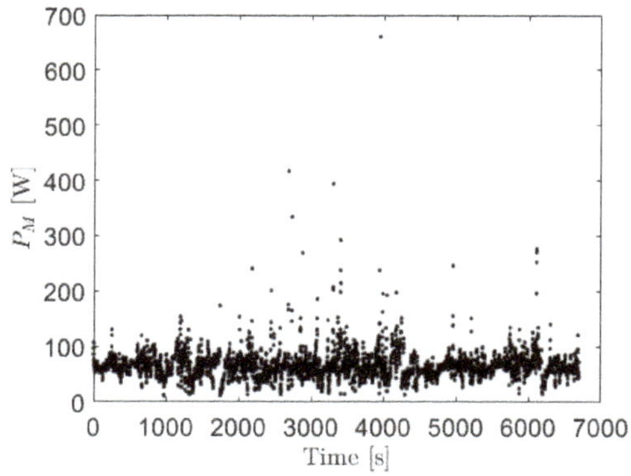

Figure 4. Time histories of the sUAV motor power.

5.2. Control Law Computation

`Cython` was used for control law computations as it is more efficient than `MATLAB` in handling nested `for` loops and two-dimensional interpolation. In our numerical experiments with dynamic programming, `Cython` was about 10 times faster than `MATLAB`.

To further speed up the value iterations, a discount factor was introduced. When testing the effect of the discount factor on the optimal policy, a zero climb angle ($\gamma = 0$) was assumed, which means that the only operating variable was the motor power. Table 2 shows the average exit time based on 100 random simulations for discount factors from 0.91 to 0.99. The stopping criterion was chosen with $\epsilon = 10^{-10}$ for all δ. Computations were performed on a Hasee K780G-i7 laptop with a CORE i7-4710MQ (2.5–3.5 GHz) processor and 24 GB of RAM.Note that the number of iterations and the computing time decrease as the discount factor decreases, but so does the exit time. The discount factor $\delta = 0.95$ was ultimately chosen as a compromise between value iteration convergence speed and solution accuracy. Figure 5 shows that the value iterations with a discount factor of $\delta = 0.95$ converge much faster than those with $\delta = 1$.

Table 2. Average exit time for different discount factor.

δ	Number of Iteration	Computing Time (min)	Exit Time with 20% Initial *SOC* (s)	Exit Time with 80% Initial *SOC* (s)
0.99	2258	830.02	6358.44	9742.99
0.97	753	100.69	6276.18	9716.86
0.95	448	58.27	6221.37	9640.24
0.93	317	39.22	6186.50	9610.22
0.91	244	30.55	6159.65	9602.55

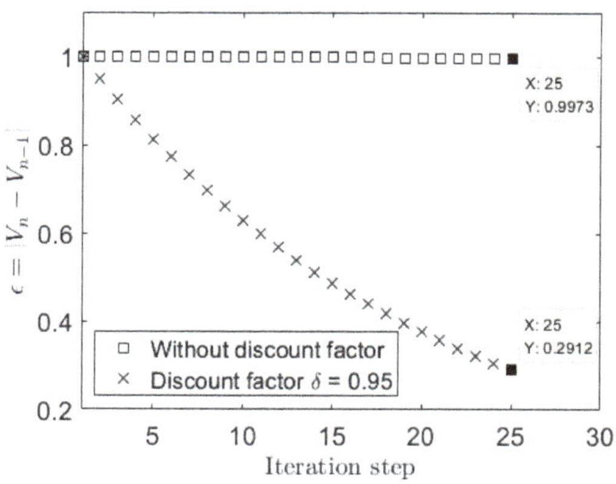

Figure 5. The effect of the discount factor in the value iteration approach.

5.3. Endurance Maximization Results

We used $\epsilon = 10^{-10}$ in the stopping criterion for the value iterations. Figure 6 illustrates the resulting control policy. Note that when SOC is low, the control policy calls for an increase in P_{FC} to charge the battery. This is reasonable given that the fuel cell cannot alone respond rapidly to fast changes in motor power request, and hence, the battery has to be charged to do so.

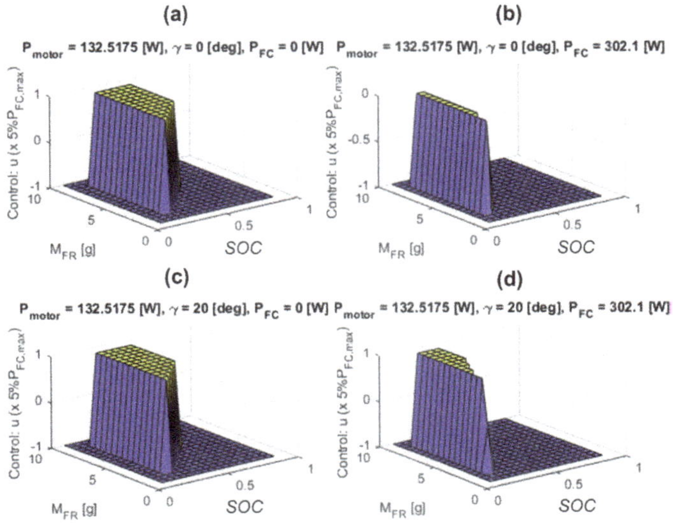

Figure 6. A cross-section of the control policy in the endurance maximization problem for $P_M = 132.52$ W and with (**a**) $\gamma = 0$ deg, $P_{FC} = 0$ W, (**b**) $\gamma = 0$ deg, $P_{FC} = 302.1$ W, (**c**) $\gamma = 20$ deg, $P_{FC} = 0$ W, and (**d**) $\gamma = 20$ deg, $P_{FC} = 302.1$ W.

The simulation results are given for three cases in Figures 7–18. The first case (Scenario I) corresponds to a higher initial SOC, and the second case (Scenario II) considers a lower initial SOC. The third scenario is for a mid-range initial SOC and is used to confirm the SOC behavior observed in the first two scenarios. In all cases, the initial fuel mass and initial fuel cell power are the same: $M_{FR,0} = 6$ g and $P_{FC,0} = 0$ W. The dashed lines in Figures 8, 12, and 16 indicate the constraints mentioned in Section 5.1. The spikes of power in Figures 7, 11, and 15 correspond to the time instants when the sUAV starts to accelerate while the positive and negative spikes of climb angle represent the time when the sUAV starts to climb or descend.

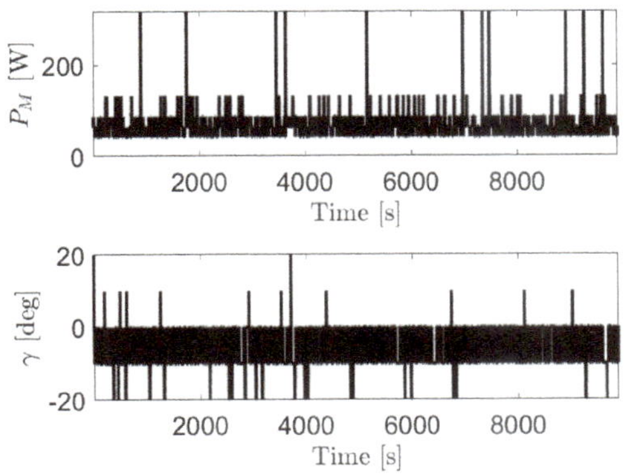

Figure 7. sUAV P_M and γ versus time, Simulation Scenario I.

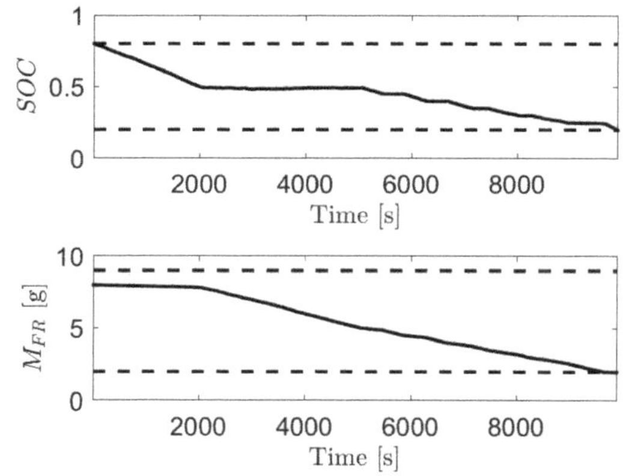

Figure 8. *SOC* and remaining M_{FR} versus time, Simulation Scenario I; dashed lines show constraints.

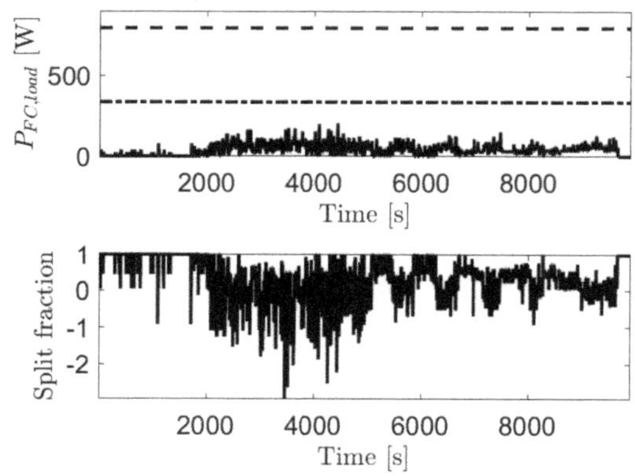

Figure 9. Fuel cell load demand power and split fraction versus time, Simulation Scenario I. The dashed and dashed-dotted lines in the top sub-figure indicate the maximum P_{FC} with $|\gamma| = 0$ deg and $|\gamma| = 20$ deg, respectively.

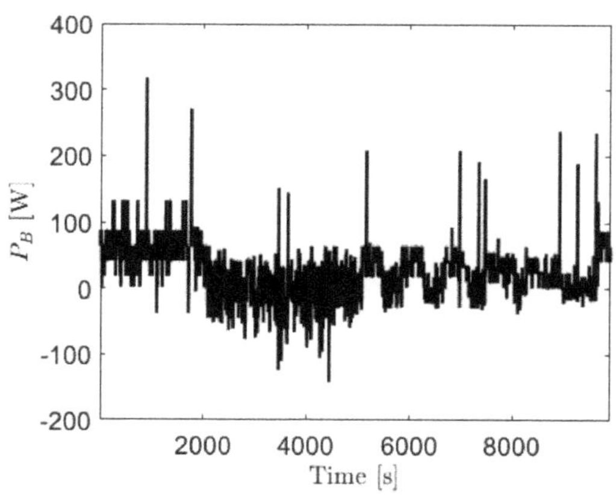

Figure 10. Battery power, Simulation Scenario I.

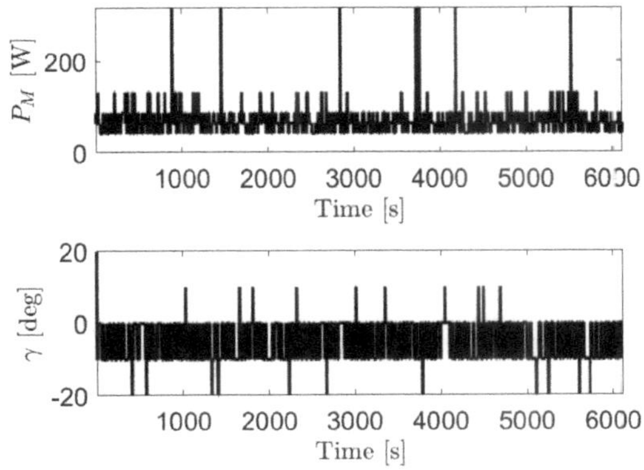

Figure 11. sUAV P_M and γ versus time, Simulation Scenario II.

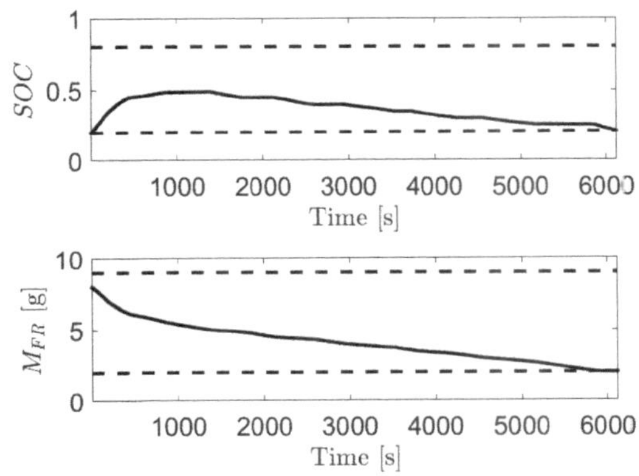

Figure 12. *SOC* and remaining M_{FR} versus time, Simulation Scenario II; dashed lines show constraints.

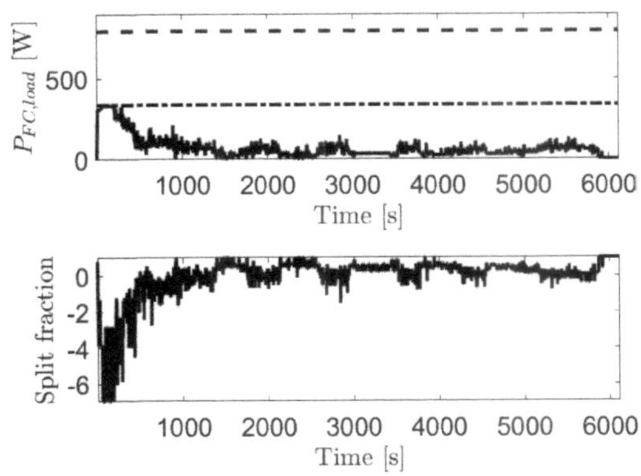

Figure 13. Fuel cell load demand power and split fraction versus time, Simulation Scenario II. The dashed and dashed-dotted lines in the top figure indicate the maximum P_{FC} with $|\gamma|$ = 0 deg and $|\gamma|$ = 20 deg, respectively.

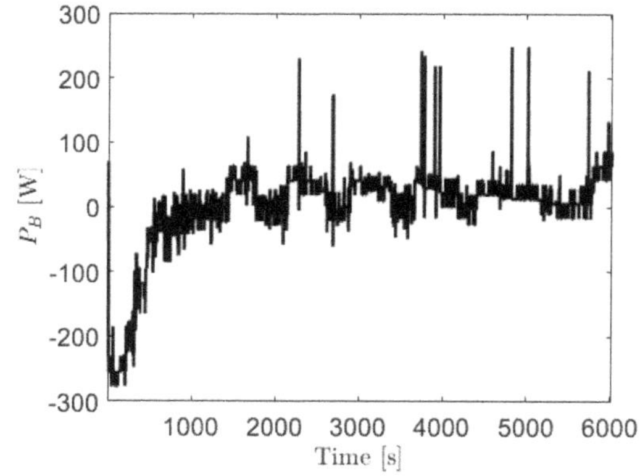

Figure 14. Battery power, Simulation Scenario II.

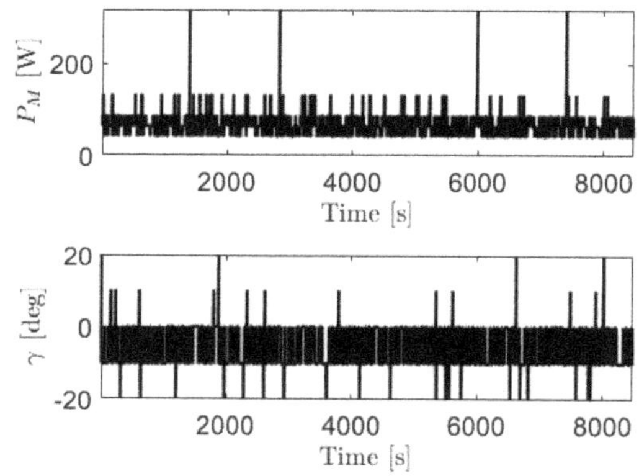

Figure 15. sUAV P_M and γ versus time, Simulation Scenario III.

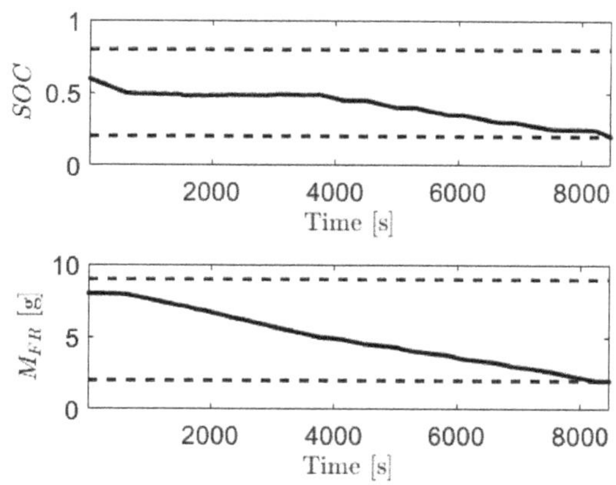

Figure 16. SOC and remaining M_{FR} versus time, Simulation Scenario III; dashed lines show constraints.

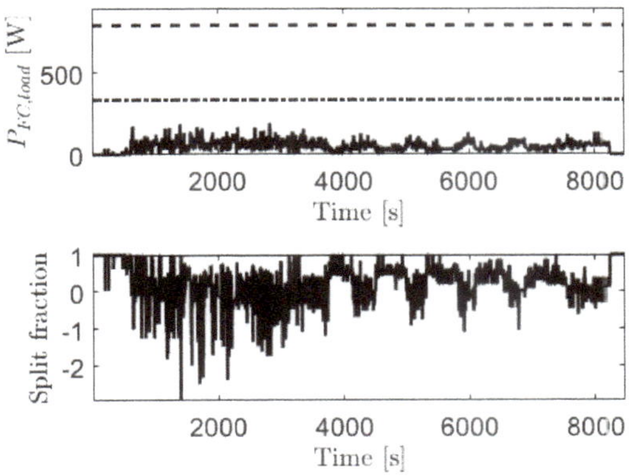

Figure 17. Fuel cell load demand power and split fraction versus time, Simulation Scenario III. The dashed and dashed-dotted lines in the top figure indicate the maximum P_{FC} with $|\gamma| = 0$ deg and $|\gamma| = 20$ deg, respectively.

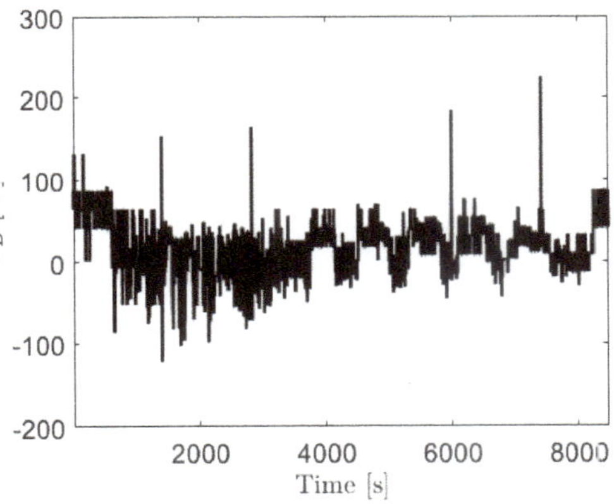

Figure 18. Battery power, Simulation Scenario III.

Figures 7–10 illustrate the closed-loop response for the first simulation scenario. The initial SOC is 0.8, and it decreases rapidly until it reaches a value of about 0.5. Then, it stays near this value between 2000 and 5000 s. Finally, when the mass of hydrogen reaches a relatively low value, the SOC starts to decrease and continues to decrease until the constraints are violated. The fuel cell load demand power keeps a relatively low value during the whole flight, and the mean value of the split fraction is negative during the 2000 to 5000 s time interval, which is the period when the SOC is kept at about 0.5.

Figures 11–13 illustrate the closed-loop response for the second simulation scenario. The initial SOC is 0.2. The battery is charged until it reaches a value of about 0.5 to enable the battery to sustain rapid propeller power fluctuations. Then, the SOC stays near that value of 0.5 between 500 and 1500 s. Finally, when the mass of hydrogen reaches a relatively low value, the SOC starts to decrease and continues to decrease until the constraints are violated. The fuel cell load demand power increases rapidly at first to charge the battery, then it keeps a relatively low value during the rest of the flight. The mean value of the split fraction is negative from the beginning to about 1500 s, which is the period when the battery is charged from $SOC = 0.2$ to about 0.5.

According to the results from Scenarios I and II, a turnpike behavior of the battery SOC is observed, with the SOC converging to about 0.5 and staying at that value for a while before decaying. To confirm this turnpike behavior, we additionally considered the responses with the developed policy to the initial SOC of 0.6. These are shown in

Figures 15–18. The exit times for Scenarios I, II, and III were, respectively, 9890 s, 6019 s, and 8478 s.

6. Conclusions

This paper considers an endurance maximization problem for a small unmanned aerial vehicle (sUAV) with a hybrid propulsion system consisting of a polymer electrolyte fuel cell and a battery, both driving an electric motor connected to a propeller. A stochastic drift counteraction optimal control (SDCOC) approach is employed to develop control policies for optimally coordinating the fuel cell and the battery while enforcing the constraints on the fuel cell power output rate of change. `Cython` is used to implement value iterations and demonstrated an order of magnitude speedup versus `MATLAB` without increasing the code complexity, due to its efficiency in handling nested `for`-loops. Additionally, the use of a discount factor is shown to significantly speed up the value iterations at the price of decreased performance. The results illustrate the effectiveness of the SDCOC strategy in regulating the charging behavior of the battery by the fuel cell to provide the capability to respond to rapidly varying motor power demand.

The proposed approach based on SDCOC is particularly suitable for handling stochastic disturbances and can be applied to sUAVs exposed to headwinds with the headwind modeled as a stochastic disturbance. Accounting for such wind disturbances, extensions to include thermal dynamics, systematic and comprehensive comparison with other energy management approaches and propulsion system choices, the systematic study of the robustness to model uncertainties, as well as actual flight experiments represent directions for continuing research. In particular, our study of the discount factor impact on the computation time and exit time suggests that the flight time is sensitive to the choice of the energy management strategy; our approach based on SDCOC is optimal in the sense of maximizing expected flight time within a stochastically modeled environment.

Author Contributions: Conceptualization, J.Z., I.K., and M.R.A.; methodology, J.Z. and I.K.; software, J.Z. and I.K.; validation, J.Z. and I.K.; formal analysis, J.Z., I.K., and M.R.A.; investigation, J.Z., I.K., and M.R.A.; resources, J.Z., I.K., and M.R.A.; data curation, J.Z., I.K., and M.R.A.; writing—original draft preparation, J.Z., I.K., and M.R.A.; writing—review and editing, I.K. and M.R.A.; visualization, J.Z.; supervision, I.K.; project administration, I.K. All authors read and agreed to the published version of the manuscript.

Funding: This research was supported in part by the U.S. National Science Foundation (NSF) under Award Number ECCS-1931738.

Institutional Review Board Statement: Not applicable.

Informed Consent Statement: Not applicable.

Data Availability Statement: Not applicable.

Conflicts of Interest: The authors declare no conflict of interest.

Abbreviations

The following abbreviations are used in this manuscript:

EPA	Environmental Protection Agency
DMFC	Direct methanol fuel cell
sUAV	Small unmanned aerial vehicle
MPC	Model predictive control
PEMFC	Polymer electrolyte membrane fuel cell
RPM	Rotations per minute
UAV	Unmanned aerial vehicle

Appendix A

The parameters of the sUAV model described in the paper are listed in Table A1.

Table A1. Parameters used in the sUAV model.

Variable	Description	Value	Unit
m	Mass of the sUAV	1.5	kg
g	Gravitational acceleration	9.81	m/s^2
S_{ref}	Wing area	0.09	m^2
C_{D0}	sUAV coefficient of drag at $\alpha = 0$	0.1038	\
K	Coefficient in Equation (3)	0.0637	\
d_P	Diameter of the propeller	0.24	m
J	Advance ratio	0.37	\
R_M	Motor resistance	0.105	Ω
$I_{M,0}$	Motor current at zero load	1.3	A
K_V	Motor speed constant	1490	RPM/V
n_{FC}	Number of single cells in series	12	\
P_0	Bias power of the fuel cell	5	W
κ_{aux}	Coefficient in Equation (12)	0.05	V
A_{FC}	Fuel cell area	200	cm^2
Δg_f	Molar specific Gibbs free energy	237.3	kJ/mol
n_e	Number of ions passed in reaction	2	\
F	Faraday constant	96,485	C/mol
$\bar{T}$	Temperature of the reaction	333.15	K
α_{FC}	Charge transfer coefficient	0.5	\
R	Universal gas constant	8.314	J/(mol · K)
R_{FC}	Ohmic resistance defined in Equation (15)	0.0024	Ω
d_0	Coefficient in Equation (16)	3e-5	V
d_1	Coefficient in Equation (16)	8	cm^2/A
k_0	Coefficient in Equation (18)	4	\
k_1	Coefficient in Equation (18)	1	\
M_h	Molecular weight of H_2	2	g/mol
n_B	Number of batteries in series	8	\
$U_{B,min}$	Open circuit voltage when $SOC = 0$	2.5	V
$U_{B,max}$	Open circuit voltage when $SOC = 1$	4.2	V
$R_{B,int}$	Battery internal resistance	0.012	Ω
C_B	Standard discharge capacity	14400	C
$I_{B,max}$	Maximum discharge current	35	A

References

1. Unmanned Aerial Vehicle (UAV) Market by Component (Hardware, Software), Class (Mini UAVs, Micro UAVs), End User (Military, Commercial, Agriculture), Type (Fixed Wing, Rotary-Wing UAVs), Capacity, and Mode of Operation—Global Forecast to 2027. Available online: https://www.meticulousresearch.com/product/unmanned-aerial-vehicle-UAV-market-5086 (accessed on 16 February 2021).
2. Kasliwal, A.; Furbush, N.J.; Gawron, J.H.; McBride, J.R.; Wallington, T.J.; De Kleine, R.D.; Kim, H.C.; Keoleian, G.A. Role of flying cars in sustainable mobility. *Nat. Commun.* **2019**, *10*, 1–9. [CrossRef] [PubMed]
3. Goyal, R. *Urban Air Mobility (uam) Market Study*; NASA: Washington, DC, USA, 2018.
4. Samy, I.; Postlethwaite, I.; Gu, D.W.; Green, J. Neural-network-based flush air data sensing system demonstrated on a mini air vehicle. *J. Aircr.* **2010**, *47*, 18–31. [CrossRef]
5. Ma, S.; Lin, M.; Lin, T.E.; Lan, T.; Liao, X.; Maréchal, F.; Yang, Y.; Dong, C.; Wang, L. Fuel cell-battery hybrid systems for mobility and off-grid applications: A review. *Renew. Sustain. Energy Rev.* **2020**, *135*, 110119. [CrossRef]
6. Lapeña-Rey, N.; Blanco, J.; Ferreyra, E.; Lemus, J.; Pereira, S.; Serrot, E. A fuel cell powered unmanned aerial vehicle for low altitude surveillance missions. *Int. J. Hydrogen Energy* **2017**, *42*, 6926–6940. [CrossRef]
7. Oh, T.H. Conceptual design of small unmanned aerial vehicle with proton exchange membrane fuel cell system for long endurance mission. *Energy Convers. Manag.* **2018**, *176*, 349–356. [CrossRef]

8. Bradley, T.; Moffitt, B.; Fuller, T.; Mavris, D.; Parekh, D. Design studies for hydrogen fuel cell powered unmanned aerial vehicles. In Proceedings of the 26th AIAA Applied Aerodynamics Conference, Honolulu, HI, USA, 18–21 August 2008; p. 6413.
9. Bradley, T.H.; Moffitt, B.A.; Mavris, D.N.; Parekh, D.E. Development and experimental characterization of a fuel cell powered aircraft. *J. Power Sources* **2007**, *171*, 793–801. [CrossRef]
10. Citroni, R.; Di Paolo, F.; Livreri, P. A novel energy harvester for powering small UAVs: Performance analysis, model validation and flight results. *Sensors* **2019**, *19*, 1771. [CrossRef] [PubMed]
11. Hochgraf, C.G.; Ryan, M.J.; Wiegman, H.L. *Engine Control Strategy for a Series Hybrid Electric Vehicle Incorporating Load-Leveling and Computer Controlled Energy Management*; SAE Technical Paper No. 1996–02-01; SAE: Detroit, MI, USA, 1996.
12. Bradley, T.; Moffitt, B.; Parekh, D.; Fuller, T.; Mavris, D. Energy management for fuel cell powered hybrid-electric aircraft. In Proceedings of the 7th International Energy Conversion Engineering Conference, Denver, CO, USA, 2–5 August 2009; p. 4590.
13. Doff-Sotta, M.; Cannon, M.; Bacic, M. Optimal energy management for hybrid electric aircraft. *arXiv* **2020**, arXiv:2004.02582.
14. Anatone, M.; Cipollone, R.; Donati, A.; Sciarretta, A. *Control-Oriented Modeling and Fuel Optimal Control of a Series Hybrid Bus*; SAE Technical Paper No. 2005–04-11; SAE: Detroit, MI, USA, 2005.
15. Dobrokhodov, V.; Jones, K.D.; Walton, C.; Kaminer, I.I. Achievable Endurance of Hybrid UAV Operating in Time-Varying Energy Fields. In Proceedings of the AIAA Scitech 2020 Forum, Orlando, FL, USA, 6–10 January 2020; p. 2197.
16. Kolmanovsky, I.V.; Lezhnev, L.; Maizenberg, T.L. Discrete-time drift counteraction stochastic optimal control: Theory and application-motivated examples. *Automatica* **2008**, *44*, 177–184. [CrossRef]
17. Balasubramanian, K.; Kolmanovsky, I.; Saha, B. Range maximization of a direct methanol fuel cell powered Mini Air Vehicle using Stochastic Drift Counteraction Optimal Control. In Proceedings of the 2012 American Control Conference (ACC), Fairmont Queen Elizabeth, MO, USA, 27–29 June 2012; pp. 3272–3277.
18. Guzzella, L.; Sciarretta, A. *Vehicle Propulsion Systems: Introduction to Modeling and Pptimization*; Springer: Berlin, Germany, 2013; Chapter 6; pp. 207–208. [CrossRef]
19. Guida, D.; Minutillo, M. Design methodology for a PEM fuel cell power system in a more electrical aircraft. *Appl. Energy* **2017**, *192*, 446–456. [CrossRef]
20. Wipke, K.; Sprik, S.; Kurtz, J.; Ramsden, T.; Ainscough, C.; Saur, G. *All Composite Data Products: National FCEV Learning Demonstration with Updates through January 18, 2012*; Technical Report, NREL/TP-5600-54021; National Renewable Energy Lab. (NREL): Golden, CO, USA, 2012.
21. Kolmanovsky, I.; Siverguina, I.; Lygoe, B. Optimization of powertrain operating policy for feasibility assessment and calibration: Stochastic dynamic programming approach. In Proceedings of the 2002 American Control Conference (ACC), Anchorage, AK, USA, 8–10 May 2002; Volume 2, pp. 1425–1430.
22. Roskam, J.; Lan, C. *Airplane Aerodynamics and Performance*; DARcorporation: Lawrence, KS, USA, 2003.
23. Hrad, P.M. *Conceptual Design Tool for Fuel-Cell Powered Micro Air Vehicles*; Technical Report; Air Force Institute of Technology: Dayton, OH, USA, 2010.
24. Rodatz, P.H. Dynamics of the polymer electrolyte fuel cell: Experiments and model-based analysis. Ph.D. Thesis, ETH, Zurich, Switzerland, 2003.
25. Specification of Product INR21700-40T. Available online: https://www.dnkpower.com/wp-content/uploads/2019/02/SAMSUNG-INR21700-40T-Datasheet.pdf (accessed on 16 February 2021).
26. Bronz, M.; Moschetta, J.M.; Brisset, P.; Gorraz, M. Towards a long endurance MAV. *Int. J. Micro Air Veh.* **2009**, *1*, 241–254. [CrossRef]
27. Murphy, O.J.; Cisar, A.; Clarke, E. Low-cost light weight high power density PEM fuel cell stack. *Electrochim. Acta* **1998**, *43*, 3829–3840. [CrossRef]
28. Quattrochi, D. Performance of Propellers. Available online: https://web.mit.edu/16.unified/www/FALL/thermodynamics/notes/node86.html (accessed on 16 February 2021).
29. Gur, O.; Rosen, A. Optimizing electric propulsion systems for UAVs. In Proceedings of the 12th AIAA/ISSMO Multidisciplinary Analysis and Optimization Conference, Victoria, BC, Canada, 10–12 September 2008; p. 5916.
30. Guzzella, L.; Amstutz, A. CAE tools for quasi-static modeling and optimization of hybrid powertrains. *IEEE Trans. Veh. Technol.* **1999**, *48*, 1762–1769. [CrossRef]
31. Mann, R.F.; Amphlett, J.C.; Hooper, M.A.; Jensen, H.M.; Peppley, B.A.; Roberge, P.R. Development and application of a generalised steady-state electrochemical model for a PEM fuel cell. *J. Power Sources* **2000**, *86*, 173–180. [CrossRef]
32. Maxoulis, C.N.; Tsinoglou, D.N.; Koltsakis, G.C. Modeling of automotive fuel cell operation in driving cycles. *Energy Convers. Manag.* **2004**, *45*, 559–573. [CrossRef]
33. Amphlett, J.; Baumert, R.; Mann, R.; Peppley, B.; Roberge, P.; Rodrigues, A. Parametric modelling of the performance of a 5-kW proton-exchange membrane fuel cell stack. *J. Power Sources* **1994**, *49*, 349–356. [CrossRef]
34. Pukrushpan, J.T.; Peng, H.; Stefanopoulou, A.G. Control-oriented modeling and analysis for automotive fuel cell systems. *J. Dyn. Syst. Meas. Control* **2004**, *126*, 14–25. [CrossRef]
35. Bansal, D.; Rajagopalan, S.; Choi, T.; Guezennec, Y.; Yurkovich, S. Pressure and air fuel ratio control of pem fuel cell system for automotive traction. In Proceedings of the IEEE Vehicle Power and Propulsion Conference, Gijón, Spain, 26–29 September 2004.
36. El-Emam, S.H.; Mousa, A.A.; Awad, M.M. Effects of stack orientation and vibration on the performance of PEM fuel cell. *Int. J. Energy Res.* **2015**, *39*, 75–83. [CrossRef]

37. Zhang, R.; Xia, B.; Li, B.; Cao, L.; Lai, Y.; Zheng, W.; Wang, H.; Wang, W.; Wang, M. A study on the open circuit voltage and state of charge characterization of high capacity lithium-ion battery under different temperature. *Energies* **2018**, *11*, 2408. [CrossRef]
38. Vahidi, A.; Greenwell, W. A decentralized model predictive control approach to power management of a fuel cell-ultracapacitor hybrid. In Proceedings of the 2007 American Control Conference (ACC), New York, NY, USA, 9–13 July 2007; pp. 5431–5437.
39. Dynkin, E.B.; Yushkevich, A.A. *Markov Processes: Theorems and Problems;* Plenum: New York, NY, USA, 1969.
40. Ross, S.M. *Introduction to Stochastic Dynamic Programming;* Academic Press: Cambridge, MA, USA, 2014.
41. U.S. Department of Energy—Fuel Economy. Detailed Test Information. Available online: http://fueleconomy.gov/feg/fe_test_schedules.shtml (accessed on 16 February 2021).
42. Schneider, A. GPS Visualizer. Available online: http://www.gpsvisualizer.com/ (accessed on 16 February 2021).
43. Wang, Y.; Zou, Y.; Henrickson, K.; Wang, Y.; Tang, J.; Park, B.J. Google Earth elevation data extraction and accuracy assessment for transportation applications. *PLoS ONE* **2017**, *12*, e0175756. [CrossRef] [PubMed]

energies

Article

Estimation of Tire Normal Forces including Suspension Dynamics

Louis Filipozzi [1,*], Francis Assadian [1], Ming Kuang [2], Rajit Johri [2] and Jose Velazquez Alcantar [2]

1 Department of Mechanical and Aerospace Engineering, University of California, Davis, CA 95616, USA; fassadian@ucdavis.edu
2 Advanced Research and Engineering, Ford Motor Company, Dearborn, MI 48126, USA; mkuang@ford.com (M.K.); rjohri@ford.com (R.J.); jvelaz42@ford.com (J.V.A.)
* Correspondence: lfilipozzi@ucdavis.edu

Abstract: Tire normal forces are difficult to measure, but information on the vehicle normal force can be used in many automotive engineering applications, e.g., rollover detection and vehicle and wheel stability. Previous papers use algebraic equations to estimate the tire normal force. In this article, the estimation of tire normal force is formulated as an input estimation problem. Two observers are proposed to solve this problem by using a quarter-car suspension model. First, the Youla Controller Output Observer framework is presented. It converts the estimation problem into a control problem and produces a Youla parameterized controller as observer. Second, a Kalman filter approach is taken and the input estimation problem is addressed with an Unbiased Minimum Variance Filter. Both methods use accelerometer and suspension deflection sensors to determine the vehicle normal force. The design of the observers is validated in simulation and a sensitivity analysis is performed to evaluate their robustness.

Keywords: normal force estimation; unbiased minimum variance estimation; controller output observer; youla parameterization

Citation: Filipozzi, L.; Assadian, F.; Kuang, M.; Johri, R.; Velazquez Alcantar, J. Estimation of Tire Normal Forces including Suspension Dynamics. *Energies* **2021**, *14*, 2378. https://doi.org/10.3390/en14092378

Academic Editors: Wiseman Yair and Francesco Bellotti

Received: 3 March 2021
Accepted: 20 April 2021
Published: 22 April 2021

1. Introduction

The automotive industry has made significant improvements in vehicle safety and driving performance in the last decades thanks to active control systems. These control systems rely on the measurement and estimation of several parameters and signals such as the wheel slip, sideslip angle. Tire normal forces can also be used to improve the vehicle safety and performance. Indeed, the longitudinal and lateral tire forces are coupled with the wheel loads [1]. Cho [2] showed that the estimate of tire forces, including tire loads, can be used to implement Global Chassis Control (GCC) systems and to further improve vehicle stability. Another practical application of tire normal force estimation is roll-over avoidance and understeering or oversteering prevention [3].

Tire loads are permanently changing when the vehicle is moving. Loads are transferred between wheels during accelerating, braking, and cornering. The position of the center of gravity, the road grade and irregularities on the road also impact the distribution of wheel normal forces making the estimation task a complex problem [4]. Due to the lack of low-cost sensors to measure the vehicle vertical force, a common approach is to consider the normal forces as constant parameters or to use an algebraic expression based on the vehicle longitudinal and lateral accelerations [5]. These open-loop estimation schemes, albeit simple, are not able to give a precise representation of the normal forces.

Doumiati et al. [6,7] provided a cascaded observer to estimate the tire normal forces. The first step of the algorithm estimates the lateral load transfer using a linear Kalman filter from the suspension deflection and accelerometer measurements and an estimate of the vehicle mass. A second observer is then used to infer the normal tire force from the lateral load transfer and a formulation of the normal force algebraic expression with coupling between longitudinal and lateral acceleration. This yields an Extended Kalman Filter. Jiang [8] extended the application of this estimation framework by adding the vehicle pitch dynamics to take in account the road angle and the road irregularities.

Ozkan [9] used a Controller Output Observer (COO) to estimate lateral and normal tire forces. The estimation model is a half-car model which includes the vehicle lateral, heave, roll, and yaw motion. The model does not include pitch dynamics. Moreover, it assumes a constant vehicle velocity and a perfectly flat road. The COO has been successfully used in other automotive applications, e.g., to estimate the vehicle states [10] or to estimate longitudinal tire forces [11].

The goal of the paper is to provide a framework capable of delivering a reliable estimate of the wheel normal forces using low-calibration estimation methods. Contrary to previous work, the estimation of the wheel force will not be derived from algebraic expressions which estimate the normal forces from the vehicle acceleration. Instead, we intend to integrate the suspension dynamics in the normal force estimator to truly capture the tire force generation using suspension deflection sensors and accelerometers. This manuscript also illustrates the application of a recently developed estimation methodology called Youla Controller Output Observer (YCOO). This new approach is compared to the established method for signal estimation: Kalman filtering. Since the estimation problem is not formulated as a state estimation but an input estimation problem, the Unbiased Minimum Variance Filter (UMVF) is used in place of the standard state-estimation Kalman Filter. The comparison between the two estimation methods is based on several criteria: estimation performance, robustness to uncertainties, and ease of design. The performance is evaluated based on simulation results. In these simulations, we cover the major ways normal force can be generated (load transfer, disturbance from the road profile, and inclined road). The robustness is analyzed by conducting a sensitivity study of the suspension parameters, by introducing discrepancies between the estimation model and the simulation model, and by introducing noise in the measurements. Finally, by considering the different requirements to implement the YCOO and UMVF and the different approaches used by each observer (the UMVF uses a stochastic approach in the time domain while the YCOO relies on a deterministic approach in the frequency domain), we aim to highlight that the YCOO estimation framework is easily implementable with a low-calibration burden to guarantee robustness.

The following section analyzes a vehicle model and provides a model that can be used for closed-loop estimation. Sections 3 and 4 introduce the YCOO and UMVF estimation frameworks. As the normal force estimation is dependent on the vehicle mass estimation problem, Section 5 explains the computation of the vehicle mass estimate, which is used by the two frameworks. Finally, both the YCOO and UMVF are tested in simulation.

2. System Modeling

The generation of vertical forces of a rigid vehicle is linked with the compliance of the suspensions [12]. To develop the estimation frameworks and evaluate the effectiveness, a vehicle model with realistic heave, roll, and pitch dynamics is indispensable. Fortunately, the literature is filled with such vehicle models. Shim [13] described a 14 degrees of freedom vehicle model and validated it against the commercial vehicle models Carsim and ADAMS/Car. Figure 1 shows a schematic of the vehicle model. The degrees of freedom are the longitudinal, lateral, heave, roll, pitch, and yaw motion of the chassis, the vertical dynamics of each unsprung masses, and the wheels spin. The model assumes rigid bodies for the sprung and unsprung masses and neglects the compliance between the chassis and the unsprung mass in the vehicle longitudinal and lateral directions.

The model described in Figure 1 gives an accurate representation of the vehicle. However, it is highly complex which limits its usage to implement vehicle observers. Before beginning the observer development, a linear system analysis of the vehicle is performed. Its purpose is to determine the cross-coupling effects between the input–output pairs to facilitate the observer design. The inputs of the model are the tire loads and the outputs are the suspension deflections.

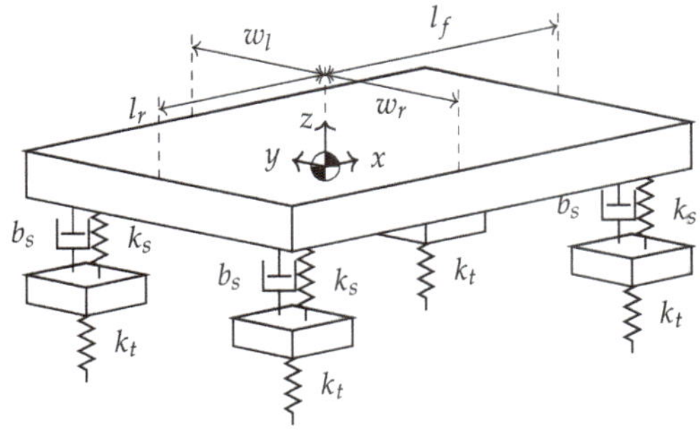

Figure 1. Fourteen degrees of freedom vehicle model. The x, y, and z axes indicate, respectively, the longitudinal, lateral, and vertical directions of the vehicle. Parameters k_s, b_s, and k_t denote the suspension stiffness, damping, and the tire stiffness.

The coupling between input-output pairs of the plant G is analyzed using the Relative Gain Array (RGA) $\Lambda = (G(0)^{-1})^T \times G(0)$ where $G(0)$ is the plant gain [14]. Since RGA is a linear analysis tool, the model mapping the tire loads f_{FLz}, f_{FRz}, and f_{RLz}, f_{RRz} to the suspension deflections q_{FLs}, q_{FRs}, q_{RLs}, and q_{RRs} must first be linearized. The operating point is chosen to be a steady-state cornering such that the vehicle velocity is $v_x = 90\,\mathrm{km\,h^{-1}}$ and the vehicle lateral acceleration is $a_y = 0.4\,\mathrm{g}$ (i.e., the vehicle lateral velocity is $v_y = 0.2\,\mathrm{m\,s^{-1}}$; the heave velocity is $v_z = 0\,\mathrm{m\,s^{-1}}$; and the roll, pitch, and roll angular velocities are $wx = wy = 0\,\mathrm{rad\,s^{-1}}$ and $w_z = 0.15\,\mathrm{rad\,s^{-1}}$). The matrices Λ_1 and Λ_2 correspond, respectively, to the RGA of a vehicle without an anti-roll bar and with an anti-roll bar on the rear axle. The inputs are $u^T = \begin{bmatrix} f_{FLz} & f_{FRz} & f_{RLz} & f_{RRz} \end{bmatrix}$.

$$\Lambda_1 = \begin{bmatrix} 0.9560 & -0.0527 & 0.0474 & 0.0493 \\ -0.0527 & 0.9604 & 0.0452 & 0.0471 \\ 0.0474 & 0.0452 & 0.9497 & -0.0423 \\ 0.0493 & 0.0471 & -0.0423 & 0.9459 \end{bmatrix}, \Lambda_2 = \begin{bmatrix} 0.9560 & -0.0527 & 0.0474 & 0.0493 \\ -0.0527 & 0.9604 & 0.0452 & 0.0471 \\ 0.0478 & 0.0457 & 1.2118 & -0.3053 \\ 0.0488 & 0.0466 & -0.3043 & 1.2089 \end{bmatrix} \quad (1)$$

The matrix Λ_1 is almost equivalent to an identity matrix which indicates that there is almost no coupling between the four corners of the vehicle without anti-roll bar. Concerning the vehicle equipped with anti-roll bar, the two axles decoupled as the bottom left and upper right 2×2 submatrices of Λ_2 are almost zero. However, the anti-roll bar introduces a coupling between the left and right rear normal loads f_{RLz} and f_{RRz}, as can be seen in the bottom right corner of Λ_2.

We assume that the vehicle is not equipped with an anti-roll bar. Thus, the wheel load of each corner of the vehicle can be estimated individually. A simpler model (Figure 2) is introduced to model the suspension of each wheel of the vehicle. The estimation model is a quarter-car model [15] with two inputs: a force applied on the sprung mass to represent the load transfer and another force which represents the tire load. In practice, the anti-roll bar of the vehicle should be considered and the coupling between the left and right normal forces should not be ignored. This coupling would appear as a load transfer from one side to the other when the two wheel loads are not equal, e.g., during cornering. The anti-roll bar would prevent the use of a quarter-car model and require a half-car model.

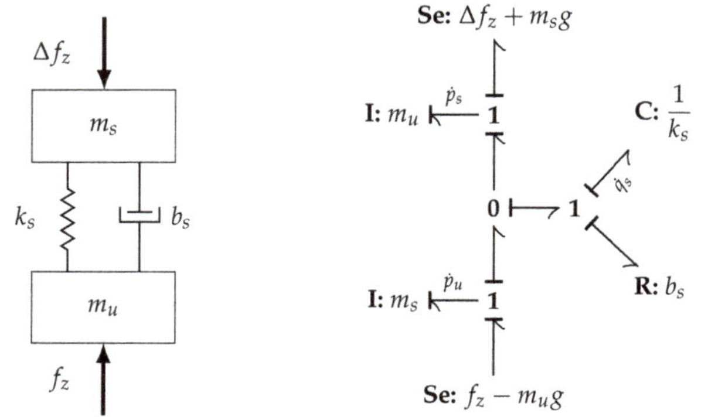

Figure 2. Quarter-car model of the suspension with wheel normal force and load transfer as inputs.

The bond graph of the model in Figure 2 yields the equation of motion of the quarter-car model

$$\dot{p}_s = k_s q_s + b_s \dot{q}_s - m_s g - \Delta f_z \tag{2}$$

$$\dot{p}_u = -k_s q_s - b_s \dot{q}_s - m_u g + f_z \tag{3}$$

$$\dot{q}_s = \frac{p_u}{m_u} - \frac{p_s}{m_s} \tag{4}$$

where p_s and p_u are the sprung mass and unsprung mass momentum, q_s is the suspension deflection, f_z is the wheel normal load, Δf_z represents the load transfer applied to the wheel, k_s is the suspension stiffness, b_s is the damper coefficient, and m_s and m_u are the sprung and unsprung masses of the corner of the vehicle. The states of the model are $x^T = \begin{bmatrix} p_s & p_u & q_s \end{bmatrix}$. It is assumed that the vehicle is equipped with suspension deflection sensors and accelerometers on the sprung mass, hence the measurements are $y^T = \begin{bmatrix} \frac{\dot{p}_s}{m_s} & q_s \end{bmatrix}$. The model inputs are $u^T = \begin{bmatrix} f_z & \Delta f_z \end{bmatrix}$.

The equation of motions of the quarter-car model (Equations (2)–(4)) are not linear but affine. The constant term due to gravity can be eliminated by translating the states, inputs, and outputs of the system as follows

$$\tilde{x}^T = \begin{bmatrix} p_s & p_u & q_s - \frac{m_s g}{k_s} \end{bmatrix}, \quad \tilde{u}^T = \begin{bmatrix} f_z - (m_s + m_u) g & \Delta f_z \end{bmatrix}, \text{ and } \tilde{y}^T = \begin{bmatrix} \frac{\dot{p}_s}{m_s} - g & q_s - \frac{m_s g}{k_s} \end{bmatrix} \tag{5}$$

3. Controller Output Observer

The Youla Controller Output Observer, based on the COO framework [9], is a model-based estimation technique that uses a controller to minimize the error between the measurement and the virtual measured signals of an estimation model. Contrary to the Luenberger observer, the YCOO does not assume that all system inputs are known; it is therefore well suited for an input estimation problem. Instead of designing a static gain controller via pole placement similarly to the Luenberger observer, the YCOO uses a dynamic controller designed with Youla parameterization [16]. This technique allows including information about the sensor dynamics and its noise content in the frequency domain to ensure good robustness and performance. A block diagram of the estimation concept is given in Figure 3. Measurements y are fed to the YCOO to provide an estimate $\hat{u}$ of the signal u. The YCOO is decomposed into two components: an estimation model $\hat{G}_p$ that maps the estimated signal $\hat{u}$ to the virtual measurement $\hat{y}$ and a controller G_c that is responsible to follow those measurements [17].

The transfer function from the the true signal u and the estimated signal $\hat{u}$ is given by $(I + L_u)^{-1} G_c G_p$ where $L_u = G_c \hat{G}_p$ is the return ratio. If there is no discrepancy between the estimated plant and the actual one ($G_p = \hat{G}_p$), then this transfer function corresponds to the closed-loop transfer function T_u. If the plant has multiplicative uncertainty such that $G_p = \hat{G}_p(I + \Delta)$, the relation becomes $\hat{u} = (T_u + Y\Delta)u$ where Y is the Youla transfer function. This shows that the YCOO relies on an accurate model of the system. Indeed, to guarantee good tracking of the measured quantities, $T_u \approx I$ is needed at low frequencies. This condition on T_u also constrains the gain of Y to the inverse of the plant gain at low frequencies since $T_u = YG_p$. If the model is not correct, however, the term $Y\Delta$ introduces a steady-state error that cannot be compensated by the observer unless the plant gain is very high at low-frequency. Moreover, for the nominal system $G_p = \hat{G}_p$, the transfer function mapping the sensor noise to the estimation error is also given by Y. Thanks to its loop shaping approach, the YCOO directly addresses the trade-off between noise rejection, bandwidth, and robustness to high-frequency multiplicative uncertainties. Indeed, a higher bandwidth would increase the gain of Y at higher frequency, making the estimation more sensitive to noise and less robust to multiplicative uncertainties.

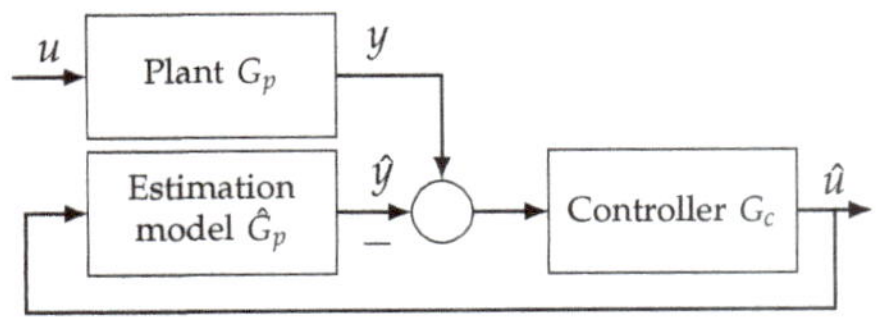

Figure 3. Block diagram of the YCOO estimation concept.

The quarter-car model described in the last section is used as the estimation model $\hat{G}_p$. The Youla parameterization technique is applied to design the controller G_c from the estimation model. The plant model can be written as a transfer function $\hat{G}_p = \frac{P}{\delta}$ mapping the signals $\hat{u}$ to $\hat{y}$.

$$\hat{G}_p = \begin{bmatrix} \frac{\ddot{p}_s}{f_z} & \frac{\ddot{p}_s}{\Delta f_z} \\ \frac{q_s}{f_z} & \frac{q_s}{\Delta f_z} \end{bmatrix} = \frac{1}{m_s m_u s^2 + (m_s + m_u) b_s s + (m_s + m_u) k_s} \begin{bmatrix} k_s + b_s s & -(m_u s^2 + b_s s + k_s) \\ m_s & m_u \end{bmatrix} \quad (6)$$

The first step in deriving a controller using the Youla parameterization technique is to find the Smith–McMillan form M_P of the plant G_p such that $M_P = U_L G_p U_R$. The Smith–McMillan form [18] is useful in multi-variable control as it gives a realization of the plant in a basis where the plant is decoupled (i.e., its transfer function matrix is a diagonal matrix). The poles and transmission zeros of the Smith–McMillan form correspond to the poles and zero of the original system and the unimodular matrices U_L and U_R describe the transformation from the original basis to the basis used by the Smith–McMillan form. The Smith–McMillan form of the plant and its unimodular matrices U_L and U_R are

$$M_P = \begin{bmatrix} \frac{1}{m_s m_u s^2 + (m_s + m_u) b_s s + (m_s + m_u) k_s} & 0 \\ 0 & \frac{1}{m_s m_u} \end{bmatrix}, \; U_L = \begin{bmatrix} 0 & 1 \\ m_u & m_u s^2 + b_s s + k_s \end{bmatrix}, \; U_R = \begin{bmatrix} 0 & \frac{1}{m_s m_u} \\ \frac{1}{m_u} & -\frac{1}{m_u^2} \end{bmatrix} \quad (7)$$

The controller is designed such that the decoupled system is a second-order Butterworth filter of unit gain with additional poles to make the controller proper, see Equation (8). The damping ratio ζ is set to $\frac{1}{\sqrt{2}}$ as it offer good trade-off between fast transient and small oscillations. A large enough bandwidth is necessary for the wheel load estimate to be used by the control system. Moreover, the frequency response of a suspension mapping the road disturbance to tire force is shaped as a band-pass filter [12] whose high cutoff frequency is the wheelhop frequency (typically located at 10 Hz). Thus, to capture the tire force response, the closed-loop bandwidth should be faster than the wheelhop frequency. The controller is designed such that $\zeta = 1/\sqrt{2}$ and the bandwidth of the closed-loop system is 30 Hz. Singular values of the closed-loop transfer function and of the controller are given in Figure 4. At frequencies below the bandwidth, T_u is 0 dB and S_u has low gain, ensuring a good tracking. At higher frequencies, the gain of T_u decreases to reject sensor noise and make the estimate robust against high-frequencies model mismatch.

$$M_T = \frac{\omega_0^2}{s^2 + 2\zeta\omega_0 s + \omega_0^2} \frac{1}{(\tau s + 1)^2} \tag{8}$$

Let M_Y such that $M_T = M_Y M_P$, the closed-loop transfer function and the controller transfer function matrix is obtained from the following equations which are derived from [17]

$$T_u = U_R M_T U_R^{-1} \tag{9}$$
$$S_u = I - T_u \tag{10}$$
$$Y = U_R M_Y U_L \tag{11}$$
$$G_c = S_u^{-1} Y \tag{12}$$

This yields the controller

$$G_c = \frac{\omega_0^2}{(s^2 + 2\zeta\omega_0 s + \omega_0^2)(\tau s + 1)^2 - \omega_0^2} \begin{bmatrix} m_u & m_u s^2 + b_s s + k_s \\ -m_s & k_s + b_s s \end{bmatrix} \tag{13}$$

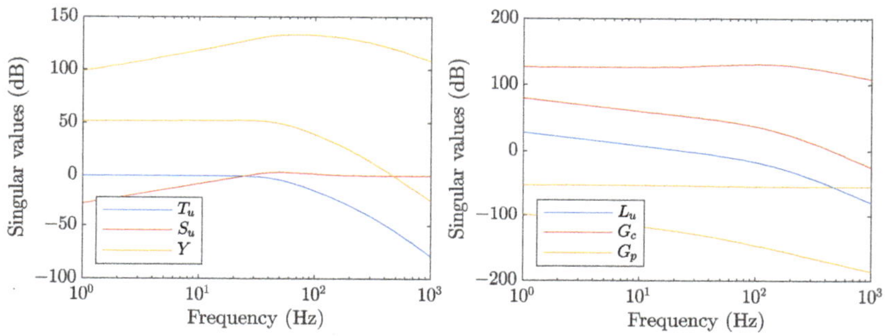

Figure 4. Singular values of the closed-loop transfer function T_u, S_u, and Y and of the return ratio L_u, G_c, and G_p.

The transfer function from the measured signal y to the estimated input $\hat{u}$ is given

$$Y = \frac{\omega_0^2}{(\tau s + 1)^2 (s^2 + 2\zeta\omega_0 s + \omega_0^2)} \begin{bmatrix} m_u & m_u s^2 + b_s s + k_s \\ -m_s & b_s s + k_s \end{bmatrix} \tag{14}$$

Hence, the estimate of the normal force given by the YCOO is

$$\hat{f}_z(s) = \left[m_u s \times \left(\frac{p_s(s)}{m_s} + s q_s(s) \right) + b_s s q_s(s) + k_s q_s(s) \right] \frac{\omega_0^2}{(s^2 + 2\zeta\omega_0 s + \omega_0^2)(\tau s + 1)^2} \tag{15}$$

The same equation can be obtained by combining (3) and (4) and by adding a filter with unit gain. Moschuk et al. [19] patented a concept to estimate wheel normal force using only suspension deflection sensors. The invention uses derivative filters to compute the suspension deflection velocity and the unsprung mass velocity (assuming the sprung mass vertical acceleration is null). It then uses damper and spring force maps to compute the tire normal force. Writing the derivative filter as F_d, the estimation in Reference [19] is

$$\hat{f}_z(s) = m_u F_d(F_d(q_s)) + b_s(F_d(q_s)) + k_s(q_s) \tag{16}$$

This is similar to the estimate of the wheel normal force given by the YCOO in (15).

4. Unbiased Minimum Variance Filtering

The Unbiased Minimum Variance Filter is a variation of the Kalman filter for systems with unknown inputs. It gives an unbiased (zero-mean error) estimate of the model states and unknown inputs [20] with the assumption that the system is strongly observable. Consider the discrete Linear Time-Invariant (LTI) system

$$\begin{aligned} x_{k+1} &= Ax_k + Bu_k + He_k \\ y_k &= Cx_k + Du_k + Ge_k \end{aligned} \tag{17}$$

where x_k are the model states, u_k the known inputs, and e_k the unknown inputs. The states and inputs of the estimation model are $x_k^T = \begin{bmatrix} p_s & p_u & q_s - \frac{m_s g}{k_s} \end{bmatrix}$, $u_k^T = \varnothing$, and $e_k^T = \begin{bmatrix} f_z - (m_s + m_u)g & \Delta f_z \end{bmatrix}$. The LTI system (17) is strongly observable if the matrix Ψ has full column rank [21] (G_d is the G matrix considering only feedthrough unknown inputs, i.e., with all zero-columns removed).

$$\Psi = \begin{bmatrix} C & G & & & \\ CA & CH & \ddots & & \\ \vdots & \vdots & \ddots & G & \\ CA^{n-1} & CA^{n-2}H & \dots & CH & G_d \end{bmatrix} \tag{18}$$

Since the first column of Ψ corresponds to the observability matrix, observability is a necessary condition for strong observability. Unfortunately, the system representing the quarter-car model with unknown force inputs (Equations (2)–(4)) is not observable. Indeed, similarity transformation shows that the state associated to the direction $(m_s\tilde{p}_s + m_u\tilde{p}_u)$ does not produce any observable output. The system is reduced to eliminate the unobservable states. Moreover, it is necessary to use additional measurements to make the system strongly observable. The measured signals used by the UMVF is $\tilde{y}^T = \begin{bmatrix} \frac{\dot{p}_s}{m_s} - g & q_s - m_s g/k_s & \dot{q}_s \end{bmatrix}$. Note that suspension deflection sensors such as linear variable transformers which defines an electrical signal based on the position of an objected it is connected to can only measure deflection [22]. The measurement of the suspension relative velocity $\dot{q}_s$ needed by the UMVF requires differentiating the signal q_s, which requires additional signal processing.

Similar to the Kalman filter, the estimated states is computed in two steps. First, the estimated signals are computed based on the plant model.

$$\hat{x}_{k+1|k} = A\hat{x}_{k|k} + Bu_k \tag{19}$$

$$P_{k+1|k} = AP_{k|k}A^T + Q_k \tag{20}$$

Second, the gain L_{k+1} is computed to guarantee an unbiased estimate of the model states.

$$\tilde{R}_{k+1} = CP_{k+1|k}C^T + R_{k+1} \tag{21}$$

$$\Phi_{k+1} = \begin{bmatrix} -G & CH \end{bmatrix} \tag{22}$$

$$\Omega_{k+1} = \begin{bmatrix} 0_{n\times p} & H \end{bmatrix} - P_{k+1|k}C^T\tilde{R}_{k+1}^{-1}\Phi_{k+1} \tag{23}$$

$$L_{k+1} = P_{k+1|k}C^T\tilde{R}_{k+1}^{-1} - \Omega_{k+1}(\Phi_{k+1}^T\tilde{R}_{k+1}^{-1}\Phi_{k+1})^{-1}\Phi_{k+1}^T\tilde{R}_{k+1}^{-1} \tag{24}$$

$$\hat{x}_{k+1|k+1} = \hat{x}_{k+1|k} + L_{k+1}(y_{k+1} - C\hat{x}_{k+1|k} - Du_{k+1}) \tag{25}$$

$$P_{k+1|k+1} = L_{k+1}\tilde{R}_{k+1}L_{k+1}^T - P_{k+1|k}C^TL_{k+1}^T - L_{k+1}CP_{k+1|k}^T + P_{k+1|k} \tag{26}$$

with n the number of states and p the number of unknown input e_k. Given an unbiased estimate $\hat{x}_{k|}$, Palanthandalam-Madapusi [23] showed that an unbiased estimate of unknown inputs can be obtained from the following equations

$$\hat{e}_k = H^{\dagger} L_{k+1}(y_{k+1} - C\hat{x}_{k+1|k} - Du_{k+1}) \tag{27}$$

$$\hat{e}_k = G^{\dagger}(y_k - C\hat{x}_{k|k} - Du_k) \tag{28}$$

where † denotes the Moore–Penrose pseudo-inverse. Computing the unknown inputs using Equations (27) and (28) guarantees that $E[\hat{e}_k] = G^{\dagger}GE[e_k]$ and $E[\hat{e}_k] = H^{\dagger}HE[e_k]$, respectively. If H or G have full column rank, then this translates to $E[\hat{e}_k] = E[e_k]$. In the general case where both H and G are not full column rank, it is necessary to combine Equations (27) and (28) to compute an unbiased estimate $\hat{e}_k$. In this article, we propose to compute the unknown input by solving a linear system. Let V_H^T and V_G^T be the matrices of left eigenvectors associated to non-zero eigenvalues of $H^{\dagger}H$ and $G^{\dagger}G$. The unknown input is the solution of:

$$\begin{bmatrix} V_H^T \\ V_G^T \end{bmatrix} \hat{e}_k = \begin{bmatrix} V_H^T H^{\dagger} L_{k+1}(y_{k+1} - C\hat{x}_{k+1|k} - Du_{k+1}) \\ V_G^T G^{\dagger}(y_k - C\hat{x}_{k|k} - Du_k) \end{bmatrix} \tag{29}$$

Without loss of generality, we can assume that rank $\begin{bmatrix} H^T & G^T \end{bmatrix} = p$, and it is possible to obtain at least p left eigenvectors of $H^{\dagger}H$ and $G^{\dagger}G$ associated to non-zero eigenvalues. Therefore, the matrix $\begin{bmatrix} V_H^T \\ V_G^T \end{bmatrix}$ has full column rank. Taking the mean of (29) yields

$$\begin{bmatrix} V_H^T \\ V_G^T \end{bmatrix} E[\hat{e}_k] = \begin{bmatrix} V_H^T \\ V_G^T \end{bmatrix} E[e_k] \tag{30}$$

Since the matrix has full column rank, this guarantee that $E[\hat{e}_k] = E[e_k]$, i.e., $\hat{e}_k$ is an unbiased estimate of e_k.

5. Vehicle Mass Estimation

Both observers require knowledge of the vehicle mass and the location of the center of gravity. Indeed, the estimate of signals $\tilde{u}$ from the observer, as given in (5), directly depends on the vehicle mass. The static wheel load must be added to this estimate to compute the wheel load. This section presents a simple algorithm to obtain the static load of each wheel. More elaborate algorithms could be applied [24].

Algebraic expressions can be used to describe the wheel load distribution in quasi-steady-state.

$$f_{ijz} = f_{ijz}^0 \pm \Delta f_j^x a_x \pm \Delta f_i^y a_y,\ (i,j) \in \{F, L\} \times \{L, R\} \tag{31}$$

where the static load and the load transfer terms are

$$f_{ijz}^0 = \frac{mg(L - l_j)(W - w_i)}{LW} \tag{32}$$

$$\Delta f_j^x = \frac{mh(W - w_j)}{LW} \tag{33}$$

$$\Delta f_i^y = \frac{mh(L - l_i)}{LW} \tag{34}$$

Variables w_L and w_R denote the distance from the center of gravity of the vehicle to the left and ride sides; $W = w_L + w_R$ is the track width; l_F and l_R denote the distance from the center of mass to the front and rear axles; $L = l_F + l_R$ is the wheelbase; m is the total vehicle mass; g is the acceleration of gravity; h is the height of the center of mass; and a_x and a_y are the longitudinal and lateral acceleration due to vehicle acceleration, which also include the gravity component on the vehicle longitudinal and lateral axes.

The vehicle mass estimation algorithm is run when the vehicle is not moving. The wheel load f_{ijz} corresponds to the force produced by the suspension deflection ignoring the weight of the unsprung mass. Hence, at steady-state, the wheel load in Equation (31) is replaced by $f_{ijz} = k_s q_{ijs}$. This yields

$$k_s q_{ijz} = f^0_{ijz} \pm \Delta f^x_j a_x \pm \Delta f^y_i a_y \tag{35}$$

where a_x and a_y are the longitudinal and lateral accelerations measured by the sensors and correspond to the acceleration of gravity in those directions if the vehicle is parked on a slope.

This corresponds to a system of four equations with four unknowns, m, l_F, w_L, and h (replacing l_R and w_R by $l - l_F$ and $w - w_L$ with l the vehicle wheelbase and w the axle track width). Thus, the position of the center of mass and the vehicle mass can be obtained when the vehicle is not moving.

6. Simulation Results

In this section, the YCOO and UMVF observers developed in Section 3 and 4 are tested in simulation. The full-car model shown in Figure 1 is assumed to represent the actual vehicle dynamics and the measured ground-truth signals are extracted from the 14 degrees of freedom vehicle model. The driving scenarios presented aim to cover all possible ways to redistribute tire loads, i.e., longitudinal or lateral load transfer, road irregularities, and sloped road. The results are also compared to the algebraic expression for normal force.

Figure 5a shows the estimates during a braking step of 3000 N m at 1 s from an initial velocity of 90 km h^{-1}. The two observers provide better estimates than the algebraic expression which suffers from a steady-state error. Moreover, the two observers are intentionally not initialized; both observers converge in approximately 0.1 s. Figure 5b shows the estimate during a double lane change maneuver with a constant velocity of 90 km/h and with maximum lateral acceleration of 0.6 g. Both estimators provide a good estimate of the tire vertical force, whereas the estimation from algebraic expression does not capture the transient response. Figure 5a,b validates the two estimators for situations where the load transfer is due to longitudinal or lateral acceleration.

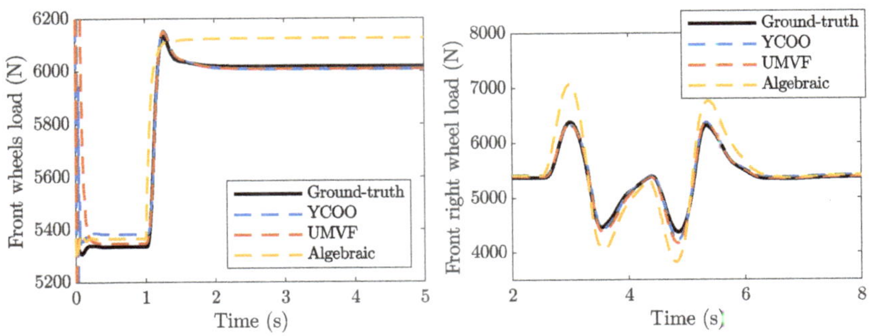

(**a**) Braking step of 3000 N m.

(**b**) Double lane change maneuver with 0.6 g maximum lateral acceleration.

Figure 5. Vertical tire force estimation on maneuver with longitudinal and lateral only acceleration.

Figure 6 shows the estimate during a bounce sine sweep test. The vehicle velocity is maintained at 20%. The road profile corresponds to sinusoidal bumps of decreasing wavelength with decreasing amplitude. The minimum wavelength is 1.6 m. Thus, the road excites the suspension over the frequency range 0 Hz to 3.5 Hz. The YCOO and the UMVF are able to estimate the wheel loads. Both observers reproduce the frequency response of the suspension: the wheel load amplitude increases when the road excitation get closer to the the suspension frequency (1 Hz obtained when $t \approx 13$ s) and remains constant at frequencies between the suspension and wheelhop frequencies. Since the longitudinal and lateral accelerations during this maneuver are almost zero, the algebraic expression is not able to provide an accurate estimation of the wheel loads. The effect of road slope on the estimation scheme is investigated in Figure 7. The vehicle is driven from a flat road to a slope of 20% gradient. The YCOO and the UMVF capture the load transfer due to the road gradient and provide a good estimate during transient.

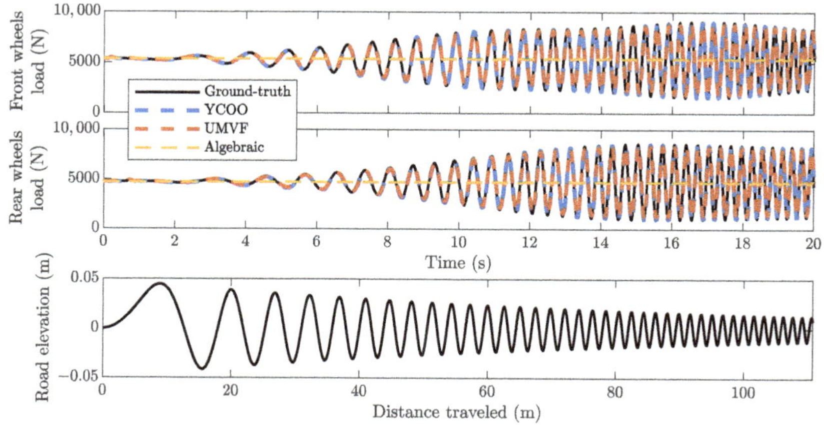

Figure 6. Vertical tire force estimation during a bounce sine sweep test. Vehicle speed is constant at $20\,\mathrm{km\,h^{-1}}$. The minimum wavelength is 1.6 m at $t = 20\%$. The bottom figure shows the road profile.

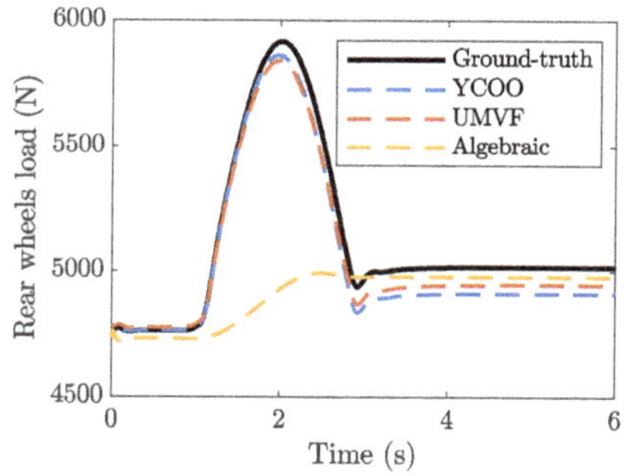

Figure 7. Vertical tire force estimation when driving on a 20 % slope.

It is not practical to assume that the estimation model is a perfect representation of the actual suspension. Figure 8 evaluates the sensitivity of the wheel load estimation against the suspension stiffness. Uncertainties over this parameter result in an offset between the real and estimated wheel load. This is due to the wrong calibration of the mass estimation strategy. The load transfer estimate also suffers from uncertainties in the suspension stiffness. Indeed, without any uncertainty, both observers yield a correct load transfer of 700 N, but with a 50% stiffer suspension the load transfer estimate is only 450 N. The robustness against the damping coefficient b_s is investigated in Figure 9a. The YCOO and the UMVF provide the same estimate, thus only the estimate given by the UMVF is shown in Figure 9a. The estimation is not robust against the damping coefficient in the transient but it does not affect the steady-state estimation. Similarly, nonlinearities in the damper map affect the transient of the wheel load estimate when the suspension operates in the region approximated by the linear damper map. The linear and nonlinear damper maps are given in Figure 9b.

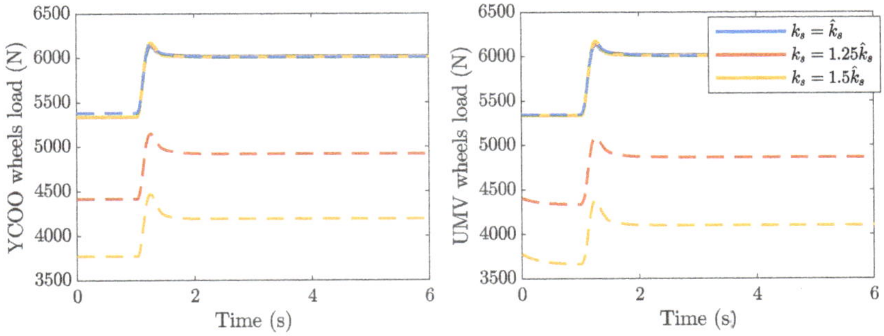

Figure 8. Robustness against suspension stiffness during a braking maneuver. Solid lines show the ground-truth signals and dashed lines show the estimated ones.

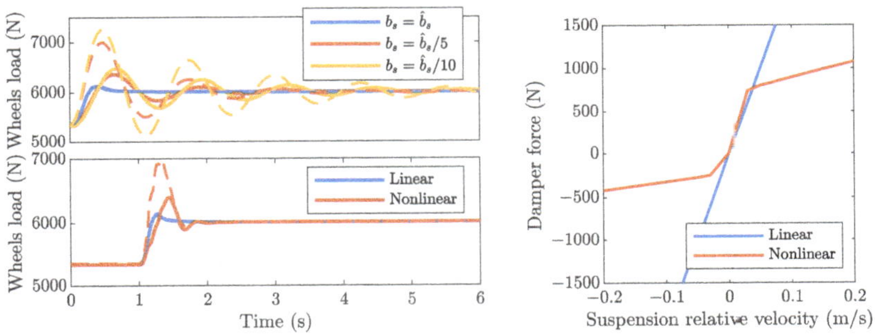

(a) Robustness against damping coefficient (**top**) and damping map (**bottom**).

(b) Suspension linear and nonlinear damper map.

Figure 9. Robustness against uncertainties in the damping map during a braking maneuver. Solid lines show the ground-truth signals and dashed lines show the estimated ones.

Finally, Figure 10 shows the estimated signals obtained with the YCOO and the UMVF when Gaussian white noise of time correlation 10 ms and of power spectral density 10^{-4} and 10^{-9} is, respectively, added to the sprung mass vertical acceleration and to the suspension deflection measurements. The YCOO offers better noise rejection than the UMVF.

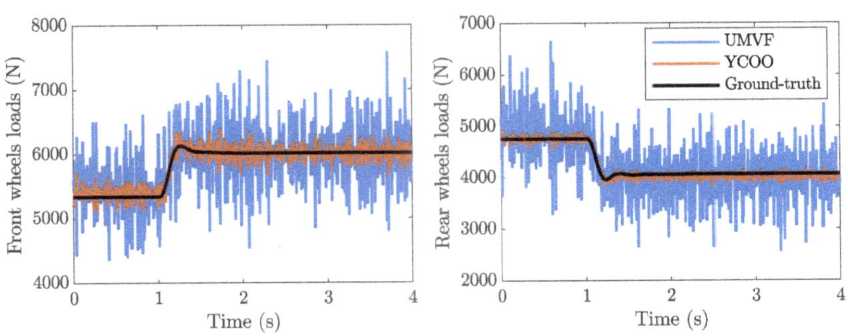

Figure 10. Estimation with noisy measurements.

7. Conclusions

The estimation of the wheel loads is formulated as an input estimation problem. A quarter-car model with load transfer and tire normal force is used as the estimation model. Two observers are designed. A Youla controller is designed to minimize the error between the measurement and the estimated model output in the Youla Controller Output Observer framework. Similarly, unobservable states of the quarter-car model are removed to use the Unbiased Minimum Variance Filter. Since the design of the YCOO is based on the frequency domain and the closed-loop transfer functions T_u, S_u, and Y, it directly addresses the trade-off among noise rejection, high bandwidth, and good robustness when tuning the observer.

Both observers were tested in simulation and provide good estimates as long as the model possesses a good enough representation of the suspension. Moreover, the anti-roll bar introduces coupling between the two wheels of the same axle. In this case, the quarter-car estimation model cannot be used and should be replaced by a half-car model. Despite using different approaches to solve the input estimation problem and different design methods, both controllers give similar performance and robustness, but the proposed YCOO provides better noise rejection than the UMVF. The YCOO provides a much simpler structure and observer tuning than the UMVF as it does not require the system to be observable and requires only two measurement when the UMVF needs a third measurement with additional signal processing.

Author Contributions: Conceptualization, L.F. and F.A.; methodology, L.F., F.A. and J.V.A.; software, L.F.; validation, L.F. and F.A.; formal analysis, L.F., F.A., and J.V.A.; supervision: M.K. and R.J.; writing—original draft preparation, review and editing: L.F. and F.A.; funding acquisition: F.A. and M.K. All authors have read and agreed to the published version of the manuscript.

Funding: This research was funded under grant Ford Motor Company: AFFORDM.

Institutional Review Board Statement: Not applicables.

Informed Consent Statement: Not applicable.

Acknowledgments: The authors thank Ford Motor Company for its support of this work.

Conflicts of Interest: The authors declare no conflict of interest.

Abbreviations

The following abbreviations are used in this manuscript:

COO	Controller Output Observer
GCC	Global Chassis Control
LTI	Linear Time-Invariant
RGA	Relative Gain Array
UMVF	Unbiased Minimum Variance Filter
YCOO	Youla Controller Output Observer

References

1. Chen, W.; Xiao, H.; Wang, Q.; Zhao, L.; Zhu, M. *Integrated Vehicle Dynamics and Control*; John Wiley & Sons Singapore Pte. Ltd.: Singapore, 2016.
2. Cho, W.; Yoon, J.; Yim, S.; Koo, B.; Yi, K. Estimation of Tire Forces for Application to Vehicle Stability Control. *IEEE Trans. Veh. Technol.* **2010**, *59*, 638–649.
3. Alatorre, V.A.; Victorino, A.; Charara, A. Estimation of Wheel-Ground Contact Normal Forces: Experimental Data Validation. *IFAC PapersOnLine* **2017**, *50*, 14843–14848. [CrossRef]
4. Doumiati, M.; Charara, A.; Victorino, A.; Lechner, D. *Vehicle Dynamics Estimation Using Kalman Filtering: Experimental Validation*; ISTE; John Wiley and Sons Inc.: London, UK; Hoboken, NJ, USA, 2013.
5. Rajamani, R. *Vehicle Dynamics and Control*; Mechanical Engineering Series; Springer: Berlin/Heidelberg, Germany, 2011.
6. Doumiati, M.; Victorino, A.; Charara, A.; Baffet, G.; Lechner, D. An Estimation Process for Vehicle Wheel-Ground Contact Normal Forces. *IFAC Proc. Vol.* **2008**, *42*, 7110–7115. [CrossRef]
7. Doumiati, M.; Victorino, A.; Charara, A.; Lechner, D. Virtual Sensors, Application to Vehicle-Tire Road Normal Forces for Road Safety. In Proceedings of the American Control Conference, San Diego, CA, USA, 17 June 2009; pp. 3337–3343. [CrossRef]

8. Jiang, K.; Pavelescu, A.; Charara, A.; Victorino, A. Estimation of vehicle's vertical and lateral tire forces considering road angle and road irregularity. In Proceedings of the 2014 17th IEEE International Conference on Intelligent Transportation Systems, ITSC 2014, Qingdao, China, 8–11 October 2014; doi:10.1109/ITSC.2014.6957714. [CrossRef]
9. Ozkan, B.; Margolis, D.; Pengov, M. The Controller Output Observer: Estimation of Vehicle Tire Cornering and Normal Forces. *J. Dyn. Syst. Meas. Control* **2008**, *130*. [CrossRef]
10. Velazquez Alcantar, J.; Assadian, F.; Kuang, M. Vehicle Velocity State Estimation using Youla Controller Output Observer. In Proceedings of the 2018 Annual American Control Conference (ACC), Milwaukee, WI, USA, 27–29 June 2018; Volume 2018; pp. 2587–2592.
11. Velazquez Alcantar, J.; Assadian, F. Longitudinal Tire Force Estimation Using Youla Controller Output Observer. *IEEE Control Syst. Lett.* **2018**, *2*, 31–36. [CrossRef]
12. Jazar, R.N. *Vehicle Dynamics: Theory and Application*; Springer: Boston, MA, USA, 2008.
13. Shim, T.; Ghike, C. Understanding the limitations of different vehicle models for roll dynamics studies. *Veh. Syst. Dyn.* **2007**, *45*, 191–216. [CrossRef]
14. Skogestad, S. *Multivariable Feedback Control: Analysis and Design*, 2nd ed.; John Wiley & Sons: Chichester, UK; Hoboken, NJ, USA, 2005.
15. Qin, Y.; Wei, C.; Tang, X.; Zhang, N.; Dong, M.; Hu, C. A novel nonlinear road profile classification approach for controllable suspension system: Simulation and experimental validation. *Mech. Syst. Signal Process.* **2019**, *125*, 79–98. [CrossRef]
16. Youla, D.; Bongiorno, J.; Lu, C. Single-loop feedback-stabilization of linear multivariable dynamical plants. *Automatica* **1974**, *10*, 159–173. [CrossRef]
17. Assadian, F.; Mallon, K. *Neoclassical Control: Control Systems and Estimation Design using the Youla Parameterization Technique*; John Wiley & Sons, Inc.: Hoboken, NJ, USA, 2021.
18. Chen, C.T. *Linear System Theory and Design*, 3rd ed.; Oxford University Press, Inc.: Oxford, MS, USA, 1998.
19. Moschuk, N.; Nardi, F.; Ryu, J.; O'Dea, K. Estimation of Wheel Normal Force and Vehicle Vertical Acceleration. U.S. Patent 8326487, 4 December 2012.
20. Gillijns, S.; De Moor, B. Unbiased minimum-variance input and state estimation for linear discrete-time systems. *Automatica* **2007**, *43*, 111–116. [CrossRef]
21. Kratz, W. Characterization of strong observability and construction of an observer. *Linear Algebra Appl.* **1995**, *221*, 31–40. [CrossRef]
22. Krauze, P.; Kasprzyk, J.; Kozyra, A.; Rzepecki, J. Experimental analysis of vibration control algorithms applied for an off-road vehicle with magnetorheological dampers. *J. Low Freq. Noise Vib. Act. Control* **2018**, *37*, 619–639. [CrossRef]
23. Palanthandalam-Madapusi, H.J.; Bernstein, D.S. Unbiased Minimum-variance Filtering for Input Reconstruction. In Proceedings of the 2007 American Control Conference, New York, NY, USA, 9–13 July 2007; pp. 5712–5717. [CrossRef]
24. Rhode, S.; Gauterin, F. Vehicle mass estimation using a total least-squares approach. In Proceedings of the 2012 15th International IEEE Conference on Intelligent Transportation Systems, Anchorage, AL, USA, 16–19 September 2012; pp. 1584–1589.

energies

Article

Optimization-Driven Powertrain-Oriented Adaptive Cruise Control to Improve Energy Saving and Passenger Comfort

Pier Giuseppe Anselma [1,2]

1 Department of Mechanical and Aerospace Engineering (DIMEAS), Politecnico di Torino, 10129 Torino, Italy; pier.anselma@polito.it
2 Center for Automotive Research and Sustainable Mobility (CARS), Politecnico di Torino, 10129 Torino, Italy

Abstract: Assessing the potential of advanced driver assistance systems requires developing dedicated control algorithms for controlling the longitudinal speed of automated vehicles over time. In this paper, a multiobjective off-line optimal control approach for planning the speed of the following vehicle in adaptive cruise control (ACC) driving is proposed. The implemented method relies on the principle of global optimality fostered by dynamic programming (DP) and aims to minimize propelling energy consumption and enhance passenger comfort. The powertrain model and onboard control system are integrated within the proposed car-following optimization framework. The retained ACC approach ensures that the distance between the following vehicle and the preceding vehicle is always maintained within allowed limits. The flexibility of the proposed method is demonstrated here through ease of implementation on a wide range of powertrain categories, including a conventional vehicle propelled by an internal combustion engine solely, a pure electric vehicle, a parallel P2 hybrid electric vehicle (HEV) and a power-split HEV. Moreover, different driving conditions are considered to prove the effectiveness of the proposed optimization-driven ACC approach. Obtained simulation results suggest that up to 22% energy-saving and 48% passenger comfort improvement might be achieved for the ACC-enabled vehicle compared with the preceding vehicle by implementing the proposed optimization-driven ACC approach. Engineers may adopt the proposed workflow to evaluate corresponding real-time ACC approaches and assess optimal powertrain design solutions for ACC driving.

Keywords: adaptive cruise control; automated driving; energy-saving; fuel-saving; optimal control; passenger comfort

Citation: Anselma, P.G. Optimization-Driven Powertrain-Oriented Adaptive Cruise Control to Improve Energy Saving and Passenger Comfort. *Energies* **2021**, *14*, 2897. https://doi.org/10.3390/en14102897

Academic Editor: Francis F. Assadian

Received: 13 April 2021
Accepted: 13 May 2021
Published: 17 May 2021

1. Introduction

Automated and connected mobility is currently forecasted reshaping public and private transportation over the next few decades [1–4]. Remarkable benefits could be achieved in general through implementing automated mobility, including enhancing passenger comfort, reducing energy consumption for propulsion, enhancing traffic management, and improving road safety, among others [5]. This technological advance demands developing effective and flexible numerical tools for controlling and designing automated vehicles [6–11].

Automated driving, as fostered by the different communication technologies (e.g., vehicle-to-vehicle, vehicle-to-infrastructure, vehicle-to-pedestrian, vehicle-to-grid, vehicle-to-device), represents an extension of advanced driver assistance systems (ADASs). Examples for ADASs currently implemented in road vehicles include cruise control (CC), where the vehicle is controlled to travel at constant longitudinal speed over time, and adaptive cruise control (ACC), where the longitudinal speed of the vehicle is controlled to vary over time according to the measured distance from the vehicle ahead. In an ACC driving scenario, the following vehicle (named hereafter as the following vehicle) typically exploits data from the preceding vehicle (named hereafter as the preceding vehicle), which can be either human-operated or automated. ACC systems use sensors, such as radar, Light Detection and Ranging (LIDAR) or cameras to identify and monitor the preceding vehicle for measuring its current distance and speed difference [12]. Current ACC systems are calibrated to regulate the longitudinal speed of the vehicle to maintain a constant headway from the preceding vehicle [13,14]. Avoidance of front-end collision between the preceding vehicle and the following vehicle can be ensured in this way. However, when propulsion and brake systems are controlled with the exclusive aim of maintaining a constant time-headway (or distance) from the preceding vehicle, it cannot be guaranteed that the ACC-enabled vehicle achieves improved performance in terms of energy economy or passenger comfort as an example [15]. New possibilities and challenges open up in this framework concerning the development of ACC approaches that can vary both the following vehicle's longitudinal speed and distance from the preceding vehicle over time regardless of the longitudinal speed of the preceding vehicle. Improving various predefined performance metrics for the preceding vehicle can be set as a control target for such an enhanced ACC system.

Literature regarding control approaches for the following vehicle's speed exploiting information coming from the preceding vehicle in automated driving can be divided between single-powertrain-based approaches and multiple-powertrain-based approaches. Single-powertrain-based ACC approaches can focus either on conventional vehicles (CVs), hybrid electric vehicles (HEVs) or battery electric vehicles (BEVs) as examples. Concerning CVs powered solely by an internal combustion engine (ICE), Lang et al. [16] in 2013 discussed a control logic aiming to minimize fuel consumption while neglecting gear shifting. He and Orosz [17] in 2017 compared feedback-based and rolling horizon optimal control-based as cooperative cruise control approaches minimizing fuel consumption. The same authors extended a fuel-optimal longitudinal speed controller to the case of heavy-duty trucks exploiting information coming from multiple vehicles ahead through vehicle-to-vehicle (V2V) communication [18]. As concerns HEVs, a recurrent research topic involves developing velocity predictors that can improve the energy management strategy of the following vehicle through the information coming from the preceding vehicle. Different categories of longitudinal speed regulation logics have been developed in the literature (e.g., heuristic, instantaneous optimization, machine learning), and various HEV powertrain layouts have been considered, such as power-split [19], parallel P0 [20], parallel P2 [21] and series-parallel P1P4 [22] as an example. Regarding BEVs, the author of this paper proposed an optimal off-line velocity controller based on dynamic programming (DP) capable of minimizing the energy consumption of the following vehicle [23]. Recently, Koch et al. [24] focused on battery-electric buses and implemented DP while assuming ideal V2I communication and a dedicated traveling road lane to generate energy-efficient driving profiles. The same authors recently proposed an algorithm validated using DP that allows the simultaneous optimization of speed profile and powertrain operation to compare different BEV powertrain architectures [25].

Regarding multiple-powertrain-based control approaches for the following vehicle in car-following scenarios, in 2018, Tate et al. [26] considered different automated driving scenarios by generating the related vehicle speed profiles with a heuristic approach according to engineering experience. Both a CV and a BEV layout were retained, and considerable reductions in greenhouse gas emissions were suggested, especially in the BEV case, thanks to implementing car-following automated driving. Plum et al. [27] in 2018 investigated a CV, an HEV and a BEV powertrain layout while considering a model predictive acceleration controller that exploited information coming from traffic light schedules and the preceding vehicle. The HEV powertrain layout was demonstrated, achieving a greater portion of up to 27.7% energy savings. Nevertheless, the controller was specifically calibrated for a limited number of predefined inner-city driving conditions. Recently, Spano et al. [28] considered a CV and an HEV and estimated the fuel consumption reduction capability at different levels of automated driving using a heuristic approach.

In general, reviewed ACC approaches for enhancing energy-saving of automated road vehicles in car-following scenarios are usually developed and calibrated ad hoc for specific powertrain configurations in terms of architecture and component sizes. Extending the reviewed approaches to reduce the energy consumption of vehicles embedding different powertrain types and component sizes might, in turn, require thorough and time-consuming re-calibration procedures. To the best of the author's knowledge, developing a control approach for the longitudinal speed of automated vehicles in car-following scenarios that can easily adapt to foster energy-saving of different powertrain layouts and component sizes still represents an open research question. To overcome the highlighted research gap, this paper aims to present a new multiobjective optimization-driven ACC algorithm that can easily estimate energy savings and passenger comfort improvements for various powertrain categories when traveling as a following vehicle in car-following scenarios. The proposed ACC approach relies on DP as a widely employed off-line control algorithm capable of identifying the optimal global solution for the considered control problem [29]. Energy consumption minimization and passenger comfort enhancement are considered as conflicting optimization targets for the proposed car-following controller. The ease of adaptability of the discussed approach is suggested through its efficient implementation retaining a CV powertrain, a BEV powertrain, a single-motor parallel HEV powertrain and a dual-motor power-split HEV powertrain. In all the presented cases, only the objective function considered in DP needs to be adapted to the given powertrain category, yet the proposed workflows can be straightforwardly applied considering different component sizes for each propulsion system category. Our results demonstrate the potential of the proposed approach for effectively and easily determining optimization-driven speed profiles over time for the following vehicle in car-following scenarios. Engineers may adopt the proposed optimization-driven ACC approach to evaluate the performance of corresponding real-time ACC approaches and to improve powertrain design methodologies considering enhanced ACC driving. The remainder of this paper is as follows: the considered vehicle powertrain layouts and the related modeling approach are first illustrated. The mathematical formulation of the car-following driving problem is then discussed, and the proposed algorithm is presented. Results are presented over different driving conditions, and conclusions are given.

2. Vehicle Powertrains

This section aims at describing the considered vehicle powertrain architectures. The adopted numerical modeling approaches find discussion as well. In this paper, a CV powertrain layout, a BEV powertrain layout, a parallel P2 HEV (P2 HEV) powertrain layout and a power-split HEV (PS HEV) powertrain layout are retained. The corresponding schematic diagrams are illustrated in Figure 1, while detailed discussion for each powertrain architecture is reported in the follow-up of this section.

2.1. CV Powertrain

For the CV powertrain layout illustrated in Figure 1a, the vehicle is propelled by an ICE alone. An automated manual transmission (AMT) is embedded capable of shifting gear according to the ICE speed and the torque request coming from the driver following a dedicated control logic. In general, a quasi-static modeling approach is implemented here in deriving speeds and torques of power components directly from the vehicle speed profile over time for the considered drive cycle [30]. The torque requested by the driver at the driven wheels T_{wheels} can particularly be evaluated following Equation (1) [31]:

$$T_{wheels} = \left(F_{roll} + F_{misc} + F_{aero} + m_{veh_{eq}} \cdot \ddot{x}\right) \cdot r_{dyn} \tag{1}$$

where F_{aero}, F_{misc} and F_{roll} represent resistive load elements corresponding to the aerodynamic drag, miscellaneous elements, such as road slope and side forces as an example, and rolling resistance, respectively. $\ddot{x}$ is the vehicle acceleration, while r_{dyn} and $m_{veh_{eq}}$, respectively represent the wheel's dynamic radius and the vehicle mass, including the equivalent mass of the rotational elements. Subsequently, the rotational speed ω_{ICE} and the requested ICE torque T_{ICE} can be computed as a function of the gear engaged j following Equations (2) and (3), respectively [32]:

$$\omega_{ICE} = \frac{\dot{x}}{r_{dyn}} \cdot \tau_{diff} \cdot \tau_{AMT}(j) \tag{2}$$

$$T_{ICE} = \frac{T_{wheels}}{\tau_{diff} \cdot \tau_{AMT}(j) \cdot \eta_{TR}^{sign(T_{wheels})}} \tag{3}$$

where $\dot{x}$ is the vehicle speed in meters per second, τ_{diff} and τ_{AMT} represent gear ratios for the differential and the instantaneous gear j engaged in the AMT, respectively. η_{TR} is the efficiency of the transmission system, and it is powered to the sign of the torque at the wheels to account for both vehicle accelerating and braking cases.

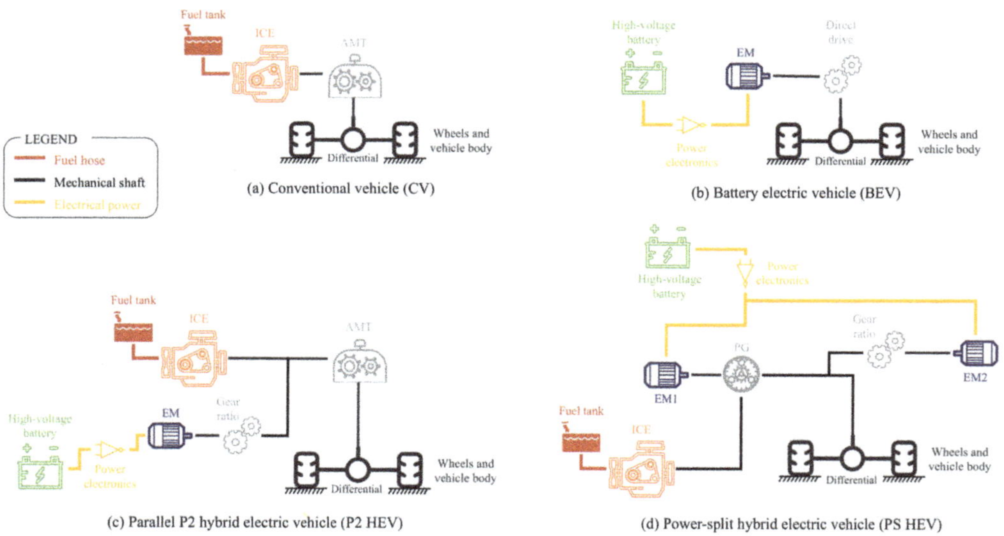

Figure 1. Schematic diagrams of the retained vehicle powertrain architectures, including (**a**) conventional vehicle (CV) powertrain; (**b**) battery electric vehicle (BEV) powertrain; (**c**) parallel P2 hybrid electric vehicle (P2 HEV) powertrain; (**d**) power-split hybrid electric vehicle (PS HEV) powertrain.

Once ω_{ICE} and T_{ICE} are determined, the instantaneous rate of fuel consumption can be determined by interpolating in a two-dimensional lookup table with speed and torque of the ICE as independent variables. As concerns selecting the gear in the AMT, a common approach implemented here refers to determining the engaged gear number according to a pre-calibrated two-dimensional lookup table with vehicle speed and driver torque demand as an independent variable [33].

2.2. BEV Powertrain

In the BEV powertrain layout illustrated in Figure 1b, the fuel tank, ICE and AMT of the CV powertrain are replaced with a high-voltage battery, an electric motor (EM) and a direct drive, respectively. Finally, power electronics enable the proper operation of the electric powertrain components. In this framework, speed ω_{EM} and torque T_{EM} of the EM can be evaluated at each time step following the same procedure illustrated for the CV case in Equations (1)–(3). A gear shift logic does not need implementation in this case, given the embedment of a direct drive. The EM electrical losses $loss_{EM}$, including inverter losses as well, can be evaluated in this case by interpolating in a two-dimensional lookup table with ω_{EM} and T_{EM} as independent variables. Finally, the battery state-of-charge (SOC) over time can be evaluated as a function of the requested battery power P_{batt} by adopting an equivalent circuit approach and following Equations (4)–(6) [34].

$$SOC(t) = \int_{t_0}^{t_{end}} \dot{SOC}[P_{batt}(t), SOC(t)]\, dt \tag{4}$$

with:

$$P_{batt}(t) = \omega_{EM}(t)\cdot T_{EM}(t) + loss_{EM}[\omega_{EM}(t), T_{EM}(t)] + loss_{aux}(t) \tag{5}$$

$$\dot{SOC}[P_{batt}(t), SOC(t)] = \frac{V_{OC}[SOC(t)] - \sqrt{\{V_{OC}[SOC(t)]\}^2 - 4\cdot R_{IN}[SOC(t)]\cdot P_{batt}(t)}}{2\cdot R_{IN}[SOC(t)]}\cdot\frac{n_P}{Ah_{batt}\cdot 3600} \tag{6}$$

where $\dot{SOC}$, t_0 and t_{end} are the instantaneous rate of SOC, the initial time instant and the final time instant of the drive cycle, respectively. $loss_{aux}$ is the power requested by the accessories (e.g., air conditioning, lubrication), and it is modeled as having a constant value in this work. R_{IN} and V_{OC} represent the internal resistance and the open-circuit voltage of the battery pack, as obtained by interpolating in 1D lookup tables with SOC ad independent variables. n_P is the number of cells in parallel according to the battery pack layout, while Ah_{batt} represents the battery pack energy capacity in ampere-hours. The factor of 3600 is considered here to convert energy units in ampere-seconds.

2.3. P2 HEV Powertrain

The parallel P2 HEV powertrain illustrated in Figure 1c represents a combination of the CV and the BEV architectures discussed above. In a parallel P2, the EM is placed downstream of the ICE output shaft before the AMT input shaft. A gear ratio between the EM and the AMT input shaft is considered in this case, while a clutch connection is included between the ICE and the EM to allow disengaging the ICE and avoiding its dragging effect in pure electric operation. Notably, P2 represents one of the potentially most efficient options among the parallel HEV powertrain architectures [35].

In a P2 HEV powertrain, the torques of ICE and EM are additive and follow Equation (7) [32]:

$$T_{wheels} = (T_{ICE} + T_{EM}\cdot\tau_{GR})\cdot\tau_{diff}\cdot\tau_{AMT}(j)\cdot\eta_{TR}^{sign(T_{wheels})} \tag{7}$$

where τ_{GR} is the gear ratio between EM output shaft and AMT input shaft. In a quasi-static modeling approach, controlling either T_{ICE} or T_{EM} allows automatically determining the value of the other variable.

Hybrid electric or pure electric operation are distinguished in this case by the value of T_{ICE} being positive or equal to zero, respectively. As a typical approach, a rule-based energy management strategy (EMS) is implemented here for the P2 HEV powertrain involving two decision steps, as reported in Figure 2 [36,37]. The first step relates to determining the ICE status (i.e., on or off), while the second step involves deciding the power split between ICE and EM in case the hybrid operation is selected.

For determining the ICE status, a set of rules is retained considering the current value of vehicle speed, the current value of battery SOC, the required output power and the current ICE status [38]. The rules constituting the control logic for the ICE status are reported in Table 1. Particularly, the ICE is controlled to be activated in the case at least one of the following criteria are met: (1) the vehicle speed exceeds a predefined threshold $\dot{x}_{thre}$, (2) the power demand is above a certain limit $P_{OUT\,thre}$, or (3) the battery SOC falls below a certain value SOC_{thre}. On the other hand, if none of the three conditions is met, the ICE is controlled to be deactivated. To reduce the frequency of ICE activation and de-activation events, different values for the discussed threshold variables are considered according to the current ICE status following $\dot{x}_{thre_{OFF}} < \dot{x}_{thre_{ON}}$, $P_{OUT\,thre_{OFF}} < P_{OUT\,thre_{ON}}$ and $SOC_{thre_{OFF}} > SOC_{thre_{ON}}$, respectively.

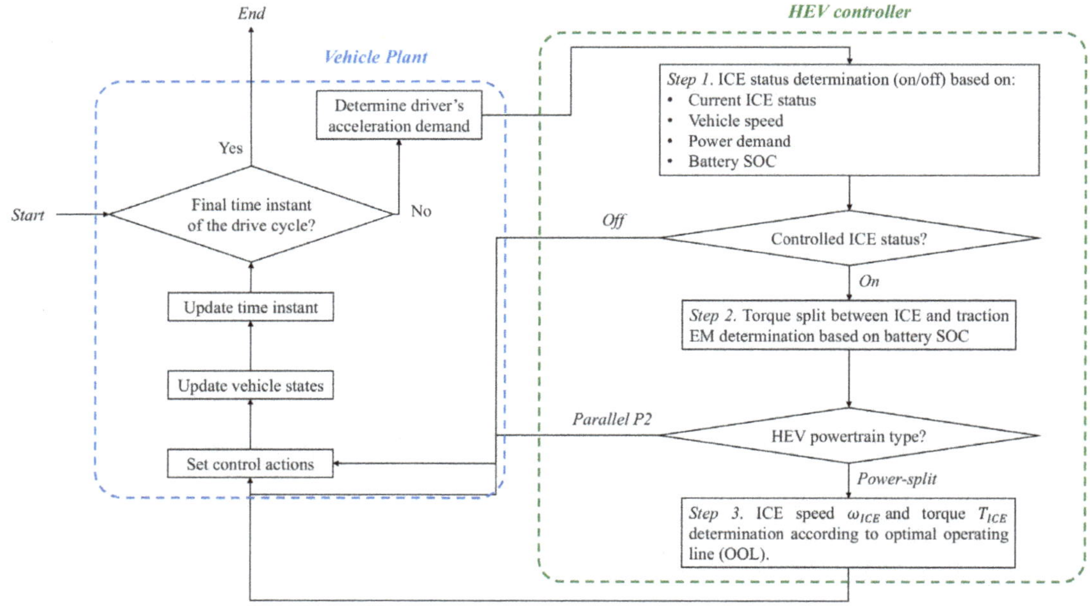

Figure 2. Flowchart of the rule-based hybrid supervisory controller implemented in the parallel P2 HEV and the power-split HEV.

Table 1. Control logic for determining the ICE status.

	Controlled ICE Status: ON	Controlled ICE Status: OFF
Current ICE status: ON	$(\dot{x} \geq \dot{x}_{thre_{OFF}}) \mid (P_{OUT} \geq P_{OUT_{thre_{OFF}}}) \mid (SOC \leq SOC_{thre_{OFF}})$	$(\dot{x} < \dot{x}_{thre_{OFF}})$ && $(P_{OUT} < P_{OUT_{thre_{OFF}}})$ && $(SOC > SOC_{thre_{OFF}})$
Current ICE status: OFF	$(\dot{x} \geq \dot{x}_{thre_{ON}}) \mid (P_{OUT} \geq P_{OUT_{thre_{ON}}}) \mid (SOC \leq SOC_{thre_{ON}})$	$(\dot{x} < \dot{x}_{thre_{ON}})$ && $(P_{OUT} < P_{OUT_{thre_{ON}}})$ && $(SOC > SOC_{thre_{ON}})$

When the hybrid operation is selected, the controlled ICE mechanical power $P_{ICE_{control}}$ can then be determined through interpolation in a one-dimensional lookup table having battery SOC as the independent variable. A tuning process for this table was carried out beforehand to simultaneously prevent excessive charge depletion of the battery and guarantee enhanced HEV fuel economy capability. Since the speeds of both the ICE and the EM are determined by the current vehicle speed and the gear engaged, the corresponding torques can be computed from $P_{ICE_{control}}$ and the resulting value of EM power.

2.4. Power-Split HEV Powertrain

The PS HEV powertrain architecture shown in Figure 1d comes from the industrial state-of-the-art, and it characterizes the embedment of two EMs [39]. EM2 represents the main traction motor, and it operates either to propel the HEV or to recover electrical energy in braking events. On the other hand, ICE, EM1 and the differential input shaft are mechanically connected to the carrier, the sun gear and the ring gear of a planetary gearset (PG). This mechanical device allows decoupling the ICE speed from the current vehicle speed, allowing the HEV powertrain to operate as an electrically variable transmission (eVT) [40]. In this framework, EM1 mainly operates as an electrical generator while ensuring reaction torque for the ICE torque being delivered to the wheels.

For the illustrated PS HEV, the same workflow as Figure 2 is implemented in the supervisory control logic as the one illustrated in Section 2.3 for the P2 HEV. However, other than determining the ICE status and the ICE mechanical power, a third control action needs achievement in this case related to controlling ω_{ICE} and T_{ICE}. To this end, once the value of $P_{ICE_{control}}$ has been determined at the previous step of the logic, ω_{ICE} and T_{ICE} may be selected as the solution of the optimization process linked to the optimal operating line (OOL) of the ICE as reported in Equations (8) and (9) [41]:

$$OOL(\omega_{ICE}, T_{ICE}) = \text{argmin}\left[\dot{m}_{fuel}(\omega_{ICE}, T_{ICE})\right] \tag{8}$$

subject to:

$$\omega_{ICE} \cdot T_{ICE} = P_{ICE_{control}} \tag{9}$$

The resulting values of speed and torque for the ICE should particularly guarantee to deliver the requested value of ICE mechanical power and minimize the fuel consumption rate at the same time. The illustrated approach thus fosters efficient ICE operation. Once ω_{ICE} and T_{ICE} are determined, speeds and torques of both the EMs can be found following the kinematic and dynamic relationships for the PG-based PS HEV powertrain that are reported in Equations (10) and (11), respectively [34]:

$$\begin{bmatrix} \omega_{EM1} \\ \omega_{EM2} \end{bmatrix} = \begin{bmatrix} -i_{PG} & i_{PG}+1 \\ \tau_{GR} & 0 \end{bmatrix} \begin{bmatrix} \omega_{IN_{FD}} \\ \omega_{ICE} \end{bmatrix} \tag{10}$$

$$\begin{bmatrix} T_{EM1} \\ T_{EM2} \end{bmatrix} = \begin{bmatrix} 0 & -\frac{1}{i_{PG}+1} \\ \frac{1}{\tau_{GR}} & -\left(\frac{i_{PG}}{i_{PG}+1}\right) \cdot \frac{1}{\tau_{GR}} \end{bmatrix} \begin{bmatrix} \frac{T_{IN_{FD}}}{\eta_{TR}^{\text{sign}(T_{wheels})}} \\ T_{ICE} \end{bmatrix} \tag{11}$$

$\omega_{IN_{FD}}$ and $T_{IN_{FD}}$ represent the rotational speed and the torque at the input shaft of the final drive, respectively. i_{PG} and τ_{GR} are the gear ratios of the PG and the transfer gearset between EM2 and the input shaft of the final drive, respectively.

3. Multiobjective Optimal Car-Following Driving Problem

In this section, the multiobjective optimal car-following driving problem under consideration is discussed. This relates to effectively plan the following vehicle's speed profile over time in the car-following driving scenario illustrated in Figure 3. In the highlighted scenario, the following vehicle receives at each time instant information from the preceding vehicle, including its position x_{prec}, velocity $\dot{x}_{prec}$, and acceleration $\ddot{x}_{prec}$, respectively. The flow of information between the two vehicles is supposed ideal and instantaneous, and a given value of inter-vehicular distance (IVD) results from the positions, speeds and accelerations of both the preceding vehicle and the following vehicle at each time instant. Two optimization targets are considered here for the following vehicle, respectively, related to the propelling energy minimization and the passenger comfort enhancement. The mathematical formulation corresponding to the optimal car-following driving problem under analysis was derived, and it is reported in Equations (12)–(16):

$$\text{argmin} J_{follow}\left(\ddot{x}_{follow}, t\right) = \int_{t_0}^{t_{end}} \alpha_{energy} \cdot P_{traction-follow}(t) + (1-\alpha_{energy}) \cdot \left|\ddot{x}_{follow}(t)\right| dt \tag{12}$$

subject to:

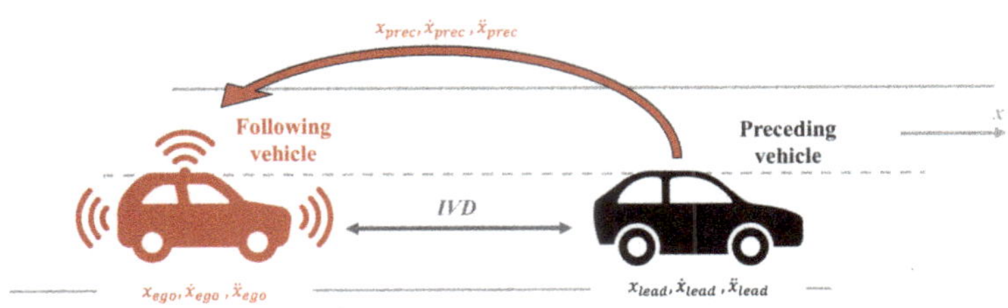

Figure 3. Schematic of the retained car-following driving scenario highlighting the information flow from the preceding vehicle to the following vehicle.

IVD constraints:

$$\left[x_{prec}(t) - x_{follow}(t)\right] \leq IVD_{MAX}(t) \tag{13}$$

$$\left[x_{prec}(t) - x_{follow}(t)\right] \geq IVD_{safety}(t) \tag{14}$$

Powertrain constraints:

$$P_{traction-follow}(t) = LHV_{fuel} \cdot \left[\dot{m}_{fuel}(t) + m_{fuel_{crank}} \cdot start_{ICE}(t)\right] - (V_{OC} \cdot Ah_{batt} \cdot 3600) \cdot S\dot{O}C \tag{15}$$

$$P_{traction-follow}(t) \leq P_{traction-follow-MAX}\left(t,\ \dot{x}_{follow}(t)\right) \tag{16}$$

where J_{follow} represents the following vehicle's cost function that needs to be minimized in the considered optimization problem. Its value represents an integration of the instantaneous cost terms throughout the entire drive cycle from the initial time instant t_0 to the final time instant t_{end}. $\ddot{x}_{follow}$ is the following vehicle's longitudinal acceleration, and it represents the only control variable affecting the overall cost function J_{follow} over time. The following vehicle's longitudinal speed trajectory can be obtained by integrating $\ddot{x}_{follow}$ over time. $P_{traction-follow}$ is the tractive power used for propelling the following vehicle, related to fuel or electricity or both of them depending on the given powertrain technology. Other than propelling energy reduction, enhancement of the ride's comfort represents a crucial potential of automated driving. To foster this aspect, several motions, path, and velocity planners proposed in the literature integrate as objective reducing variables related to vehicle longitudinal acceleration, lateral acceleration and yaw rate as an example [42,43]. Following a similar approach, reducing the following vehicle's longitudinal acceleration over the entire simulated drive cycle is considered here as an optimization target in J_{follow} to enhance the passenger comfort. In most cases, reducing the overall energy used for propulsion and improving the passenger comfort may represent contrasting objectives. For example, in electrified vehicles, several fluctuations can be observed in the longitudinal speed when optimizing for energy consumption solely [23]. A trade-off between the two optimization targets retained might, therefore, be implied. In this framework, α_{energy} represents a weighting coefficient that can be tuned to give more emphasis either on the fuel economy enhancement or on the passenger comfort improvement. Two categories of constraints are considered for the optimal car-following driving optimization problem illustrated in Equation (12). They, respectively, relate to IVD and the following vehicle's powertrain, and they find illustration in the follow-up of this section.

3.1. Constraints on the IVD

Looking at Equations (13) and (14), the IVD is defined according to x_{prec} and x_{follow}, which represent the preceding vehicle's position and the following vehicle's position, respectively. The instantaneous value of IVD is thus set in Equations (13) and (14) to be always below IVD_{MAX} and above IVD_{safety} over time. IVD_{MAX} represents the maximum distance that the following vehicle can attain from the preceding vehicle. Here, the value of IVD_{MAX} is made varying over time depending on the current road type. As a general consideration, the IVD might be limited in urban areas to ease traffic flow and to reduce general road occupancy. This correlates well with reduced values of vehicle speed generally associated with urban driving. Contrarily, enhanced road surface availability and generally higher values of vehicle speed might lead to allow a higher value of IVD_{MAX} in extra-urban and highway driving conditions. In light of these considerations, when extra-urban and highway driving conditions are encountered, the maximum achievable value of IVD is assumed here to be 300 m. This hypothesis stems from the current approximate range of V2V communication, which might contribute to long-distance preceding-vehicle detection [44]. On the other hand, the allowed value of IVD_{MAX} is assumed to be reduced to 100 m in urban driving conditions to limit road occupancy.

In Equation (14), IVD_{safety} stands for the minimum safety IVD. Its value is obtained by interpolating in a two-dimensional lookup table with the following vehicle's speed and relative vehicle speed between the preceding vehicle and the following vehicle as independent variables. The lookup table is shown in Figure 4, and it is derived from a time-headway numerical braking model proposed in [45]. The interested reader can consult [23] for further information regarding this procedure.

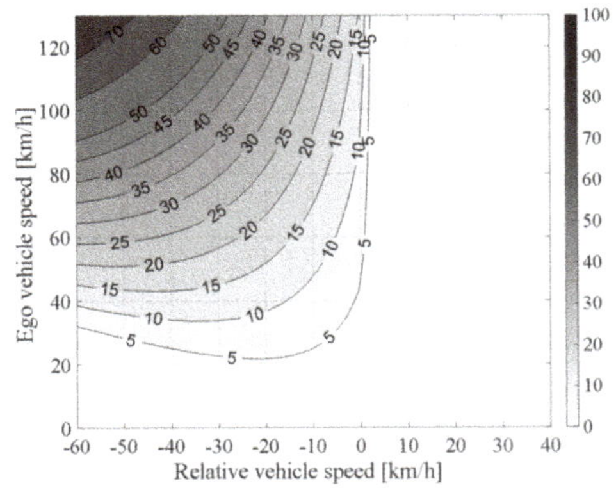

Figure 4. Minimum safety IVD as a function of the following vehicle's speed and the relative vehicle speed between the preceding vehicle and the following vehicle.

3.2. Constraints on the Following Vehicle Powertrain

The following vehicle's powertrain constraints are considered in the optimal car-following driving problem illustrated in Equations (15) and (16). These allow identifying energy-saving oriented $\ddot{x}_{follow}$ control solutions that are tailored according to the given powertrain layout and control logic for the following vehicle. First, the instantaneous following vehicle's tractive power $P_{traction-follow}$, expressed in watts, is evaluated as a function of fuel and electrical energy consumption. In Equation (15), $\dot{m}_{fuel}$ and $m_{fuel_{crank}}$ represent the instantaneous fuel rate (as obtained by interpolating in the empirical lookup table as a function of ICE speed and torque) and the mass of fuel needed to crank the ICE in grams, respectively. $start_{ICE}$ represents a binary variable detecting ICE activations over time, and its value is set to 1 in those time instants in which the sign of the fuel consumption is positive, while it was zero in the previous time instant. LHV_{fuel} is the lower heating value of the fuel and corresponds to 43,700 J/g here.

$\dot{m}_{fuel}$ and $m_{fuel_{crank}}$ are considered only when the powertrain layout includes an ICE, i.e., for CVs and HEVs. On the other hand, the battery power term is retained only when a high-voltage battery is embedded in the powertrain layout, i.e., for HEVs and BEVs. In Equation (15), the battery power term includes the SOC rate $S\dot{O}C$ as given by Equation (6) and the term $(V_{OC}\cdot Ah_{batt}\cdot 3600)$ that represents the battery energy in watt-seconds. The negative sign of the battery SOC variation relates to negative values of $S\dot{O}C$ corresponding to battery charge depletion. Especially for HEVs, this formulation for the first powertrain constraint allows obtaining the same unit of measure (i.e., watts) between the fuel chemical power term and the battery electrical power term. A proper balance between the usage of the two energy sources can be achieved in this way. At each time instant, both $\dot{m}_{fuel}$, $start_{ICE}$ and $S\dot{O}C$ related to the following vehicle can be evaluated within the control optimization process according to numerical models of powertrain layout and related onboard control logic as it has been detailed in the previous section for each retained powertrain category. Tailoring the propelling energy consumption minimization according to the specific powertrain layout of the following vehicle can be achieved in this way.

The final powertrain constraint reported in Equation (16) involves limiting the tractive power of the following vehicle within the instantaneous maximum limit represented by $P_{traction-follow-MAX}$. This term is given as a function of the instantaneous maximum tractive power of each power component embedded in the retained following vehicle's powertrain layout, and it typically depends on the current value of the following vehicle's speed $\dot{x}_{follow}$.

4. Solving the Optimal Car-Following Driving Problem

A flexible and effective approach is presented in this section for solving the multiobjective optimal car-following driving problem illustrated in the previous section. Here, the considered control problem is solved off-line, i.e., exploiting the knowledge of the speed profile over time for the entire drive cycle of the preceding vehicle a priori. Dynamic programming (DP) is considered in this framework as a widely known optimization algorithm capable of returning the optimal global solution for the control problem under analysis [46–50]. The operating principle of DP involves exhaustively sweeping discretized values of control variables and state variables at each time instant of the retained control problem. Thus, the optimal sequence of control actions is identified by minimizing the overall value of a predefined cost function [51–53]. To find the global optimal control trajectory, the DP workflow is iterated at each time instant backwardly in the retained drive cycle from t_{end} to t_0. The DP adaption presented by Sundstrom and Guzzella [54] is retained here as a control workflow for all the considered powertrains. The state-space X and the control space U considered here are reported in Equation (17), and they relate to the DP algorithm implemented to solve the optimal car-following driving problem illustrated in Equation (12) [23].

$$X = \begin{Bmatrix} x_{prec} - x_{follow} \\ \dot{x}_{follow} \end{Bmatrix}, \; U = \left\{ \ddot{x}_{follow} \right\} \tag{17}$$

The state-space includes the IVD and the following vehicle's speed $\dot{x}_{follow}$. The IVD is particularly considered to ensure compliance over time with the corresponding optimization constraints reported in Equations (13) and (14). The following vehicle's longitudinal speed is retained to evaluate its trajectory over time by integrating the following vehicle's longitudinal acceleration, representing the only control variable included in U.

In the follow-up of this section, the workflow for assessing the capability of the introduced DP as ACC algorithm for car-following scenarios is discussed for all the four vehicle powertrain architectures illustrated in Section 2. As introduced earlier, energy-saving and passenger comfort improvement compared with the preceding vehicle are retained as evaluation metrics for the following vehicle's performance. Passenger comfort is particularly measured here in terms of the root-mean-square (RMS) of the longitudinal vehicle acceleration throughout the given driving mission. The value of RMS for vehicle acceleration indeed represents a common index for evaluating the quality of passengers' ride perception [55,56].

4.1. CV Powertrain

The flowchart for validating the proposed ACC algorithm for the CV powertrain is illustrated in Figure 5a, together with the corresponding flowchart considering the BEV powertrain in Figure 5b. Once the vehicle data, control logic and the drive cycle under analysis are defined, an online simulation is performed first for the preceding vehicle, following the exact speed profile over time for the given input cycle. Focusing on the CV flowchart reported in Figure 5a, the gear engaged is determined at each time instant according to the considered shift logic, while the fuel consumption can be computed according to values for speed and torque of the ICE following Equations (1)–(3). Then, the following vehicle's case is retained while sweeping different values of α_{energy} ranging from 0 to 1. The following vehicle's longitudinal velocity is planned off-line following the illustrated DP approach and integrating the controlled value of longitudinal acceleration $\ddot{x}_{follow}$ over time. An online simulation is subsequently performed to double-check the fuel consumption and the passenger comfort resulting from the following vehicle's driving solution obtained from DP. Evaluating the multiobjective optimal Pareto front when minimizing both the fuel consumption and the RMS of the vehicle acceleration for the following vehicle is allowed in this way. Moreover, the performance of the preceding vehicle that follows the drive cycle speed profile under analysis can be benchmarked with the correspondingly obtained following-vehicle control solutions.

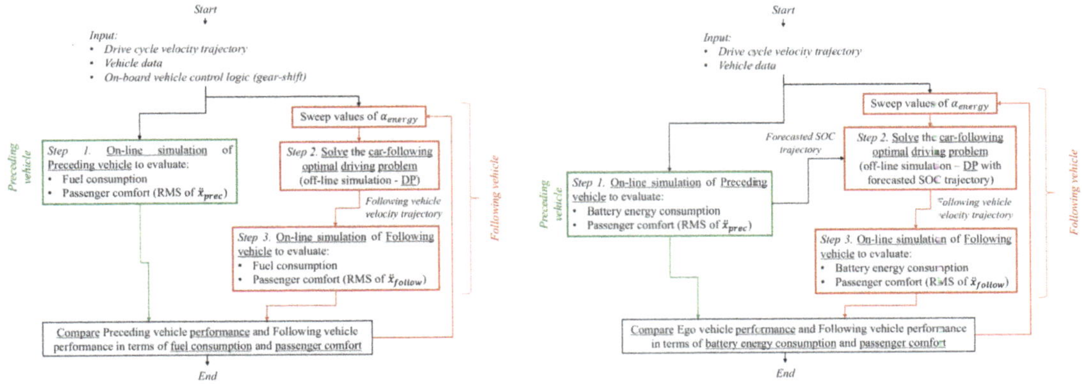

(a) Conventional vehicle (CV) (b) Battery electric vehicle (BEV)

Figure 5. Flowchart of the validation methodology for the proposed optimal V2V vehicle velocity planner considering a CV powertrain (**a**) and a BEV powertrain (**b**).

4.2. BEV Powertrain

As regards the BEV powertrain, it should be noted that the SOC variation modeled as in Equation (6) and embedded in the car-following optimal driving problem illustrated in Equation (12) depends on the current value of SOC. Accounting for the current value of SOC over time when solving the optimal car-following driving problem with DP would, in turn, require an additional state variable. Nevertheless, DP is notably affected by the curse of dimensionality depending on the size of both control and state sets. In this framework, not only is the DP typically associated with a remarkable computational cost but this latter has been demonstrated exponentially increasing as a function of the number of control variables and state variables considered [32]. To avoid this drawback, an alternative approach is proposed here. Once the preceding vehicle has been simulated online in the retained drive cycle, the obtained SOC trajectory over time is used as information when running the DP for the following vehicle's velocity planning in Figure 5b. Assuming an equal SOC value for both the preceding vehicle and following vehicle at the beginning of the drive cycle, this approach provides to the DP an estimation of the punctual SOC value for the following BEV at each time instant. The forecasted SOC value is used in turn to evaluate the SOC variation of the following vehicle as a function of its controlled acceleration. It is true that at each time instant, the following vehicle's SOC may be slightly higher than the corresponding preceding vehicle's SOC as a result of the energy savings achieved thanks to the optimal velocity planner. However, this small difference between SOC values is usually observed having little impact on the final solution of the optimization process. To answer this concern anyway, step 3 of the discussed validation methodology in Figure 5b involves performing an online simulation of the following BEV operation the following the longitudinal velocity planned earlier by DP. Evaluating the exact SOC variation over time for the following vehicle's performance of the given drive cycle in a car-following scenario is achieved in this way. Results obtained for the following vehicle by varying α_{energy} can be benchmarked with the corresponding the preceding vehicle's performance for the BEV layout as well.

4.3. P2 HEV and PS HEV Powertrains

The flowchart of the validation methodology for the proposed ACC optimization-driven approach is illustrated in Figure 6 for the P2 HEV and the PS HEV powertrains.

The first step aims at assessing the operation of the preceding vehicle in the retained drive cycle using an online simulation, which accounts for the HEV onboard control logic. For the P2 HEV, control variables at each time step include the gear engaged and the torque split between ICE and EM. On the other hand, ICE speed and ICE torque represent the variables to be controlled at each time instant for the PS HEV. In both cases, the fuel consumption and the battery SOC trajectory can be evaluated, and this latter is used as input to the following vehicle's DP-based velocity planner for estimating the punctual SOC value at each time instant of the optimization process as it was described above for the BEV powertrain. In this case, the estimated value of SOC allows not only calculating the SOC variation, but it also impacts at each time instant on the HEV control logic influencing the power split between ICE and EMs as it has been described in Sections 2.3 and 2.4. DP is subsequently executed in the second step of Figure 6 to identify the optimal control pattern according to the drive cycle, the HEV powertrain, the related control logic and the weighted objective function under analysis. The third step then aims at online simulating the following vehicle's operation according to the longitudinal velocity profile over time identified at the previous step. Furthermore, the flowchart illustrated in Figure 6 can be repeated several times by varying the value of α_{energy} and benchmarking obtained results with the corresponding the preceding vehicle's performance.

5. Results

This section aims at presenting obtained results for the proposed optimization-driven ACC approach. The performance of the implemented algorithm is particularly evaluated for all the retained powertrain categories in a different drive cycle, such as the urban dynamometer driving schedule (UDDS), the worldwide harmonized light vehicle test procedure (WLTP), the highway federal test procedure (HWFET) and the US06 supplemental procedure (US06). Energy savings and comfort improvement are quantified for the ACC-enabled following vehicle by benchmarking with the performance of the preceding vehicle that embeds the same powertrain. The preceding vehicle reproduces the vehicle speed profile provided over time for the given driving mission, while the following vehicle reproduces the longitudinal speed profile provided off-line using DP and solving the car-following optimal driving problem illustrated in Section 3.

Table 2 reports the vehicle and powertrain data considered in all performed simulations and related to each retained powertrain category. The vehicle body is the same for all cases, and it was retained from [57]. The same ICE is embedded in CV, P2 HEV and PS HEV powertrain layouts, and it refers to a 1.2-L spark-ignition engine. ICE data, including an efficiency map, were generated according to the methodology implemented in Amesim® software (version 2020.1, Siemens PLM, Camberley, UK) and discussed in [58]. A 6-gear AMT layout is considered for the CV, and the P2 HEV, with corresponding gear-shift logic, developed the following approach illustrated in [33]. Efficiency maps and operational data of interior permanent magnet synchronous machines were generated for the EMs according to the corresponding methodology implemented in Amesim® software and described in [59]. Battery SOC dependent parameters (i.e., voltage and resistance) related to 42 kWh and 2 kWh high-voltage battery packs were derived for the BEV and the HEV layouts, respectively. To this end, the electrical storage system sizing tool implemented in Amesim® described in [60] was used. The HEV control logics were developed taking inspiration from [38], and they were tuned following a trial-and-error procedure to achieve charge-sustained operation while reducing fuel consumption. Finally, a constant auxiliary loss power of 500 W is considered for BEV, P2 HEV and PS HEV powertrains.

The effectiveness of the proposed optimization-driven ACC approach for the following vehicle is assessed here by performing several simulations of the following methodologies illustrated in Figures 5 and 6. Each described powertrain category was simulated in all the considered drive cycles. In this case, six values are retained for α_{energy} in each performed evaluation corresponding to (0.01 0.2 0.4 0.6 0.8 0.99) to assess a wide range of optimization targets obtained as a combination of energy minimization objective and comfort improvement objective. The simulation time step is set to 1 s. The initial IVD between the preceding vehicle and the following vehicle is set to be 50 m, and the final value is constrained to be below 50 m to ensure that the two vehicles travel the same mileage. The initial battery SOC is assumed to be 95% and 70% for the BEV powertrain and the two HEV powertrains, respectively.

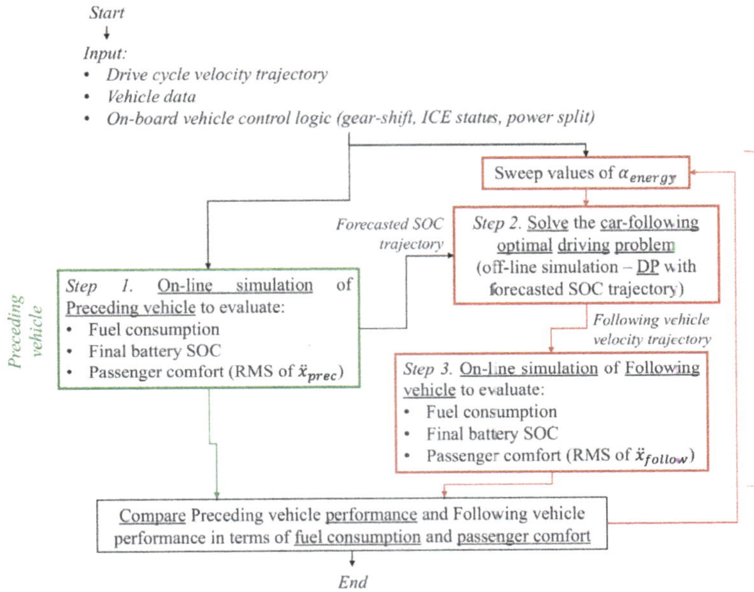

Figure 6. Flowchart of the validation methodology for the proposed optimal V2V vehicle velocity planner considering P2 HEV and PS HEV powertrains.

Table 2. Vehicle and powertrain parameters.

Component	Parameter	CV	BEV	P2 HEV	PS HEV
Vehicle body	Mass (kg)	1248	1248	1248	1248
	RL_A (N)	143	143	143	143
	RL_B (N/(m/s))	0.9	0.9	0.9	0.9
	RL_C (N/(m/s^2))	0.44	0.44	0.44	0.44
	Wheel dynamic radius (m)	0.273	0.273	0.273	0.273
Transmission	AMT gear ratios (-)	(3.58; 2.06; 1.40; 1; 0.71; 0.58)	-	(3.58; 2.06; 1.40; 1; 0.71; 0.58)	-
	PG ratio (-)	-	-	-	2.6
	EM to FD ratio (-)	-	2.3	2	1.26
	FD ratio (-)	4.1	3.4	4.1	3.27
	Efficiency (-)	0.9	0.9	0.9	0.85
ICE	Displacement (l)	1.2	-	1.2	1.2
	Max power	89 kW @ 4000 rpm	-	89 kW @ 4000 rpm	89 kW @ 4000 rpm
	Max torque	230 Nm @ 2000 rpm	-	230 Nm @ 2000 rpm	230 Nm @ 2000 rpm
EM1	Max power	-	95 kW @ 14,500 rpm	26 kW @ 9000 rpm	26 kW @ 9000 rpm
	Max torque	-	147 Nm @ (0–4800 rpm)	118 Nm @ (0–2000 rpm)	118 Nm @ (0–2000 rpm)
EM2	Max power	-	-	-	45 kW @ 9000 rpm
	Max torque	-	-	-	204 Nm @ (0–2000 rpm)
Battery pack	Nominal capacity (Ah)	-	115.5	6.5	6.5
	Voltage (V)	-	364	310	310

The Pareto fronts for the obtained results in terms of energy consumption and RMS of the vehicle acceleration are shown in Appendix A from Figures A1–A4 for each retained powertrain category in all the considered drive cycles. Both preceding vehicle and following vehicle's simulation cases are illustrated. Particularly for the following vehicle, different control solutions are obtained by sweeping various values of α_{energy}, i.e., the weighting coefficient between energy economy target and passenger comfort target. The fuel consumption and the battery energy consumption are reported as energy economy metrics in the x-axis for the CV and the BEV powertrain layouts, respectively. On the other hand, to achieve a fair comparison for the HEV powertrains, the amount of fuel needed to recharge the battery up to the initial SOC value (70% in this case) needs to be calculated and added to the previously calculated fuel consumption. A method to evaluate this additional fuel consumption was retained from literature and applied to the specific HEV architectures under analysis [61]. A parameter named equivalent fuel consumption (EFC) can be obtained in this way that accounts for both the fuel consumed by the ICE throughout the simulation of the considered drive cycle and for the amount of fuel representative of the net battery SOC increase or decrease throughout the simulated drive cycle. The EFC is thus used as an evaluation metric for the HEV energy economy assessed in the x-axis of the Pareto fronts reported in Figures A3 and A4. On the other hand, the RMS of the following vehicle's acceleration is reported in the y-axis of all Pareto fronts displayed from Figures A1–A4 as an indicator of the level of passenger comfort. In general, both lower values for both propelling energy consumption and RMS of the vehicle acceleration can be obtained for the following vehicle compared with the preceding vehicle in all Pareto fronts displayed in Figures A1–A4. The capability of the proposed optimization-driven ACC approach to effectively improve the following vehicle's performance both in terms of energy consumption and passenger comfort may be suggested in this way.

For each evaluation case, corresponding to a given powertrain architecture and a predefined drive cycle, two suboptimal control solutions for the following vehicle were identified corresponding to energy-saving maximization (i.e., "Opt_energy") and passenger comfort improvement maximization (i.e., "Opt_comfort"). "Opt_energy" and "Opt_comfort" control solutions for the following vehicle relate to maximum and minimum set values for α_{energy}, i.e., 0.99 and 0.01, respectively. The Pareto fronts for the following-vehicle control solutions embedding the CV powertrain layout displayed in Figure A1 are characterized by a regular trend for all the four drive cycles under consideration. In particular, progressive reduction in fuel consumption can be achieved at the expense of increasing the RMS of the vehicle acceleration when gradually increasing the value of α_{energy}. Similar behavior can be observed for the BEV layout and the P2 HEV layout in Figures A2 and A3, respectively, even with different slopes for the Pareto fronts. For example, a Pareto front characterized by large steepness can be observed for the BEV layout in HWFET in Figure A2c, suggesting that only marginal improvement can be obtained in the overall battery energy consumption by increasing α_{energy} for the retained highway driving conditions. As concerns passenger comfort, larger improvement can be achieved within the following-vehicle control solutions of the BEV layout in HWFET. This relates to the corresponding RMS of the following vehicle's acceleration decreasing from 0.30 m/s^2 to 0.27 m/s^2 when decreasing the value of α_{energy}. On the other hand, the Pareto front for the P2 HEV layout in UDDS shown in Figure A3b characterizes for its reduced steepness. In this case, a larger variation in EFC can be observed when varying the value of α_{energy} among the following-vehicle control solutions, while narrower variation is obtained in the RMS value of the following vehicle's acceleration. As a result, urban driving conditions are found promising for potentially reducing the EFC of the retained P2 HEV layout, shifting from 4.59 L/100 km to 4.33 L/100 km as the value of α_{energy} gradually increases. As regards the PS HEV layout, the corresponding Pareto fronts are shown in Figure A4 are characterized by a less conventional trend than the remaining three powertrain layouts. Indeed, only a few following-vehicle control solutions are located on the dual-objective optimal front for EFC and RMS of vehicle acceleration, while the remaining ones are located in non-optimal regions. This effect is particularly emphasized for WLTP, and US06 drive cycles in Figure A4a,d since the optimized Pareto front is represented by only "Opt_energy" and "Opt_comfort" control solutions. For both WLTP and US06, increasing the value of α_{energy} from 0.01 to 0.8 for the PS HEV following-vehicle control appears to bring overall improvements both in terms of energy consumption and passenger comfort. As a consequence, the "Opt_energy" control solution in both Figure A4a,d corresponds to a value of 0.8 for α_{energy} instead of 0.01 as for the remaining cases under analysis. A possible explanation for this behavior of the following-vehicle control performance for the retained PS HEV powertrain layout may relate to the proposed optimization-drive ACC approach being capable of further improving the passenger comfort as well when focusing on EFC improvement. The opposite behavior can be observed for UDDS and HWFET in Figure A4b,c, respectively, since the related "Opt_comfort" control solution corresponds to a value of 0.6 for α_{energy} instead of 0.99 as for the remaining considered cases. For these drive cycles, further optimizing the following PS HEV control solution for EFC reduction may indeed bring to excessively irregular trends of following-vehicle velocity solutions, thus in turn involving increased values for the EFC.

Results displayed in Figures A1–A4 in Appendix A demonstrate how the proposed optimization-driven ACC approach preserves the engineer's freedom when selecting the control solution being oriented either to energy-saving, comfort improvement or a blending of the two objectives. Energy consumption and the RMS of vehicle acceleration are reported in Table 3 for each suboptimal control solution identified (both "Opt_energy" and "Opt_comfort"), together with the corresponding percentage of improvement compared with the related preceding vehicle's case. In general, obtained results suggest how enhancing energy-saving and improving passenger comfort can be achieved for the following vehicle in car-following automated driving conditions implementing the proposed control approach. Particularly, energy savings vary depending on the drive cycle and on the powertrain type from 1.8% to even 22.1% for the PS HEV in US06. On the other hand, the RMS of the vehicle acceleration can be reduced from 0.4% up to even 48.2% for the BEV powertrain in UDDS.

The energy-saving potential and the passenger comfort improvement potential, respectively related to "Opt_energy" and "Opt_comfort" control solutions, are highlighted in Figure 7 for each drive cycle and for each retained powertrain category. As shown in Figure 7a, the PS HEV powertrain is suggested to achieve the best energy-saving potential in most driving conditions. This relates to the effectiveness of the proposed optimization-driven ACC approach and in part to the considered PS HEV generally exhibiting a slightly larger EFC compared with both CV and P2 HEV, thus increasing the energy-saving potential achievable using the ACC technology. On the other hand, the BEV powertrain exhibits the lowest energy-saving capability in WLTP, HWFET and US06. As regards the remaining powertrain categories, Figure 7a suggests how their ranking in terms of energy-saving capability varies across different driving conditions, making it impractical to establish a general hierarchy for the retained powertrains. As a common trend, the energy-saving potential might be considerably reduced when only highway driving conditions are encountered, such as in HWFET. The reduced speed variation over time in highway driving conditions might reduce the energy-saving opportunity achievable by the optimization-driven ACC approach in this framework.

Focusing on the passenger comfort enhancement displayed in Figure 7b, urban driving conditions, such as in UDDS, are suggested to offer the highest potential for improvement by implementing the proposed optimization-driven ACC approach. This relates to urban driving generally distinguishing for frequent vehicle start–stop events and sudden acceleration and deceleration events. In this framework, it was possible to reduce the RMS of the vehicle acceleration from 38.7% for the P2 HEV to 48.2% for the BEV for the optimal control solution provided by DP. On the other hand, as has been observed for energy-saving, highway driving conditions are suggested to exhibit the lowest potential for comfort improvement. A clear trend cannot be observed for the rank of powertrain categories even considering passenger comfort improvement, thus opening up the need for a dedicated evaluation of each given case.

Time-series of the results for the suboptimal control solutions obtained in WLTP both as the preceding vehicle and following vehicle are reported in Appendix B from Figures A5–A12. The trend of vehicle speed, gear engaged, IVD, fuel consumption and battery SOC is particularly displayed over time when applicable. Regarding the IVD, the reader can see how the proposed optimization-driven following vehicle's ACC approach always respects the upper and lower limits retained. Vehicle operations associated with the "Opt_Energy" suboptimal control solutions in Figures A5–A8 are distinguished by considerable savings in fuel and electrical energy for the following vehicle. On the other hand, improved uniformity in the following vehicle's speed trend over time can be observed for the "Opt_Comfort" suboptimal control solutions in Figures A9–A12.

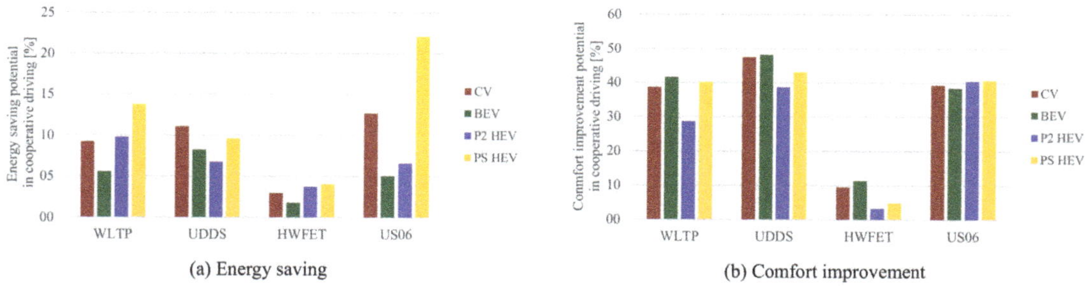

Figure 7. Overall energy-saving (**a**) and comfort improvement (**b**) potential for the following vehicle compared with the preceding vehicle when being controlled by the proposed V2V optimal velocity planner for all retained powertrain categories.

6. Conclusions

This paper proposes a multiobjective off-line optimization-driven ACC approach for car-following automated driving scenarios that can flexibly adapt to different powertrain categories. A CV powertrain, a BEV powertrain, a parallel P2 HEV powertrain and a PS HEV powertrain are considered as test cases, and their numerical model was presented along with the related onboard control strategies. The optimal problem for car-following driving has then been outlined. Propelling energy-saving and passenger comfort improvement were selected as the two optimization targets when controlling the following vehicle's longitudinal acceleration throughout a given drive cycle. Dedicated constraints were integrated for the maximum and minimum achievable values of IVD, along with specific powertrain-related constraints. An optimization-driven control solution for the presented car-following driving problem can be obtained by implementing a DP technique. Simulation results obtained in different driving conditions highlight the potential of the proposed ACC approach in identifying improved control solutions for the following vehicle in terms of energy-saving and passenger comfort considering a wide range of powertrain categories. Up to 22.1% energy-saving and up to 48.2% reduction in the RMS of the vehicle acceleration were demonstrated by the following-vehicle led using the proposed approach compared with the preceding vehicle, depending on the tuning performed for the two optimization targets.

In general, the illustrated approach preserves the engineer's freedom to select the weights for energy-saving and passenger comfort improvement for the following vehicle's operation. The obtained optimization-driven results might be used to benchmark different ACC approaches in this way. Moreover, the proposed approach could pave the way for developing real-time-capable control algorithms for the following vehicle in car-following scenarios that mimic optimal control actions forecasted by the introduced off-line optimization-driven approach. Furthermore, improving the fidelity level for the modeling approach might be achieved in terms of powertrain, vehicle dynamics, and ACC sensing using radar, LIDAR or cameras. For example, adaptations in the onboard control logic for gear-shifting, ICE activation and power split could be examined to further enhance the powertrain efficiency when traveling as a following vehicle in car-following driving. Finally, the optimization-drive approach could be extended considering multiple ACC-enabled vehicles traveling behind the preceding vehicle.

Table 3. Results for the preceding and following vehicles.

Drive Cycle	Vehicle	Parameter	CV	BEV	P2 HEV	PS HEV
WLTP	Preceding	Fuel consumption/battery energy consumption/EFC	4.27 L/100 km	15.39 kWh/100 km	4.74 L/100 km	5.45 L/100 km
		RMS (acceleration) (m/s^2)	0.53	0.53	0.53	0.53
	Following–Opt_energy	Fuel consumption/battery energy consumption/EFC	3.87 L/100 km (−9.2%)	14.54 kWh/100 km (−5.5%)	4.27 L/100 km (−9.8%)	4.70 L/100 km (−13.7%)
		RMS (acceleration) (m/s^2)	0.47 (−10.9%)	0.52 (−2.5%)	0.51 (−4.5%)	0.33 (−38.7%)
	Following–Opt_comfort	Fuel consumption/battery energy consumption/EFC	4.12 L/100 km (−3.5%)	14.59 kWh/100 km (−5.2%)	4.35 L/100 km (−8.2%)	4.71 L/100 km (−13.6%)
		RMS (acceleration) (m/s^2)	0.33 (−38.6%)	0.31 (−41.6%)	0.38 (−28.7%)	0.32 (−40.2%)
UDDS	Preceding	Fuel consumption/battery energy consumption/EFC	7.50 L/100 km	10.80 kWh/100 km	4.65 L/100 km	4.46 L/100 km
		RMS (acceleration) (m/s^2)	0.62	0.62	0.62	0.62
	Following–Opt_energy	Fuel consumption/battery energy consumption/EFC	6.67 L/100 km (−11.1%)	9.91 kWh/100 km (−8.2%)	4.33 L/100 km (−6.8%)	4.03 L/100 km (−9.6%)
		RMS (acceleration) (m/s^2)	0.61 (−2.7%)	0.58 (−7.9%)	0.55 (−11.3%)	0.46 (−27.0%)
	Following–Opt_comfort	Fuel consumption/battery energy consumption/EFC	7.15 L/100 km (−4.6%)	9.98 kWh/100 km (−7.6%)	4.59 L/100 km (−1.2%)	4.11 L/100 km (−7.8%)
		RMS (acceleration) (m/s^2)	0.33 (−47.5%)	0.32 (−48.2%)	0.38 (−38.7%)	0.36 (−43.1%)
HWFET	Preceding	Fuel consumption/battery energy consumption/EFC	3.82 L/100 km	15.22 kWh/100 km	4.11 L/100 km	4.55 L/100 km
		RMS (acceleration) (m/s^2)	0.30	0.30	0.30	0.30
	Following–Opt_energy	Fuel consumption/battery energy consumption/EFC	3.71 L/100 km (−3.0%)	14.95 kWh/100 km (−1.8%)	3.95 L/100 km (−3.8%)	4.37 L/100 km (−4.0%)
		RMS (acceleration) (m/s^2)	0.28 (−6.2%)	0.30 (−0.4%)	0.29 (−2.5%)	0.29 (−3.2%)
	Following–Opt_comfort	Fuel consumption/battery energy consumption/EFC	3.76 L/100 km (−1.8%)	14.95 kWh/100 km (−1.8%)	3.95 L/100 km (−3.7%)	4.37 L/100 km (−4.0%)
		RMS (acceleration) (m/s^2)	0.27 (−9.5%)	0.27 (−11.3%)	0.29 (−3.3%)	0.28 (−4.8%)
US06	Preceding	Fuel consumption/battery energy consumption/EFC	4.54 L/100 km	19.23 kWh/100 km	4.78 L/100 km	7.30 L/100 km
		RMS (acceleration) (m/s^2)	0.99	0.99	0.99	0.98
	Following–Opt_energy	Fuel consumption/battery energy consumption/EFC	3.97 L/100 km (−12.7%)	18.26 kWh/100 km (−5.0%)	4.46 L/100 km (−6.6%)	5.69 L/100 km (−22.1%)
		RMS (acceleration) (m/s^2)	0.84 (−14.4%)	0.94 (−4.2%)	0.94 (−4.3%)	0.58 (−40.4%)
	Following–Opt_comfort	Fuel consumption/battery energy consumption/EFC	4.08 L/100 km (−10.3%)	18.30 kWh/100 km (−4.9%)	4.51 L/100 km (−5.6%)	5.70 L/100 km (−21.9%)
		RMS (acceleration) (m/s^2)	0.60 (−39.3%)	0.61 (−38.4%)	0.59 (−40.4%)	0.58 (−40.7%)

Funding: This research was funded by the Doctoral School of Politecnico di Torino.

Conflicts of Interest: The author declares no conflict of interest.

Abbreviations

ACC	Adaptive cruise control
AMT	Automated manual transmission
BEV	Battery electric vehicle
CACC	Cooperative adaptive cruise control
CC	Cruise control
CV	Conventional vehicle
DP	Dynamic programming
EFC	Equivalent fuel consumption
EM	Electric motor
eVT	Electrically variable transmission
HEV	Hybrid electric vehicle
HWFET	Highway federal test procedure
ICE	Internal combustion engine
IVD	Inter-vehicular distance
LIDAR	Light Detection and Ranging
OOL	Optimal operating line
P2 HEV	Parallel P2 hybrid electric vehicle
PG	Planetary gearset
PS HEV	Power-split hybrid electric vehicle
SOC	State-of-charge
UDDS	Urban dynamometer driving schedule
US06	US06 supplemental procedure
V2V	Vehicle-to-vehicle
WLTP	Worldwide harmonized light-vehicle test procedure

Appendix A. Optimal V2V Driving Pareto Fronts

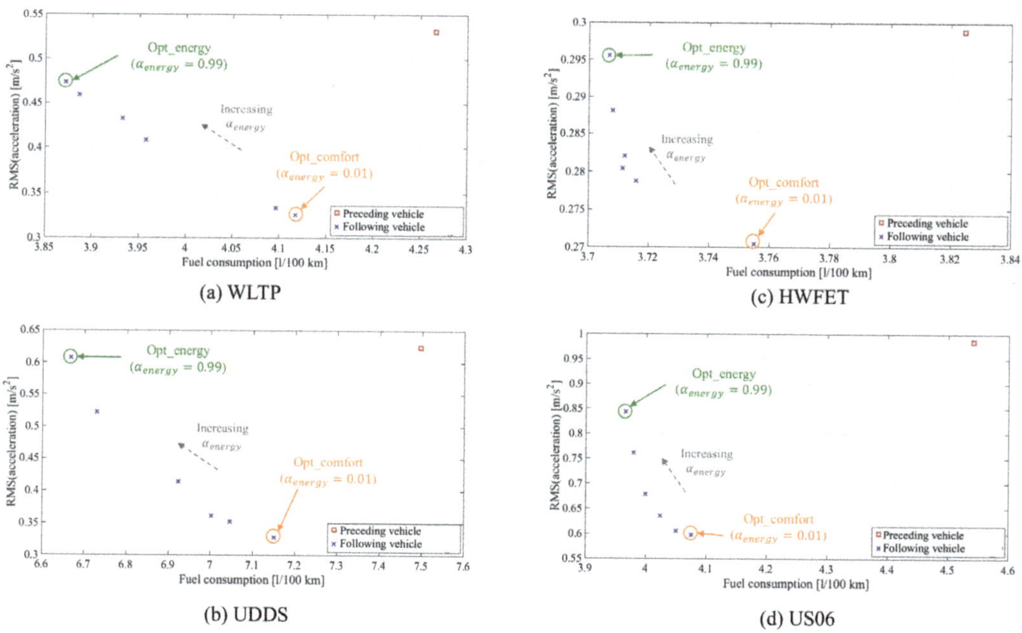

Figure A1. Pareto fronts for fuel consumption and RMS of the vehicle acceleration in WLTP (**a**), UDDS (**b**), HWFET (**c**) and US06 (**d**)—CV powertrain.

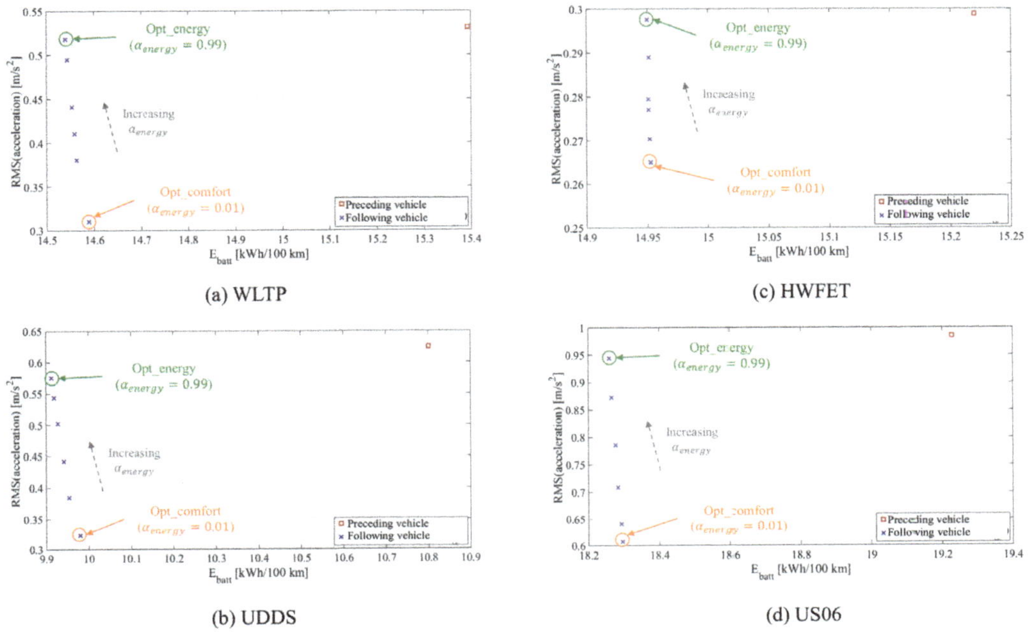

Figure A2. Pareto fronts for fuel consumption and RMS of the vehicle acceleration in WLTP (**a**), UDDS (**b**), HWFET (**c**) and US06 (**d**)—BEV powertrain.

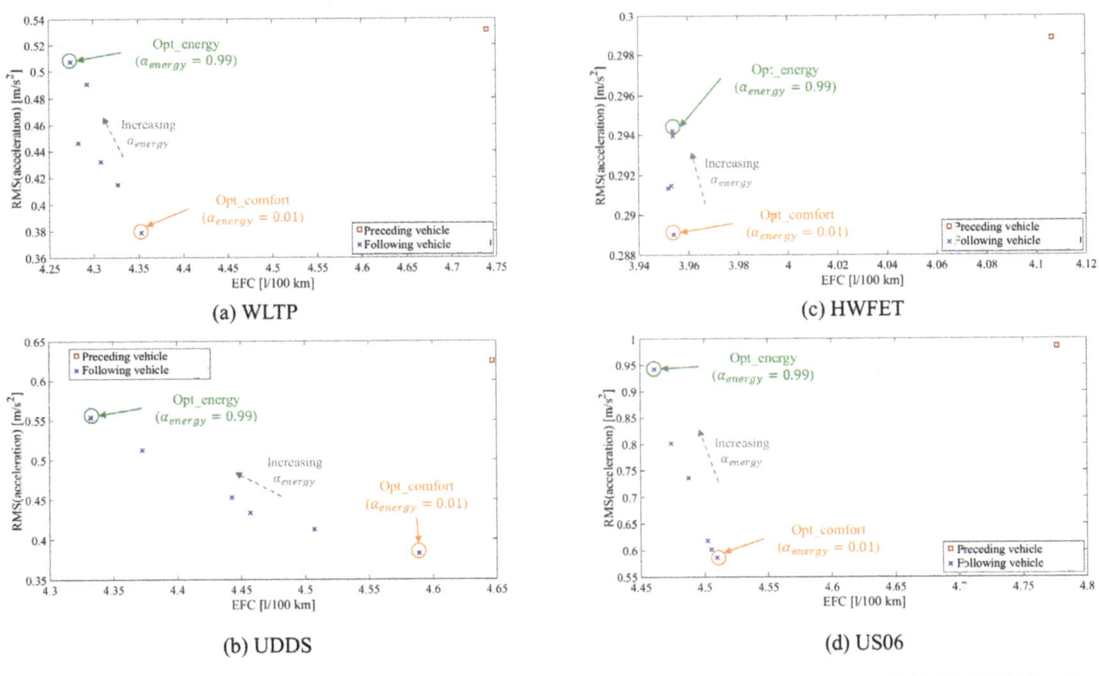

Figure A3. Pareto fronts for fuel consumption and RMS of the vehicle acceleration in WLTP (**a**), UDDS (**b**), HWFET (**c**) and US06 (**d**)—P2 HEV powertrain.

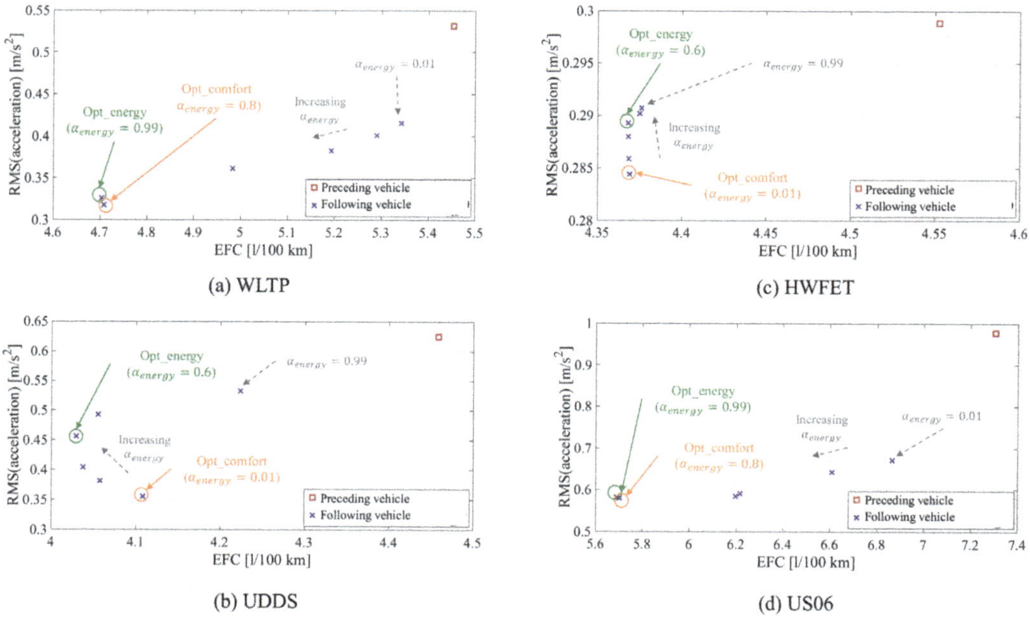

Figure A4. Pareto fronts for fuel consumption and RMS of the vehicle acceleration in WLTP (**a**), UDDS (**b**), HWFET (**c**) and US06 (**d**)—PS HEV powertrain.

Appendix B. Time-Series of Suboptimal Control Solutions in WLTP

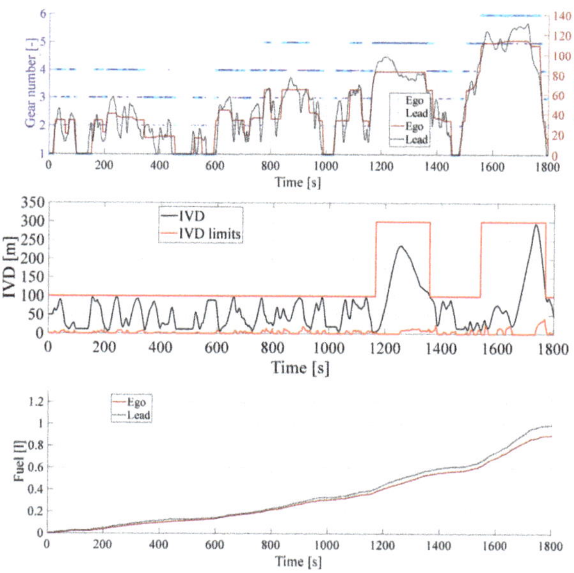

Figure A5. Time-series for the simulation results of the CV powertrain in WLTP both as the preceding vehicle and the following vehicle for the "Opt_Energy" suboptimal control solution.

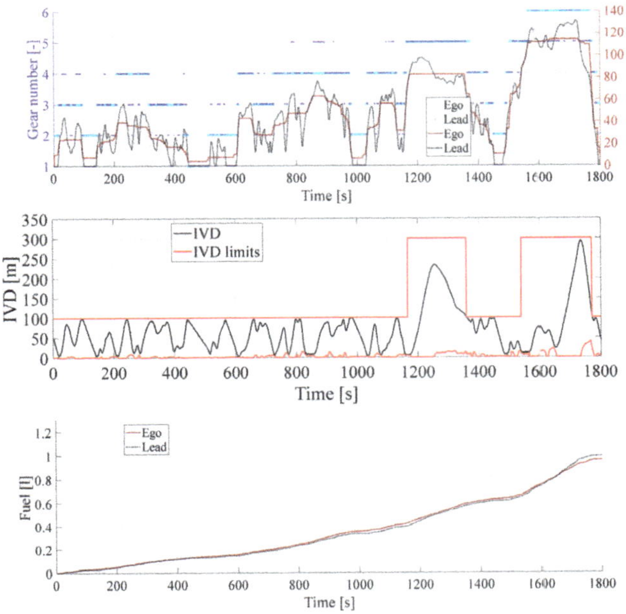

Figure A6. Time-series for the simulation results of the CV powertrain in WLTP both as the preceding vehicle and the following vehicle for the "Opt_Comfort" suboptimal control solution.

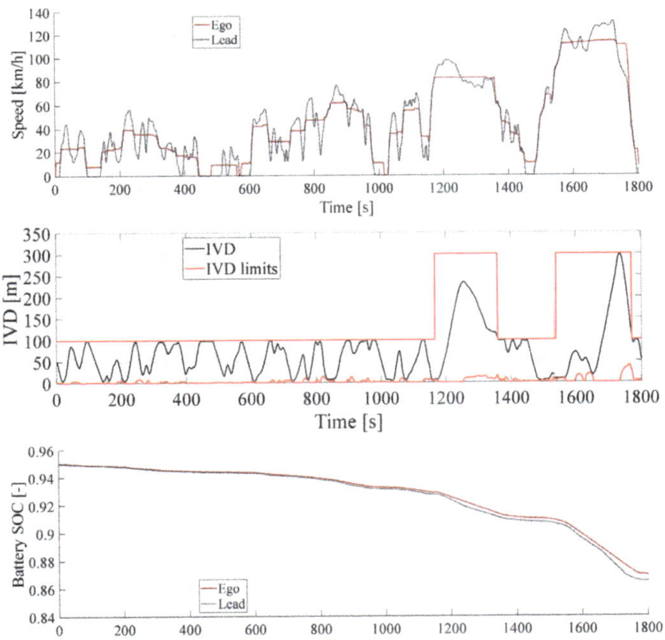

Figure A7. Time-series for the simulation results of the BEV powertrain in WLTP both as the preceding vehicle and the following vehicle for the "Opt_Energy" suboptimal control solution.

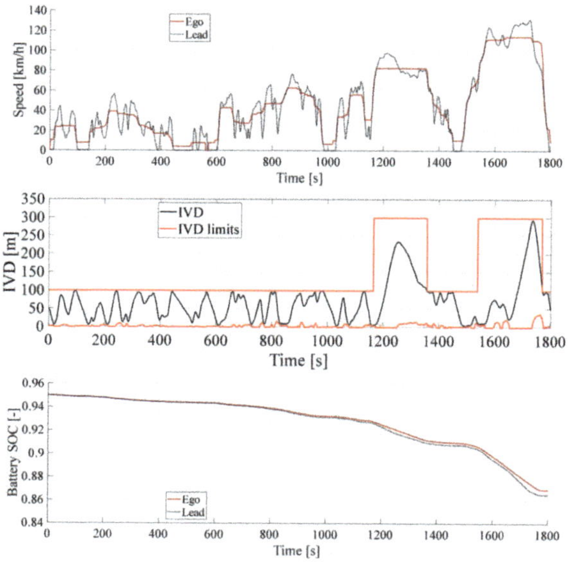

Figure A8. Time-series for the simulation results of the BEV powertrain in WLTP both as the preceding vehicle and the following vehicle for the "Opt_Comfort" suboptimal control solution.

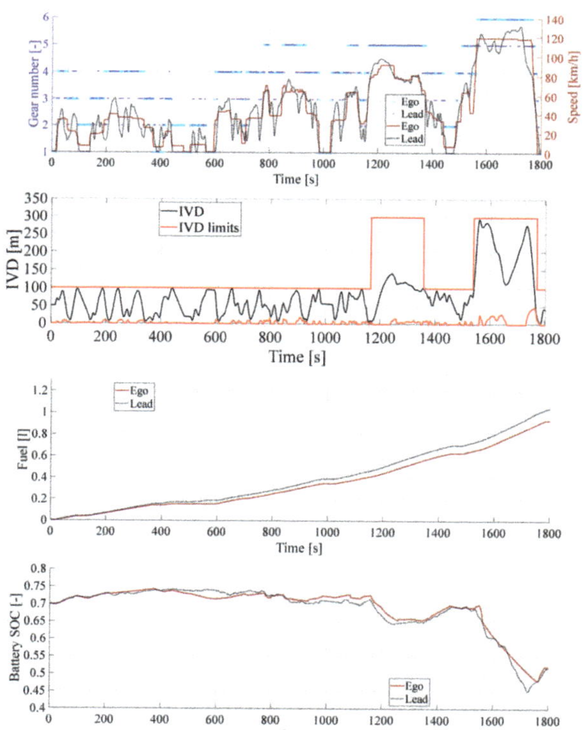

Figure A9. Time-series for the simulation results of the P2 HEV powertrain in WLTP both as the preceding vehicle and the following vehicle for the "Opt_Energy" suboptimal control solution.

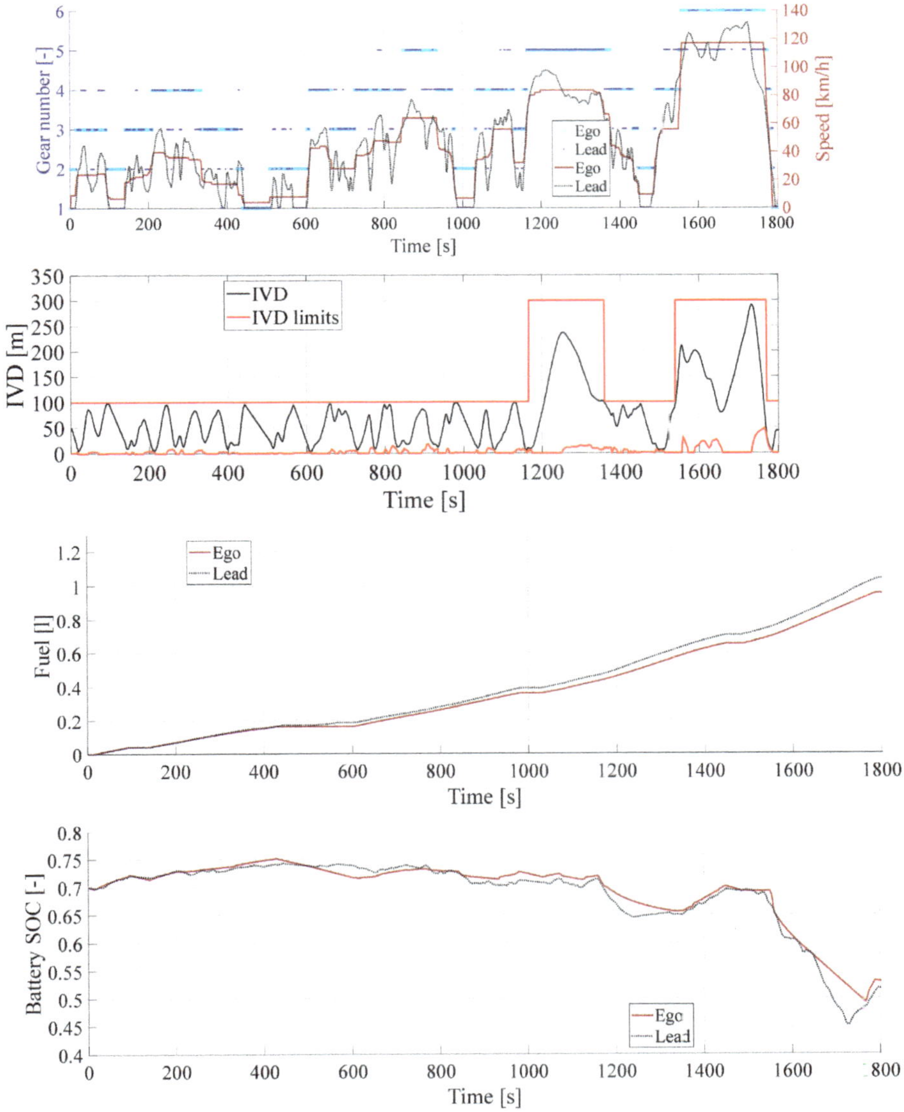

Figure A10. Time-series for the simulation results of the P2 HEV powertrain in WLTP both as the preceding vehicle and the following vehicle for the "Opt_Comfort" suboptimal control solution.

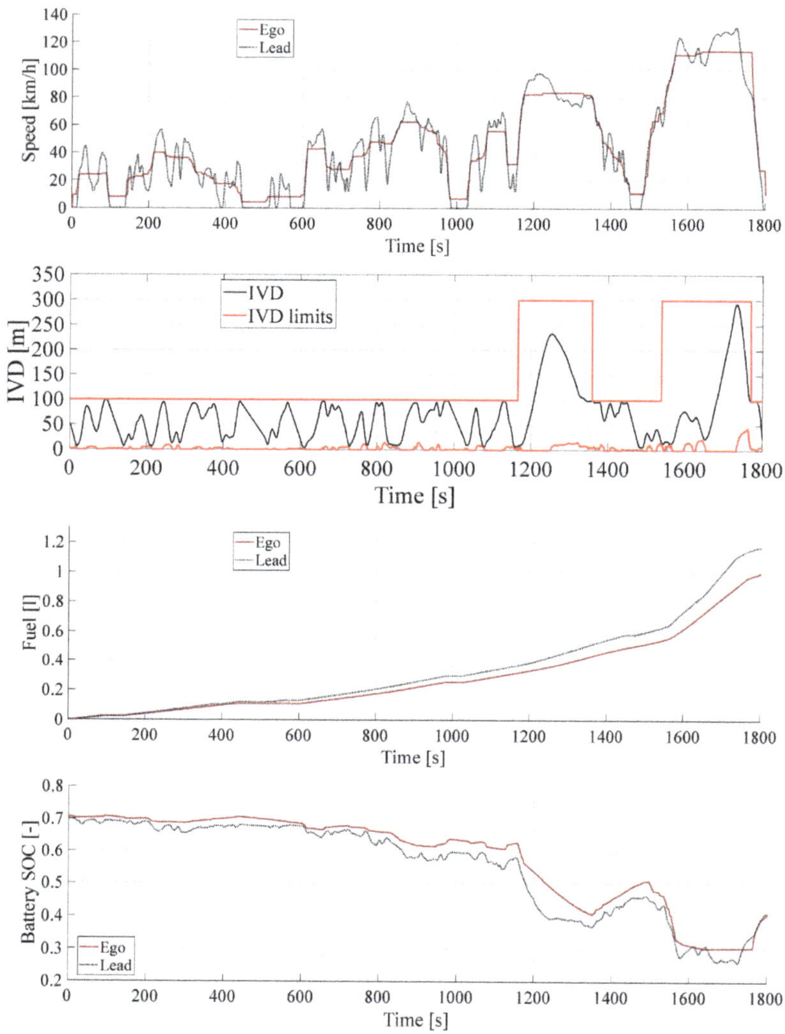

Figure A11. Time-series for the simulation results of the PS HEV powertrain in WLTP both as the preceding vehicle and the following vehicle for the "Opt_Energy" suboptimal control solution.

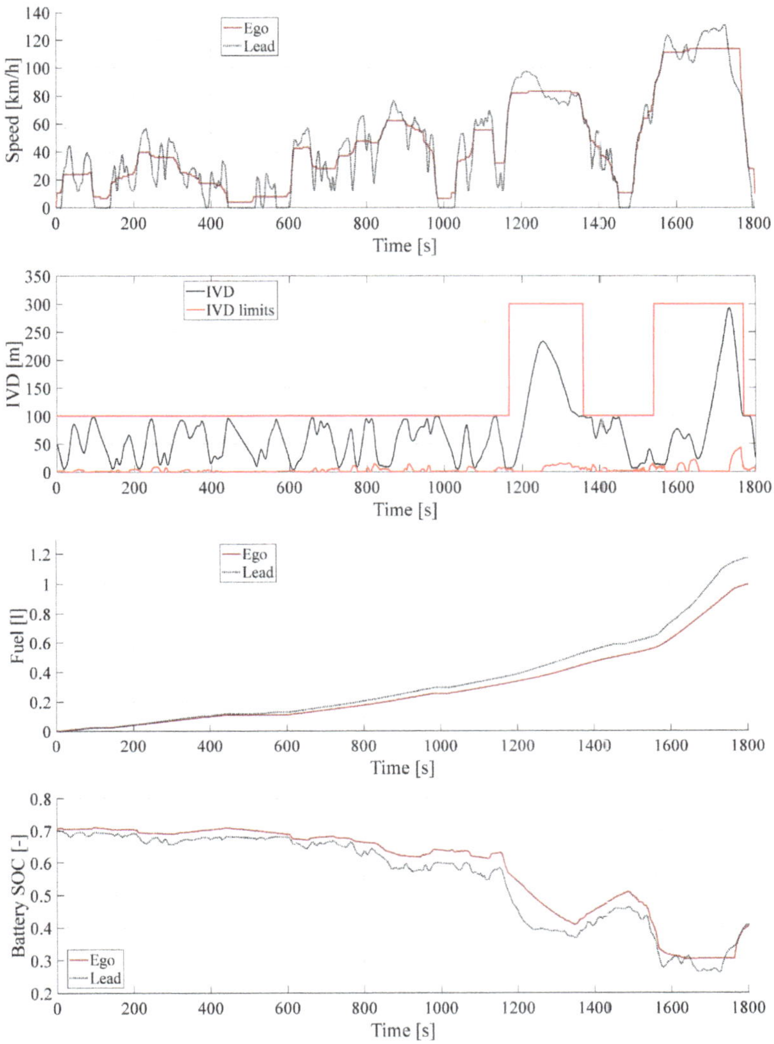

Figure A12. Time-series for the simulation results of the PS HEV powertrain in WLTP both as the preceding vehicle and the following vehicle for the "Opt_Comfort" suboptimal control solution.

References

1. Iclodean, C.; Cordos, N.; Varga, B.O. Autonomous Shuttle Bus for Public Transportation: A Review. *Energies* **2020**, *13*, 2917. [CrossRef]
2. Lim, H.S.M.; Taeihagh, A. Autonomous Vehicles for Smart and Sustainable Cities: An In-Depth Exploration of Privacy and Cybersecurity Implications. *Energies* **2018**, *11*, 1062. [CrossRef]
3. Bonfitto, A.; Feraco, S.; Amati, N.; Tonoli, A. Virtual Sensing in High-Performance Vehicles with Artificial Intelligence. In Proceedings of the International Design Engineering Technical Conferences and Computers and Information in Engineering Conference, Anaheim, CA, USA, 18–21 August 2019; Volume 59216.
4. Rios-Torres, J.; Malikopoulos, A.A. Impact of Partial Penetrations of Connected and Automated Vehicles on Fuel Consumption and Traffic Flow. *IEEE Trans. Intell. Veh.* **2018**, *3*, 453–462. [CrossRef]
5. Smith, S.W.; Kim, Y.; Guanetti, J.; Li, R.; Firoozi, R.; Wootton, B.; Kurzhanskiy, A.A.; Borrelli, F.; Horowitz, R.; Arcak, M. Improving Urban Traffic Throughput With Vehicle Platooning: Theory and Experiments. *IEEE Access* **2020**, *8*, 141208–141223. [CrossRef]
6. Ozatay, E.; Ozguner, U.; Onori, S.; Rizzoni, G. Analytical Solution to the Minimum Fuel Consumption Optimization Problem with the Existence of a Traffic Light. In *Proceedings of the ASME 2012 5th Annual Dynamic Systems and Control Conference joint with the JSME 2012 11th Motion and Vibration Conference*; ASME: Fairfield, CT, USA, 2012; pp. 837–846.

7. Wang, P.; Gao, S.; Li, L.; Sun, B.; Cheng, S. Obstacle Avoidance Path Planning Design for Autonomous Driving Vehicles Based on an Improved Artificial Potential Field Algorithm. *Energies* **2019**, *12*, 2342. [CrossRef]
8. Rizzoni, G.; Özgüner, U.; Onori, S.; Wollaeger, J.; Kumar, A.; Khayyer, P.; Ozatay, E. Big Data in Automotive Applications: Cloud Computing Based Velocity Profile Generation for Minimum Fuel Consumption. 2016. Available online: https://pdfs.semanticscholar.org/68c6/39425c042ffd2e213b5a172270aa6eb6c6fe.pdf (accessed on 22 January 2021).
9. Feraco, S.; Bonfitto, A.; Amati, N.; Tonoli, A. Combined Lane Keeping and Longitudinal Speed Control for Autonomous Driving. In Proceedings of the International Design Engineering Technical Conferences and Computers and Information in Engineering Conference, Anaheim, CA, USA, 18–21 August 2019; Volume 59216.
10. Feraco, S.; Luciani, S.; Bonfitto, A.; Amati, N.; Tonoli, A. A local trajectory planning and control method for autonomous vehicles based on the RRT algorithm. In Proceedings of the 2020 AEIT International Conference of Electrical and Electronic Technologies for Automotive (AEIT AUTOMOTIVE), Torino, Italy, 18–20 November 2020; pp. 1–6.
11. Anselma, P.G.; Belingardi, G. *Next Generation HEV Powertrain Design Tools: Roadmap and Challenges*; SAE Technical Paper 2019-01-2602; SAE International: Warrendale, PA, USA, 2019.
12. Makridis, M.; Mattas, K.; Ciuffo, B. Response Time and Time Headway of an Adaptive Cruise Control. An Empirical Characterization and Potential Impacts on Road Capacity. *IEEE Trans. Intell. Transp. Syst.* **2019**, *21*, 1677–1686. [CrossRef]
13. Makridis, M.; Mattas, K.; Ciuffo, B.; Re, F.; Kriston, A.; Minarini, F.; Rognelund, G. Empirical study on the properties of adaptive cruise control systems and their impact on traffic flow and string stability. *Transp. Res. Rec.* **2020**, *2674*, 471–484. [CrossRef]
14. Mahdinia, I.; Arvin, R.; Khattak, J.A.; Ghiasi, A. Safety, energy, and emissions impacts of adaptive cruise control and cooperative adaptive cruise control. *Transp. Res. Rec.* **2020**, *2674*, 253–267. [CrossRef]
15. He, Y.; Makridis, M.; Fontaras, G.; Mattas, K.; Xu, H.; Ciuffo, B. The energy impact of adaptive cruise control in real-world highway multiple-car-following scenarios. *Eur. Transp. Res. Rev.* **2020**, *12*, 1–11. [CrossRef]
16. Lang, D.; Stanger, T.; Del Re, L. *Opportunities on Fuel Economy Utilizing V2V Based Drive Systems*; SAE Technical Paper 2013-01-0985; SAE International: Warrendale, PA, USA, 2013.
17. He, C.R.; Orosz, G. Saving fuel using wireless vehicle-to-vehicle communication. In Proceedings of the 2017 American Control Conference (ACC), Seattle, WA, USA, 24–26 May 2017; pp. 4946–4951.
18. He, C.R.; Ge, J.I.; Orosz, G. Fuel Efficient Connected Cruise Control for Heavy-Duty Trucks in Real Traffic. *IEEE Trans. Control Syst. Technol.* **2019**, *28*, 2474–2481. [CrossRef]
19. Tunnell, J.A.; Asher, Z.D.; Pasricha, S.; Bradley, T.H. Towards Improving Vehicle Fuel Economy with ADAS. *SAE Int. J. CAV* **2018**, *1*, 81–92. [CrossRef]
20. Olin, P.; Aggoune, K.; Tang, L.; Confer, K.; Kirwan, J.; Deshpande, S.R.; Gupta, S.; Tulpule, P.; Canova, M.; Rizzoni, G. *Reducing Fuel Consumption by Using Information from Connected and Automated Vehicle Modules to Optimize Propulsion System Control*; SAE Technical Paper 2019-01-1213; SAE Technical Paper Series; SAE International: Warrendale, PA, USA, 2019.
21. Zhang, F.; Xi, J.; Langari, R. Real-Time Energy Management Strategy Based on Velocity Forecasts Using V2V and V2I Communications. *IEEE Trans. Intell. Transp. Syst.* **2016**, *18*, 416–430. [CrossRef]
22. Plianos, A.; Jokela, T.; Hancock, M. *Predictive Energy Optimization for Connected and Automated HEVs*; SAE Technical Paper 2018-01-1179; SAE International: Warrendale, PA, USA, 2018.
23. Anselma, P.G.; Belingardi, G. Enhancing Energy Saving Opportunities through Rightsizing of a Battery Electric Vehicle Powertrain for Optimal Cooperative Driving. *SAE Int. J. CAV* **2020**, *3*, 71–83. [CrossRef]
24. Koch, A.; Teichert, O.; Kalt, S.; Ongel, A.; Lienkamp, M. Powertrain Optimization for Electric Buses under Optimal Energy-Efficient Driving. *Energies* **2020**, *13*, 6451. [CrossRef]
25. Koch, A.; Bürchner, T.; Herrmann, T.; Lienkamp, M. Eco-Driving for Different Electric Powertrain Topologies Considering Motor Efficiency. *World Electr. Veh. J.* **2021**, *12*, 6. [CrossRef]
26. Tate, L.; Hochgreb, S.; Hall, J.; Bassett, M. *Energy Efficiency of Autonomous Car Powertrain*; SAE Technical Paper 2018-01-1092; SAE International: Warrendale, PA, USA, 2018.
27. Plum, T.; Wegener, M.; Eisenbarth, M.; Ye, Z.; Etzold, K.; Pischinger, S.; Andert, J. A simulation-based case study for powertrain efficiency improvement by automated driving functions. *Proc. Inst. Mech. Eng. Part D J. Automob. Eng.* **2019**, *233*, 1320–1330. [CrossRef]
28. Spano, M.; del Vescovo, D.; Sangeorzan, B.; Misul, D.; Belingardi, G. Optimizing engine downsizing and driving behaviour in conventional and hybrid powertrains for autonomous driving applications. *Int. J. Mech. Control* **2020**, *21*, 1–12.
29. Belingardi, G.; Anselma, P.G.; Demic, M. Optimization-Based Controllers for Hybrid Electric Vehicles. *Mobil. Veh. Mech.* **2018**, *44*, 53–67. [CrossRef]
30. Rizzoni, G.; Guzzella, L.; Baumann, B. Unified modeling of hybrid electric vehicle drivetrains. *IEEE/ASME Trans. Mechatron.* **1999**, *4*, 246–257. [CrossRef]
31. Anselma, P.G.; Huo, Y.; Roeleveld, J.; Emadi, A.; Belingardi, G. Rapid optimal design of a multimode power split hybrid electric vehicle transmission. *Proc. Inst. Mech. Eng. Part D J. Automob. Eng.* **2019**, *233*, 740–762. [CrossRef]
32. Anselma, P.G.; Biswas, A.; Belingardi, G.; Emadi, A. Rapid assessment of the fuel economy capability of parallel and series-parallel hybrid electric vehicles. *Appl. Energy* **2020**, *275*, 115319. [CrossRef]
33. Ayman, M.; Kim, N.; Rousseau, A. *Vehicle Modeling for Use in the CAFE Model: Process Description and Modeling Assumptions*; Argonne National Lab: Argonne, IL, USA, 2016.

34. Anselma, P.G.; Kollmeyer, P.; Belingardi, G.; Emadi, A. Multi-Objective Hybrid Electric Vehicle Control for Maximizing Fuel Economy and Battery Lifetime. In Proceedings of the 2020 IEEE Transportation Electrification Conference & Expo (ITEC), Chicago, IL, USA, 24–26 June 2020; pp. 1–6.
35. Anselma, P.G.; Belingardi, G.; Falai, A.; Maino, C.; Miretti, F.; Misul, D.; Spessa, E. Comparing Parallel Hybrid Electric Vehicle Powertrains for Real-world Driving. In Proceedings of the 2019 AEIT International Conference of Electrical and Electronic Technologies for Automotive (AEIT AUTOMOTIVE), Torino, Italy, 2–4 July 2019; pp. 1–6.
36. Bianchi, D.; Rolando, L.; Serrao, L.; Onori, S.; Rizzoni, G.; Al-Khayat, N.; Hsieh, T.-M.; Kang, P. A Rule-Based Strategy for a Series/Parallel Hybrid Electric Vehicle: An Approach Based on Dynamic Programming. In *Proceedings of the ASME 2010 Dynamic Systems and Control Conference*; ASME International: New York, NY, USA, 2010; Volume 44175, pp. 507–514.
37. Rizzoni, G.; Onori, S. Energy management of hybrid electric vehicles: 15 years of development at the Ohio State University. *Oil Gas Sci. Technol.* **2015**, *70*, 41–54. [CrossRef]
38. Biswas, A.; Anselma, P.G.; Rathore, A.; Emadi, A. Comparison of Three Real-Time Implementable Energy Management Strategies for Multi-mode Electrified Powertrain. In Proceedings of the 2020 IEEE Transportation Electrification Conference & Expo (ITEC), Chicago, IL, USA, 23–26 June 2020; pp. 514–519.
39. Kim, N.; Rousseau, A.; Rask, E. *Autonomie Model Validation with Test Data for 2010 Toyota Prius*; SAE Technical Paper 2012-01-1040; SAE International: Warrendale, PA, USA, 2012.
40. Yang, Y.; Schofield, N.; Emadi, A. Integrated Electromechanical Double-Rotor Compound Hybrid Transmissions for Hybrid Electric Vehicles. *IEEE Trans. Veh. Technol.* **2016**, *65*, 4687–4699. [CrossRef]
41. Kim, N.; Cha, S.; Peng, H. Optimal Control of Hybrid Electric Vehicles Based on Pontryagin's Minimum Principle. *IEEE Trans. Control Syst. Technol.* **2011**, *19*, 1279–1287.
42. Németh, B. Coordination of Lateral Vehicle Control Systems Using Learning-Based Strategies. *Energies* **2021**, *14*, 1291. [CrossRef]
43. Jin, X.; Yan, Z.; Yin, G.; Li, S.; Wei, C. An Adaptive Motion Planning Technique for On-Road Autonomous Driving. *IEEE Access* **2021**, *9*, 2655–2664. [CrossRef]
44. US Department of Energy. Nhtsa Issues Advance Notice of Proposed Rulemaking and Research Report on Ground-Breaking Crash Avoidance Technology: "Vehicle-To-Vehicle Communications: Readiness of V2V Technology for Application". Available online: https://www.nhtsa.gov/sites/nhtsa.dot.gov/files/documents/v2v_fact_sheet_101414_v2a.pdf (accessed on 21 January 2021).
45. Chen, C.; Lü, N.; Liu, L.; Pei, Q.-Q.; Li, X.-J. Critical safe distance design to improve driving safety based on vehicle-to-vehicle communications. *J. Cent. South Univ.* **2013**, *20*, 3334–3344. [CrossRef]
46. Bellman, R.; Lee, E. History and development of dynamic programming. *IEEE Control Syst. Mag.* **1984**, *4*, 24–28. [CrossRef]
47. Nazari, S.; Siegel, J.; Middleton, R.; Stefanopoulou, A. Power Split Supercharging: A Mild Hybrid Approach to Boost Fuel Economy. *Energies* **2020**, *13*, 6580. [CrossRef]
48. Jiang, Z.; Qiao, Y.; Chen, Y.; Ji, C. A New Reservoir Operation Chart Drawing Method Based on Dynamic Programming. *Energies* **2018**, *11*, 3355. [CrossRef]
49. An, L.N.; Tuan, T.Q. Dynamic Programming for Optimal Energy Management of Hybrid Wind–PV–Diesel–Battery. *Energies* **2018**, *11*, 3039. [CrossRef]
50. Soldo, J.; Škugor, B.; Deur, J. Synthesis of Optimal Battery State-of-Charge Trajectory for Blended Regime of Plug-in Hybrid Electric Vehicles in the Presence of Low-Emission Zones and Varying Road Grades. *Energies* **2019**, *12*, 4296. [CrossRef]
51. Delprat, S.; Lauber, J.; Guerra, T.; Rimaux, J. Control of a Parallel Hybrid Powertrain: Optimal Control. *IEEE Trans. Veh. Technol.* **2004**, *53*, 872–881. [CrossRef]
52. Pan, C.; Liang, Y.; Chen, L.; Chen, L. Optimal Control for Hybrid Energy Storage Electric Vehicle to Achieve Energy Saving Using Dynamic Programming Approach. *Energies* **2019**, *12*, 588. [CrossRef]
53. Qiao, Y.; Song, Y.; Huang, K. A Novel Control Algorithm Design for Hybrid Electric Vehicles Considering Energy Consumption and Emission Performance. *Energies* **2019**, *12*, 2698. [CrossRef]
54. Sundstrom, O.; Guzzella, L. A generic dynamic programming Matlab function. In Proceedings of the 2009 IEEE Control Applications, (CCA) & Intelligent Control, (ISIC), St. Petersburg, Russia, 8–10 July 2009; pp. 1625–1630.
55. Du, Y.; Liu, C.; Li, Y. Velocity Control Strategies to Improve Automated Vehicle Driving Comfort. *IEEE Intell. Transp. Syst. Mag.* **2018**, *10*, 8–18. [CrossRef]
56. Luciani, S.; Bonfitto, A.; Amati, N.; Tonoli, A. Model predictive control for comfort optimization in assisted and driverless vehicles. *Adv. Mech. Eng.* **2020**, *12*, 1–14. [CrossRef]
57. Anselma, P.; Kollmeyer, P.; Belingardi, G.; Emadi, A. *Multitarget Evaluation of Hybrid Electric Vehicle Powertrain Architectures Considering Fuel Economy and Battery Lifetime*; SAE Technical Paper 2020-37-0015; SAE International: Warrendale, PA, USA, 2020.
58. Alix, G.; Dabadie, J.; Font, G. *An ICE Map Generation Tool Applied to the Evaluation of the Impact of Downsizing on Hybrid Vehicle Consumption*; SAE Technical Paper 2015-24-2385; SAE International: Warrendale, PA, USA, 2015.
59. Le Berr, F.; Abdelli, A.; Postariu, D.-M.; Benlamine, R. Design and Optimization of Future Hybrid and Electric Propulsion Systems: An Advanced Tool Integrated in a Complete Workflow to Study Electric Devices. *Oil Gas Sci. Technol.* **2012**, *67*, 547–562. [CrossRef]

60. Petit, M.; Marc, N.; Badin, F.; Mingant, R.; Sauvant-Moynot, V. A Tool for Vehicle Electrical Storage System Sizing and Modelling for System Simulation. In Proceedings of the 2014 IEEE Vehicle Power and Propulsion Conference (VPPC), Coimbra, Portugal, 27–30 October 2014; pp. 1–5.
61. Ligterink, N.E.; van Mensch, P.; Cuelenaere, R.F.A.; Hausberger, S.; Leitner, D.; Silberholz, G. *Correction Algorithms for WLTP Chassis Dynamometer and Coast-Down Testing*; TNO-Report 2015 R10955; TNO: Delft, The Netherlands, 2015.

energies

MDPI

Article

A Study of Control Methodologies for the Trade-Off between Battery Aging and Energy Consumption on Electric Vehicles with Hybrid Energy Storage Systems

Kevin Mallon * and Francis Assadian

Department of Mechanical and Aerospace Engineering, University of California, Davis, CA 95616, USA; fassadian@ucdavis.edu
* Correspondence: krmallon@ucdavis.edu

Abstract: Hybrid and electric vehicle batteries deteriorate from use due to irreversible internal chemical and mechanical changes, resulting in decreased capacity and efficiency of the energy storage system. This article investigates the modeling and control of a lithium-ion battery and ultracapacitor hybrid energy storage system for an electric vehicle for improved battery lifespan and energy consumption. By developing a control-oriented aging model for the energy storage components and integrating the aging models into an energy management system, the trade-off between battery degradation and energy consumption can be minimized. This article produces an optimal aging-aware energy management strategy that controls both battery and ultracapacitor aging and compares these results to strategies that control only battery aging, strategies that control battery aging factors but not aging itself, and non-optimal benchmark strategies. A case study on an electric bus with variously-sized hybrid energy storage systems shows that a strategy designed to control battery aging, ultracapacitor aging, and energy losses simultaneously can achieve a 28.2% increase to battery lifespan while requiring only a 7.0% decrease in fuel economy.

Keywords: electric vehicle; hybrid vehicle; energy management; lithium ion; ultracapacitor; battery aging

Citation: Mallon, K.; Assadian, F. A Study of Control Methodologies for the Trade-Off between Battery Aging and Energy Consumption on Electric Vehicles with Hybrid Energy Storage Systems. *Energies* **2022**, *15*, 600. https://doi.org/10.3390/en15020600

Academic Editor: Daniel-Ioan Stroe

Received: 30 November 2021
Accepted: 11 January 2022
Published: 14 January 2022

Publisher's Note: MDPI stays neutral with regard to jurisdictional claims in published maps and institutional affiliations.

1. Introduction

Due to their low operating speeds and frequent stopping and starting, buses are a prime candidate for hybridization or electrification in the goal of reducing transportation sector emissions. The stop-and-go behavior, in particular, means that regenerative braking can recover a large portion of expended power. However, for handling bus loads and ranges, the lithium ion batteries needed for electric vehicles (EVs) and hybrid-electric vehicles (HEVs) can be prohibitively expensive and heavy [1]. Additionally, the large current spikes from acceleration and deceleration can degrade the battery, reducing range, increasing energy consumption [2,3], and, in general, adding new operational costs to such vehicles.

One possible solution to this is to use a Hybrid Energy Storage System (HESS)—a combination of lithium ion energy storage with an ultracapacitor (UC) sized to handle large charge and discharge currents—in place of standard battery energy storage. In general, lithium ion batteries have a high energy density but low power density: they can store a large amount of charge, but cannot access it quickly without degrading. Specifically, large currents to and from the battery cause its capacity to fade and internal resistance to grow. High temperatures and deep discharges also contribute to battery aging. On the contrary, ultracapacitors have a low energy density and high power density [4]. A HESS, then, allows one to obtain the efficient storage of lithium batteries while allowing an ultracapacitor to handle the large currents [4,5]. The aging of UCs does not depend on current magnitude or discharge depth, rather on time, temperature, and cell voltage [6–8]; therefore, there is not necessarily a tradeoff between battery aging and UC aging in HESSs.

Current research on HESSs considers HEV, EV, and fuel cell vehicle applications. The bulk of the literature, for instance [9–14], is concerned with the optimal sizing of the HESS so as to maximize the cost-effectiveness of such a system. However, battery aging is often not considered directly in this optimization; instead, battery aging factors such as high temperatures and currents are minimized, rather than battery aging directly, and the benefits to overall aging are only assumed [14]. Some related works that does directly address aging are described here. In reference [15], for instance, an optimal control policy is developed to control UC behavior. This policy demonstrated clear aging improvements over an uncontrolled system using passive energy management. Reference [16] used multi-objective optimization while directly incorporating an aging model and using a rule-based control system to govern energy management for a study on HESS sizing in EVs. Reference [17], likewise, carried out a parametric study on battery degradation versus UC size in EVs, using a control system based on fuzzy logic. Reference [18] considered a HESS that used lead-acid batteries rather than lithium ion, and developed an HEV energy management strategy that tuned for battery life extension. Notably, they found that, for the HESS to be cost-effective, a 50% increase in battery cycle life was required. Reference [19] compared the aging benefits of an optimally-sized HESS to the theoretical maximum benefits—battery aging reductions with an infinitely large HESS. These benefits were experimentally verified, with the developed approach decreasing battery power fade and temperature rise in lithium-ion batteries on a vehicle load profile. Most notably, references [20–22] demonstrate an optimal control strategy to directly minimize battery aging in a HESS for a plug-in HEV.

However, the literature is lacking in direct aging control for EVs, in the impact of ultracapacitor aging in the HESS, and in methods to assess the economic benefit of the HESS given UC aging. Although studies on direct aging control for HEVs do exist, for instance [2,3,23,24], EVs pose a unique control problem due to the fewer controlled variables and different component sizes. This research fills these gaps in knowledge: new energy management strategies to control battery aging and to jointly control battery aging, ultracapacitor aging, and energy losses are developed and compared to existing methods. Then, the cost/benefit analysis of a HESS that considers ultracapacitor aging is performed, and the drawbacks of overusing the ultracapacitor are discussed.

This paper begins by developing the aging models for an electric vehicle hybrid energy storage system, with an aside showing why there must necessarily be a trade-off between battery aging and energy consumption for this vehicle configuration. Next, energy management systems (EMS) for aging control are developed, including Deterministic Dynamic Programming (DDP), Stochastic Dynamic Programming (SDP), and Load-Leveling (LL). The vehicle model and aging-aware controllers are then applied to a case study of an electric bus with a HESS, simulated for the lifespan of the battery. Finally, these simulation results are analyzed, and conclusions are drawn regarding the benefits of optimal control and aging-aware control for vehicle energy management.

2. Modeling

This section develops a model for an electric vehicle with a hybrid energy storage system. Specifically, the model is of a HESS-equipped electric bus using lithium-ion batteries for energy storage and ultracapacitor modules for handling large currents, as depicted in Figures 1 and 2. The first subsection presents the overall vehicle model, including vehicle dynamics, motor efficiencies, battery dynamics, and so on. The next two subsections deal with the battery aging and ultracapacitor aging models. The final subsection discusses why, based on the provided models, aging control can have a negative impact on fuel economy.

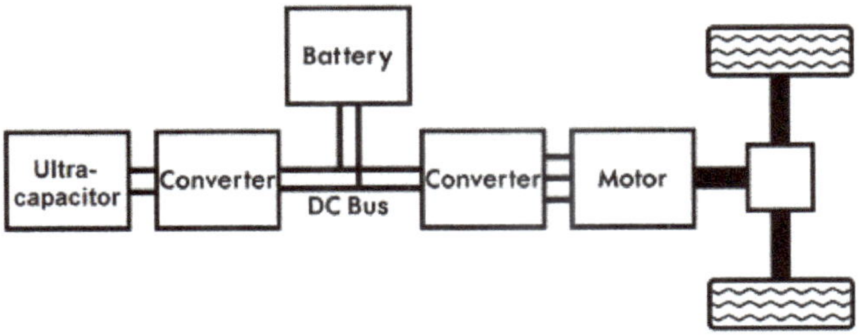

Figure 1. Powertrain of an EV with a UC.

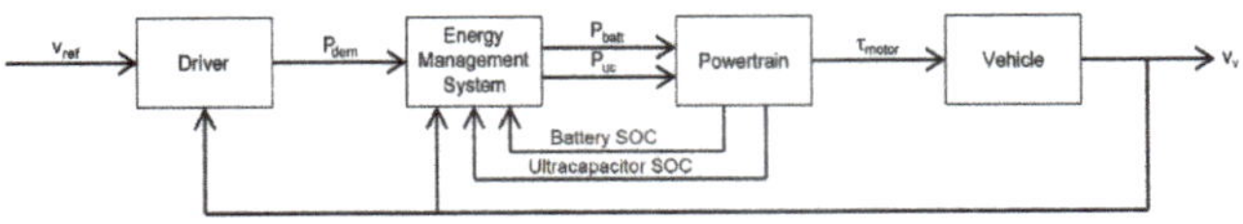

Figure 2. Block diagram for energy management of an EV with a UC.

2.1. Vehicle Modeling

2.1.1. Vehicle Dynamics

For this study, a backwards-facing quasi-static vehicle model [25] is used to represent the vehicle dynamics. In this model, it is assumed that the driver accurately follows the velocity of a given drive cycle, eliminating the need for a driver model and allowing the time-history of the electrical power demand to be calculated in advance.

This research uses a backwards-facing quasi-static vehicle model [25] to simulate the vehicle dynamics. This method assumes that the model accurately follows a specified velocity profile, which allows for calculating the acceleration and, therefore, the electrical power request can be computed in advance, eliminating the need for a driver model.

The vehicle body is affected by inertial forces, aerodynamic drag, and rolling resistance, while gravitational forces (such as those due to driving on inclines) are neglected. The drag force is given by

$$F_{drag} = \frac{1}{2}\rho A_f C_D (v_v)^2 \tag{1}$$

where ρ, A_f, C_D, and v_v are the air density, frontal area, drag coefficient, and vehicle velocity, respectively. Rolling resistance is given by

$$F_{roll} = M_v g C_R \tag{2}$$

where M_v, g, and C_R are the vehicle's total mass (including components such as the engine and generator), acceleration due to gravity, and rolling resistance coefficient. In a backwards-facing model, the acceleration and the vehicle mass determine the inertial force on the vehicle as

$$F_{inertial} = M_{eq}\frac{dv_v}{dt}. \tag{3}$$

M_{eq} is the combined bus mass and equivalent mass due to the rotational inertia of the motor and wheels

$$M_{eq} = M_v + 4J_w\left(\frac{1}{R_w}\right)^2 + J_m\left(\frac{N_{fd}N_{gb}}{R_w}\right)^2, \tag{4}$$

where J_w, R_w, J_m, N_{fd}, and N_{gb} are the rotational inertia of a single wheel, the wheel radius, the rotational inertia of the motor, the final drive ratio, and the gearbox ratio for a single-speed gearbox, respectively. The acceleration term in (3) is approximated from a given velocity profile according to

$$\frac{dv_v}{dt}(t) \approx \frac{v_v(t+\Delta t) - v_v(t-\Delta t)}{2\Delta t}. \tag{5}$$

These three forces sum together to give the tractive force on the bus.

$$F_{traction} = F_{inertial} + F_{drag} + F_{roll} \tag{6}$$

Parameter values for the vehicle model can be found in Table 1. The bus is assumed to be fully loaded and at its maximum allowable weight. The bus's physical parameters are based on the existing literature on bus simulation [26–28].

Table 1. Vehicle model physical parameters.

Parameter	Variable	Value
Vehicle Mass	M_v	18,181 kg
Frontal Area	A_f	8.02 m^2
Drag Coefficient	C_D	0.55
Roll Resistance Coefficient	C_R	0.008
Wheel Inertia	J_w	20.52 kg-m^2
Motor Inertia	J_m	0.277 kg-m^2
Wheel Radius	R_w	0.48 m
Final Drive Ratio	N_{fd}	5.1:1
Gearbox Ratio	N_{gb}	5:1
Transmission Efficiency	η_{trans}	96%

2.1.2. Transmission

Next, the vehicle velocity and tractive force are equated to motor torque and angular velocity. The motor torque is given by

$$\tau_m = \begin{cases} \left(\frac{R_w}{N_{fd}N_{gb}}F_{traction}\right)/\eta_{trans}, & F_{traction} \geq 0 \\ \left(\frac{R_w}{N_{fd}N_{gb}}F_{traction}\right) \cdot \eta_{trans}, & F_{traction} < 0 \end{cases} \tag{7}$$

η_{trans} is the transmission efficiency, represented as torque losses. The motor speed is then given by

$$\omega_m = \frac{N_{fd}N_{gb}}{R_w} v_v \tag{8}$$

The mechanical power P_{mech} needed to drive the vehicle is expressed in terms of the above torque and angular velocity.

$$P_{mech} = \tau_m \cdot \omega_m \tag{9}$$

In this formulation, P_{mech} is positive during acceleration. Parameter values for the transmission can be found in Table 1.

2.1.3. Motor and Power Electronics

The motor torque and angular velocity are used to find the motor efficiency η_{motor}, which is constrained to $0 < \eta_{motor} < 1$. The motor efficiency is determined from a static efficiency map from the National Renewable Energy Laboratory's Advanced Vehicle Simulator (ADVISOR) data library [29]. This efficiency includes both the motor itself as well as the associated power electronics. The bus model in this research uses a 250 kW AC induction motor.

Once the motor efficiency is found, it can be used to evaluate the driver's electrical power request, P_{req}.

$$P_{req} = \begin{cases} P_{mech}/\eta_{motor} & \tau_m \geq 0 \\ P_{mech} \cdot \eta_{motor} & \tau_m < 0 \end{cases} \tag{10}$$

The electrical power request is met with power from the battery P_{batt} and ultracapacitor P_{uc}. Given that this is a backwards-facing simulation, the power request must always be met.

$$P_{req} = P_{batt} + P_{uc} \tag{11}$$

In later sections, the ultracapacitor power is developed as the energy management system's controlled variable. Then, P_{batt} is dependent on P_{req} and P_{uc}, and Equation (11) is rewritten as

$$P_{batt} = P_{req} - P_{uc} \tag{12}$$

2.1.4. Battery

The EV's lithium-ion battery cells are modeled with the simple equivalent circuit shown in Figure 3, where V_{cell} is the single-cell open-circuit voltage (OCV) and R_{cell} is the single-cell equivalent series resistance [4]. The individual cells are then combined into a larger battery pack. Only one state variable is required for this model, the state of charge (SOC). The OCV and internal resistance are variable parameters dependent on SOC. The formulas for these parameters are given in [30], which develops a lithium-iron-phosphate battery from experimental data.

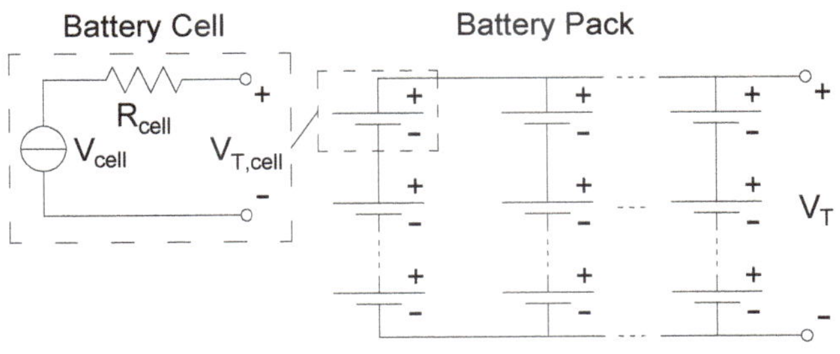

Figure 3. Battery pack equivalent circuit.

The battery pack equivalent resistance R_{eq} is given by

$$R_{eq} = R_{cell} \cdot \frac{N_{ser}}{N_{par}} \tag{13}$$

where N_{ser} and N_{par} are the number of cells in series and in parallel, respectively. The battery pack OCV is likewise given by

$$V_{ocv} = N_{ser} \cdot V_{cell} \tag{14}$$

The battery pack's terminal voltage is found from the OCV and battery power P_{batt} using the equivalent circuit in Figure 3.

$$I_{batt} = P_{batt} / V_T \tag{15}$$

$$V_T = V_{ocv} - I_{batt} \cdot R_{eq} \tag{16}$$

Then, substituting the current equation into the voltage equation and solving yields

$$V_T^2 = V_{ocv} \cdot V_T - P_{batt} \cdot R_{eq} \tag{17}$$

$$V_T = 1/2\left(V_{ocv} + \sqrt{V_{ocv}^2 - 4 \cdot P_{batt} \cdot R_{eq}} \right) \tag{18}$$

Substituting V_T into Equation (15) allows the battery current to be found explicitly. Then, integrating the battery current yields the state of charge.

$$SOC(k+1) = SOC(k) + \Delta t \cdot \frac{I_{batt}}{Q_{batt}}, \tag{19}$$

where Q_{batt} is the battery pack's charge capacity in coulombs and Δt is the integration time step.

The battery model parameters are given in Table 2. The number of cells in series ensures that the battery pack, the OCV, is in line with the requirements of [31]. The number of parallel cells was chosen so that the bus can meet the power requirements in [31,32] to drive continuously on an urban bus velocity profile for four hours. Note that Table 2 only gives the nominal values for R_{cell} and V_{cell}—in reality, these parameters vary with SOC and other operating conditions [30].

Table 2. Battery model parameters.

Parameter	Variable	Value
Battery Cells in Parallel	N_{par}	400 cells
Parallel Sets in Series	N_{ser}	100 sets
Total Charge Capacity	Q_{batt}	340 Ah
Nominal Open Circuit Voltage	V_{cell}	3.8 v
Nominal Equivalent Resistance	R_{cell}	7.5 mΩ

2.1.5. Ultracapacitor

The ultracapacitor modules are modeled as the first-order equivalent circuit shown in Figure 4. The model itself is based on the 100F ultracapacitor model derived in [33]. Ultracapacitor model parameters are given in Table 3. The ultracapacitor pack, similar to the battery pack, consists of ultracapacitors arranged in N_{pc} modules in a parallel set and N_{sc} sets in series. The number of modules is variable so that the effectiveness and cost-benefit of the HESS can be considered across a range of designs.

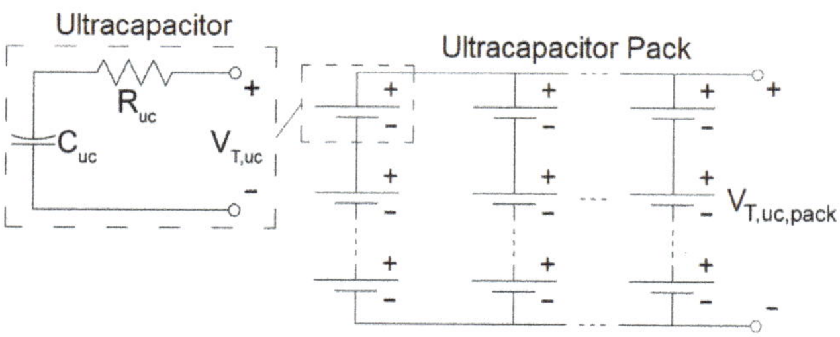

Figure 4. Ultracapacitor pack equivalent circuit.

Table 3. Ultracapacitor model parameters.

Parameter	Variable	Value
UC modules in Parallel	N_{pc}	variable
UC Parallel Sets in Series	N_{sc}	100 sets
Resistance	R_{uc}	44.3 mΩ
Capacitance	C_{uc}	105.9 F

The UC pack is connected to the DC bus through a converter, as shown in Figure 1. The converter allows the UC pack to operate independently of the DC bus voltage. The ultracapacitor pack power is indicated by P_{uc}, where P_{uc} is positive while discharging and negative while charging. Then, each individual module has power $P_{uc,module}$ given by

$$P_{uc,module} = \frac{P_{uc}}{N_{pc}N_{sc}} \tag{20}$$

For capacitor charge q_{uc} at some given power $P_{uc,module}$, the current and terminal voltage I_{uc} and $V_{T,uc}$ are found similarly to Equations (15)–(18).

$$I_{uc} = P_{uc,module}/V_{T,uc} \tag{21}$$

$$V_{T,uc} = q_{uc}/C_{uc} - I_{uc}R_{uc} \tag{22}$$

Substituting Equation (21) into Equation (22) and solving yields

$$V_{T,uc}^2 = q_{uc}/C_{uc} \cdot V_{T,uc} - P_{uc,module}R_{uc} \tag{23}$$

$$V_{T,uc} = \frac{1}{2}\left(\frac{q_{uc}}{C_{uc}} + \sqrt{\frac{q_{uc}}{C_{uc}} - 4P_{uc,module}R_{uc}}\right) \tag{24}$$

$V_{T,uc}$ can then be substituted back into Equation (21) to obtain the ultracapacitor current. Then, the state equation for the capacitor is

$$\dot{q}_{uc} = I_{uc} - \frac{q_{uc}}{R_{uc}C_{uc}} \tag{25}$$

Then, for the complete ultracapacitor pack,

$$I_{uc,pack} = I_{uc} \cdot N_{pc} \tag{26}$$

$$V_{T,uc,pack} = V_{T,uc} \cdot N_{sc} \tag{27}$$

$$R_{uc,pack} = R_{uc} \cdot \frac{N_{sc}}{N_{pc}} \tag{28}$$

where $I_{uc,pack}$ is the total current going to or from the UC pack, $V_{T,uc}$ is the terminal voltage of the overall UC pack, and $R_{uc,pack}$ is the equivalent series resistance of the entire pack.

2.2. Battery Aging Model

This research uses the cycle-life aging model presented in Reference [34], and develops it here into an aging model that can be used in dynamic control of the energy storage systems. Reference [34] models the cycle life of a battery as a function of depth of discharge DoD, charging current I_c, discharging current I_d, and temperature T.

$$CL = g(DoD, I_c, I_d, T) \tag{29}$$

This model starts as a simple curve fit of cycle life to depth-of-discharge at a reference point of $I_c = 1$ C, $I_d = 1$ C, $T = 25$ °C . This baseline cycle life is denoted as CL_{DoD}. Then,

$$CL_{DoD} = a_1 e^{a_2 \cdot DoD} + a_3 e^{a_4 \cdot DoD} \tag{30}$$

where the a_i terms are curve fit parameters. The cycle life is then obtained by modifying CL_{DoD} based on the actual operating I_c, I_d, and T.

$$CL = CL_{DoD} \cdot A_{I_d} \cdot A_{I_c} \cdot A_T \tag{31}$$

where

$$A_{I_d} = \frac{a_5 e^{a_6 \cdot I_d} + a_7 e^{a_8 \cdot I_d}}{a_5 e^{a_6} + a_7 e^{a_8}} \tag{32}$$

$$A_{I_c} = \frac{a_9 e^{a_{10} \cdot I_c} + a_{11} e^{a_{12} \cdot I_c}}{a_9 e^{a_{10}} + a_{11} e^{a_{12}}} \tag{33}$$

$$A_T = \frac{a_{13} T^3 + a_{14} T^2 + a_{15} T + a_{16}}{25^3 a_{13} + 25^2 a_{14} + 25 a_{15} + a_{16}} \tag{34}$$

where the a_i terms are, again, curve fit parameters. The a_5–a_8 parameters are found from a curve fit of cycle life to varying I_d for $I_c = 1\ C$, $T = 25^\circ$, and $DoD = 100\%$. The a_9–a_{12} parameters are found from a curve fit of cycle life to varying I_c for $I_d = 1\ C$, $T = 25^\circ$, and $DoD = 100\%$. The a_{13}–a_{16} parameters are found from a curve fit of cycle life to varying T for $I_d = 1C$, $I_c = 1\ C$, and $DoD = 100\%$.

The cycle life model in [34] assumes uniform charge and discharge cycles over the life of the battery. The Palmgren–Miner (PM) rule can be used, then, to handle the non-uniform cycles of vehicle operation. This method, originally developed for analyzing material fatigue life, has been shown to effectively approximate the battery health over non-uniform charge and discharge cycles [35–37]. Under the assumptions of this method, each charge and discharge cycle damages the battery an amount equal to the inverse of the cycle life at that cycle's operating conditions. In other words, if we assume a cycle k with depth of discharge DoD_k, charge current magnitude $I_{c,k}$, discharge current magnitude $I_{d,k}$, and temperature T_k, then the cycle life for these operating conditions is CL_k. Under the PM rule, this cycle damages the battery an amount D_k given by

$$D_k = 1/CL_k \tag{35}$$

Damage accumulates linearly for each charge and discharge cycle. Therefore, the damage from each individual cycle can be summed to find the total damage. The total damage D_{tot} through the k-th cycle is therefore

$$D_{tot}(k) = \sum_{i=1}^{k} D_i \tag{36}$$

where each D_i represents the damage from a single cycle with operating conditions DoD_i, $I_{c,i}$, $I_{d,i}$, and T_i. In this way, the damage of individual cycles with unique operating conditions is summed to obtain a total measure of battery health. Zero total damage indicates that the battery is at its beginning of life, while total damage of one indicates the battery's end of life. Battery end-of-life corresponds to a 20% capacity fade, therefore the capacity fade CF can be put in terms of the damage as

$$CF(k) = 0.2 \cdot D_{tot}(k) \tag{37}$$

The above method requires full knowledge of the charge and discharge time histories, which is not practical for use in energy management; the EMS must act at a much faster rate than the pace at which these cycles develop. It is possible, however, that the EMS could determine how a control decision might cause the damage from the current cycle to lessen or grow. For instance, imagine a battery operating at conditions of DoD_j, $I_{c,j}$, $I_{d,j}$, T_j. Then, let the energy management system make some decision that produces new operating conditions of DoD_k, $I_{c,k}$, $I_{d,k}$, T_k. Using Equations (29) and (35), the change in damage ΔD due to the EMS's decision can be computed as

$$\Delta D = D_k - D_j = \frac{1}{g(DoD_k, I_{c,k}, I_{d,k}, T_k)} - \frac{1}{g(DoD_j, I_{c,j}, I_{d,j}, T_j)} \tag{38}$$

Then, an energy management strategy could incorporate Equation (38) for a measure of potential battery damage. In this way, the strategy would try to minimize the damage from the control decision made at each time step. Note that, when controlling aging in this manner, the EMS can only be aware of the DoD up until the current point in time and can only assess damage relative to the current DoD, while the "true" aging depends on the size of the completed cycle. Despite this discrepancy, this method still proves an effective way to control battery aging, as will be shown in later sections.

The resistance growth model in Reference [34] can be treated in an identical manner to Equations (29)–(38). The capacity fade and resistance growth models both use rainflow counting, as in [38], to determine the aging from the irregular cycling operations experienced while a vehicle is in operation. For simplicity, it is assumed that the battery operates at a constant internal temperature of 35 °C.

2.3. Ultracapacitor Aging

A novel aspect of this research is that ultracapacitor aging is considered in addition to battery aging. Reference [6] provides the following model for ultracapacitor aging. This model is based on Eyring's Law, a chemical rate equation which gives an ultracapacitor lifespan based on the operating voltage and internal temperature where aging increases exponentially as the voltage and temperature increase. Then, the aging rate at an instance in time is based on the inverse of the lifespan at the given operating conditions. In this model, SoA is the state of aging that characterizes both capacitance fade and resistance growth, where a SoA of 0 indicates start-of-life and of 1 indicates end-of-life.

$$\frac{dSoA}{dt} = \frac{1}{T_{life}^{ref}} \cdot \exp\left(\ln(2)\frac{\theta_c - \theta_c^{ref}}{\theta_0}\right) \cdot \left(\exp\left(\ln(2)\frac{V - V^{ref}}{V_0}\right) + K\right) \quad (39)$$

where θ_c and V are the UC temperature and voltage, respectively, and the remaining variables (T_{life}^{ref}, θ_c^{ref}, θ_0, V^{ref}, V_0, and K) are parameters fitted to experimental data. Then, from [6], the instantaneous capacitance C_{uc} and internal resistance R_{uc} are given by

$$C_{uc} = C_{uc,0} \times (0.95 - 0.15 \cdot SoA) \quad (40)$$

$$R_{uc} = R_{uc,0} \times (1 - 0.3 \cdot SoA)^{-1} \quad (41)$$

where $C_{uc,0}$ and $R_{uc,0}$ are the initial values of C_{uc} and R_{uc}.

This model can be used in control without modification. The standard end-of-life conditions for ultracapacitors are defined similarly to batteries: when the UC capacitance has faded by 20% [6]. It is assumed that the ultracapacitor operates at a constant internal temperature of 55 °C, estimated from the operating conditions found in [39].

2.4. Aging and Fuel Economy Trade-Off

This paper analyzes the trade-off between battery aging and energy consumption in an electric vehicle with a HESS. This section briefly touches on why there must necessarily be a trade-off.

Consider the two paths in which power can flow through the HESS, shown in Figure 5. Consider the first case, where power flows primarily or entirely along the upper path, directly between the battery and the electric motor. In this case, the ultracapacitor is used marginally or not at all, meaning there is little to no change in either energy consumption or battery aging compared to a conventional EV that does not include an ultracapacitor. Next, consider a case where power flows primarily on the lower path, such that the ultracapacitor is heavily used and acts as a buffer between the battery and the electric motor. On one hand, the power flowing to or from the battery can be controlled to reduce aging factors such as large currents. On the other hand, the ultracapacitor introduces new resistances to the energy storage system as well as converter inefficiencies, resulting in increased losses. Therefore, any use of the ultracapacitor to reduce battery aging necessarily incurs new energy losses from the internal resistance of the ultracapacitor. In short, battery lifespan cannot be extended without an increase in energy consumption.

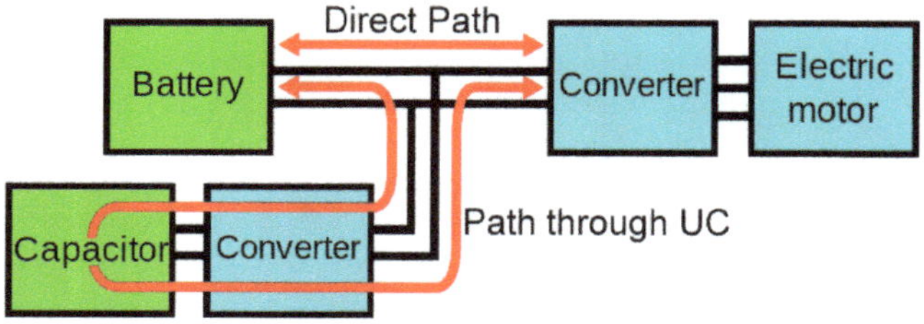

Figure 5. HESS power paths. Power on the direct path between battery and motor experiences minimal losses, while power on the path through the UC experience additional losses from the UC internal resistance.

3. Control

In order to fully investigate the benefits of aging-aware control, seven different types of energy management systems are considered:

1. DDP with Battery Aging Penalty, denoted DDP-B;
2. DDP with Energy Loss Penalty and Battery and Capacitor Aging Penalties, denoted DDP-EC;
3. DDP with Battery Power Penalty, denoted DDP-P;
4. SDP with Battery Aging Penalty, denoted SDP-B;
5. SDP with Energy Loss Penalty and Battery and Capacitor Aging Penalties, denoted SDP-EC;
6. SDP with Battery Power Penalty, denoted SDP-P;
7. Load Leveling, denoted LL.

3.1. Dynamic Programming

The first six strategies use DP to generate an optimal controller, with the first set of three using DDP and the second set of three using SDP. The development of DP for HEV energy management has been covered by a variety of literature, such as [40–43]. For both DDP and SDP, the optimization problem considers a discrete-time dynamic system

$$x(k+1) = f(x(k), u(k), w(k)) \tag{42}$$

where $x(k)$ is the state vector at time k, $u(k)$ is the control vector, and $w(k)$ is a vector of any inputs or disturbances. x, u, and w are assumed to exist in finite ranges $x \in X$, $u \in U$, and $w \in W$.

DDP uses exact knowledge of the driver behavior, including knowledge of future behavior, to minimize a given cost function over the complete driving trajectory.

$$J = \sum_{k=0}^{N} L(x(k), u(k), w(k)) \tag{43}$$

where $L(x, u, w)$ is an instantaneous cost function, and x, u, and w are the state variables, controlled variables, and system inputs, respectively. Equation (43) is minimized by solving a recursive cost-to-go function

$$V(x, N) = \min_{u \in U}\{L(x, u, w(N))\} \tag{44}$$

$$V(x, k) = \min_{u \in U}\{L(x, u, w(k)) + V(f(x, u, w(k)), k+1)\} \tag{45}$$

$$\text{for } k = N-1, \ldots, 0$$

starting at $k = N$ and working backward through time to $k = 0$. The key point of the DDP method is that, at each optimization step, the entire cost from the current time k to the final time N is minimized, not just the instantaneous cost. $V(x, k)$ is evaluated for each $x \in X$, so that $V(f(x,u,w(k)),k+1)$ can be interpolated from the prior update. The optimal control is found by a direct search of $u \in U$. Then, the optimal control u^* is given by

$$u^*(x,k) = \arg\min_{u \in U} \{L(x,u,w(k)) + V(f(x,u,w(k)),k+1)\} \quad (46)$$

Meanwhile, SDP uses a stochastic model of driver behavior to anticipate the future driver power or torque requests and minimize the expected value of a given cost function

$$J = \mathbf{E}\left[\sum_{k=0}^{N} \gamma^k L(x(k),u(k),w(k))\right] \quad (47)$$

where the function $\mathbf{E}[\cdots]$ denotes an expected value, and γ is a discount factor $0 < \gamma < 1$ that allows the cost function to converge as $k \to \infty$. Equation (47) is again minimized with a recursive cost-to-go function

$$V(x,w,N) = \min_{u \in U}\{L(x,u,w)\} \quad (48)$$

$$V(x,w,k) = \min_{u \in U}\{L(x,u,w) + \gamma \cdot \mathbf{E}[V(f(x,u,w),w,k+1)]\} \quad (49)$$

$$\text{for } k = N-1, \ldots, 0$$

where, this time, the expected future costs are considered, rather than the exact future costs.

The SDP problem can be treated as a finite horizon problem, where N is a fixed number of updates. Alternatively, it can be treated as an infinite horizon problem, where N is arbitrarily large and the updates to the cost-to-go function are carried out until the control policy converges, in other words

$$V(x,w,k) = V(x,w,k+1)\ \forall\ x \in X \text{ and } u \in U. \quad (50)$$

As noted earlier, a value of $0 < \gamma < 1$ ensures convergence of the number as $N \to \infty$ [44]. Then, the optimal control u^* is given by

$$u^*(x,w) = \arg\min_{u \in U}\{L(x,u,w) + \gamma \cdot \mathbf{E}[V(f(x,u,w),w,1)]\}. \quad (51)$$

That is, the control optimizes the final update of the cost-to-go function. Although the SDP problem is solved backwards in time like the DDP problem, the resulting control policy is both time-invariant and causal. This is because the SDP problem does not require future knowledge of w; instead, it relies on the time-invariant stochastic model.

For this research, the state variables are the ultracapacitor state of charge SOC_c and the battery depth of discharge for the current cycle DoD. The controlled variable is the power allotted to ultracapacitor P_{uc}. The driver power request P_{req} is an input to the controller. For DDP, it is a precisely known function of time, while, for SDP, the future power request is estimated from the current driver power request and the current vehicle wheel speed ω_{wh}, based on a stochastic model as described in [40].

It should be noted that, in general, dynamic programming control strategies are considered too computationally expensive to run in real time on a vehicle [5]. Instead, the control policy must be computed off-line and be implemented on the vehicle using a lookup table. This approach requires quantizing the variables into discrete grids of points; linear interpolation can then be used to find the optimal control at any given operating point. The implementation of such lookups tables has been shown to operate well in real time [45].

The three strategies employed by DDP and SDP in this research have a component of their respective instantaneous cost functions $L(x,u,w)$ to penalize battery aging and a component to penalize deviation of the UC SOC from a target value $SOC_{c,tgt} = 60\%$. The

SOC deviation penalty serves two purposes: first, it helps to maintain the UC's readiness to handle large currents. If the UC is near its maximum charge, it may be unable to accept a large charging power request, and if the UC is near its minimum, it will be unable to accept a large discharging power request, both of which can strain the battery. Keeping the charge near a central value combats this problem. Second, by varying the penalty on the deviation from the target value, the extent to which the ultracapacitor is used for aging control can be tuned, allowing for a better comparison of lifespan improvements between simulation cases. The manner in which the instantaneous cost functions penalize aging varies, as described below.

The first strategy, employed by DDP-B and SDP-B, directly penalizes battery aging according to

$$L(x,u,w) = (SOC_c - SOC_{c,tgt})^2 + Q_{1,\Delta D} \cdot \Delta D \tag{52}$$

where ΔD is the damage to the battery as a result of a given control decision, as given in Equation (38), and $Q_{1,\Delta D}$ is a tuned weighting parameter. This strategy is denoted as DP-B when referring to the DDP-B and SDP-B types together.

The second, used for DDP-EC and SDP-EC, penalizes a combination of battery aging, ultracapacitor aging rate, and electrical energy losses according to

$$L(x,u,w) = Q_{2,SOC}(SOC_c - SOC_{c,tgt})^2 + Q_{2,\Delta D} \cdot \Delta D + Q_{2,SOA} \cdot \frac{dSoA}{dt} + Q_{2,loss}E_{loss} \tag{53}$$

where E_{loss} is the energy losses from the battery and ultracapacitor, obtained from

$$E_{loss} = R_{eq}I_{batt}^2 + R_{uc,pack}I_{uc,pack}^2 \tag{54}$$

where R_{eq} is the battery pack series resistance, I_{batt} is the current through the battery, $R_{uc,pack}$ is the ultracapacitor pack series resistance, and $I_{uc,pack}$ is the current through the ultracapacitor pack, per the models presented in Section 2.1. Returning to Equation (53), the $Q_{2,i}$ terms are weighting parameters for their respective elements in the cost function. The three weighting parameters $Q_{2,\Delta D}$, $Q_{2,SOA}$, and $Q_{2,loss}$ are set according to industrial average prices for lithium-ion batteries, ultracapacitors, and energy from the electrical grid [46,47], such that the battery aging, ultracapacitor aging, and energy loss terms are all equally weighted based on their real-world values. Then, the remaining term $Q_{2,SOC}$ is used to tune the strategy. This strategy is denoted as DP-EC when referring to the DDP-EC and SDP-EC types together.

The third and final strategy does not directly penalize aging but rather penalizes large power going to or from the battery

$$L(x,u,w) = (SOC_c - SOC_{c,tgt})^2 + Q_{3,P} \cdot P_{Batt}^2 \tag{55}$$

where $Q_{3,P}$ is a tuned weighting parameter and P_{batt} is the power going to or from the battery, per Equation (12). In this way, we limit battery damage using only simple knowledge of how the battery ages—that large currents to and from the battery degrade it. Thus, we can distinguish the benefits of direct aging control in the DDP,-B, SDP-B, DDP-EC, and SDP-EC strategies from the benefits of DP control generally. This strategy is denoted as DP-P when referring to the DDP-P and SDP-P types together.

In this research, DDP is used to obtain the global-optimal control strategy for a given cost function and represents the best-case scenario for a controller type. SDP, on the other hand, represents a causal, implementable controller and offers a more realistic understanding of the capabilities of a given cost function design. Because it is causal, it is also a better comparison to the Load Leveling controller. It should be noted that it is possible to adapt the results of DDP optimization into a causal rule base; however, this method is not used in this research.

3.2. Load-Leveling

The final strategy considered is a simple method called "Load-Leveling." In this method, the battery is assigned a maximum allowable current for charging and discharging, $I_{b,max}$, which corresponds to minimum and maximum battery powers $P_{b,min}$ and $P_{b,max}$. $P_{b,min}$ provides the limit on power going into the battery while charging (negative values of P_{batt}) and $P_{b,max}$ provide the limit on discharging. Any power request from the driver that exceeds the allowable amount is handled by the ultracapacitor.

$$P_{uc} = \begin{cases} P_{req} - P_{b,max}, & P_{req} > P_{b,max} \\ P_{req} - P_{b,min}, & P_{req} < P_{b,min} \\ P_{reset}, & \text{otherwise} \end{cases} \tag{56}$$

where P_{reset} is a small amount of power from the battery used to return the ultracapacitor SOC to a target value of $SOC_{c,tgt} = 60\%$.

$$P_{reset} = \begin{cases} 13\ \text{kW}, & SOC_c > SOC_{c,tgt} \\ -13\ \text{kW}, & SOC_c < SOC_{c,tgt} \\ 0 & \text{otherwise} \end{cases} \tag{57}$$

The 13 kW value corresponds approximately to a 0.1 C battery charge or discharge rate, considered sufficiently low to not majorly affect the battery aging. The particular value of $I_{b,max}$ is varied to tune the response of the controller.

This controller serves as a lower bound for EMS performance, as it has neither an aging model nor any form of optimal control.

4. Case Study

The model and developed controllers are, in this section, used for a case-study analysis of aging-aware energy management: simulation is used to determine how the various strategies perform relative to each other.

Each strategy is simulated on the Manhattan Bus Cycle (MBC) drive cycle [48] for an array of different controller tunings.

1. DDP-B and SDP-B had the $Q_{1,\Delta D}$ parameter varied from 10^4 to 10^{10};
2. DDP-EC and SDP-EC had the $Q_{2,SOC}$ parameter varied from 10^{-4} to 10^2;
3. DDP-P and SDP-P had the $Q_{3,P}$ parameter varied from 10^{-14} to 10^{-11};
4. LL has the $I_{b,max}$ parameter varied from 2 C to 0.8 C.

The range of weights is determined by looking at orders of magnitude of the element of the cost functions. For instance, for Equation (55), the $(SOC_c - SOC_{c\ tgt})^2$ term has an order of magnitude of, at most, 10^{-2}, while the P_{batt}^2 term can have an order of magnitude of up to 10^{10}. Thus, tuning of $Q_{3,P}$ begins at $Q_{3,P} = 10^{-12}$ and is varied from that point.

Additionally, a single baseline case that does not use the ultracapacitor is simulated. This corresponds to $Q_{1,\Delta D} = 0$, $Q_{2,SOC} \to \infty$, $Q_{3,P} = 0$, or $P_{max} \to \infty$.

Simulations begin with both the battery and ultracapacitor at the beginning of their life. After each full discharge cycle, the aging of the battery and ultracapacitor are measured, and the capacity, capacitances, and resistances of the HESS are updated. For the purpose of measuring aging, the battery is assumed to recharge at a rate of 0.5 C. Simulations are then repeated until the battery reaches the end of its life, at which point the cycle life, ultracapacitor state-of-aging, and average energy consumption are measured and recorded.

The above is repeated for three HESS designs: a small ultracapacitor unit $N_{pc} = 10$, a middle-sized ultracapacitor unit $N_{pc} = 40$, and a large ultracapacitor unit $N_{pc} = 100$. Both DDP and SDP simulations are performed for the $N_{pc} = 100$ case, in order to establish that the SDP controllers will closely follow the DDP results. For the $N_{pc} = 10$ and $N_{pc} = 40$, only the causal controllers (SDP-B, SDP-EC, SDP-P, and LL) are simulated.

Energy consumption is measured in equivalent miles per gallon (MPGe), while battery aging is measured in capacity loss per mile. For ease of interpretation, the battery cycle life is converted to an approximate lifespan using an estimate of the average number of miles driven per year, $D_{avg\text{-}year}$ = 34,000 [49].

Finally, the cost-benefit of the HESS is determined. First, the value of the HESS is determined based on the industrial average price per kWh of \$300/kWH for lithium ion batteries and \$15,000/kWh ultracapacitors from [46] and on the battery and ultracapacitor size given in Section 2.1. This gives a battery value of V_{batt} = \$38,760 and an ultracapacitor value of V_{uc} = \$13,021. Additionally, an average electrical energy price of V_{nrg} = \$0.1065/kWh for the U.S. is obtained from [47]. Costs and benefits are normalized by mile driven for a fair comparison between configurations. Then, the battery cost per mile (*BCPM*) is determined from the miles driven over the life of the battery, denoted as "battery lifetime miles driven" (BLMD).

$$BCPM = \frac{V_{batt}}{BLMD} \tag{58}$$

The ultracapacitor cost per mile (*UCCPM*) is similarly determined, this time including a term for the ultracapacitor state of aging at the battery end-of-life (BEOL), SoA_{BEOL}.

$$UCCPM = SoA_{BEOL} \times \frac{V_{uc}}{BLMD} \tag{59}$$

Finally, the energy costs per mile (*ECPM*) are given as

$$ECPM = \frac{V_{nrg} \times GGE}{MPGe} \tag{60}$$

where *GGE* is the gasoline gallon equivalent to convert from gallons of gasoline to kWh, $GGE = 33.41$ kWh/gal. gasoline.

Then, the cost or benefit of the HESS can be determined by comparing the result to the nominal case where no UC is present. Letting the subscript *nom* denote the nominal case and (k) denote any particular simulation, the benefit per mile (*BPM*) is given by

$$BPM(k) = (BCPM_{nom} - BCPM(k)) - UCCPM(k) + (ECPM_{nom} - ECPM(k)) \tag{61}$$

where a positive benefit per mile indicates that value is being added to the system, while a negative value indicates that the cost of the UC outweighs the benefit it adds.

Finally, the payback time $T_{payback}$ (in years) for the HESS can be estimated from the UC value, average miles driven per year, and the benefit per mile.

$$T_{payback} = V_{uc} \times \frac{1}{BPM(k)} \times \frac{1}{D_{avg\text{-}year}} \tag{62}$$

Payback time assumes a positive benefit per mile. If the *BPM* is zero or negative, then a payback time does not exist.

5. Results

The simulation results are analyzed as follows: first, it is verified that the SDP controllers closely follow the DDP controllers. Next, the impact of aging-aware control on the causal controllers is assessed. Then, the effect of overuse of the ultracapacitor is discussed. Finally, the cost-benefit of the HESS is analyzed and discussed.

5.1. Verification of DP Controllers

First, the DDP and SDP methods are compared for the $N_{pc} = 100$ case for each of the three DP cost functions, as given in Equations (52)–(55). These simulation results are shown in Figure 6. In the case of the DP methods with an incorporated aging model, the SDP controller closely tracks the global optimal DDP controller. For the cases that use a battery aging model, the lifespan of the SDP-controlled battery is typically within 1% of the DDP result for a given MPGe, while the difference is greater for the controller that only limits battery power, especially near the peak. These results demonstrate that the causal SDP controllers are able to closely match the DDP global optima and indicate that the SDP controllers behave as intended.

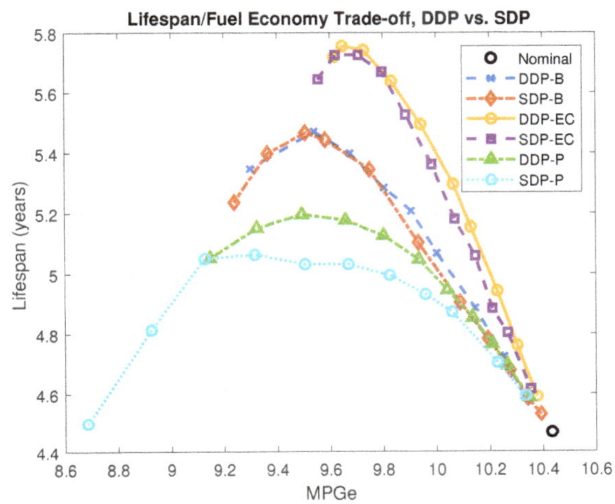

Figure 6. Energy consumption and battery aging for the DP-based methods for $N_{pc} = 100$.

The DP results are summarized in Table 4. Note that the value for "Mean Life Difference" is the average difference in lifespan of an SDP controller compared to a DDP controller at any given operating MPGe value between the nominal point and the DDP peak. Again, the key result of these data is that the SDP controllers follow the DDP controllers within 1.7%, and within 1.0% for the aging-aware control specifically. Comparisons of the different cost functions are discussed next.

Table 4. Comparison of DP Controllers for $N_{pc} = 100$.

Type	DDP Max Lifespan (Years)	MPGe at DDP Peak (MPGe)	SDP Lifespan at DDP Peak (Years)	Difference at Peak (%)	Mean Life Difference (%)
DP-B	5.47	9.54	5.46	−0.27	−0.45
DP-EC	5.76	9.65	5.73	−0.51	−0.95
DP-P	5.20	9.49	5.04	−3.11	−1.67

5.2. Effect of Aging-Aware Control

Next, all four causal strategies (SDP-B, SDP-EC, SDP-P, and Load Leveling) are compared. The $N_{pc} = 10$ case is shown in Figure 7, the $N_{pc} = 40$ case is shown in Figure 8, and the $N_{pc} = 100$ case is shown in Figure 9.

First, it can be seen that a larger HESS allows for greater improvements to battery lifespan. This is expected—the $N_{pc} = 10$ UC can only reduce current to or from the battery by approximately 0.5 C, while the largest power request from the driver corresponds to 2.5 C. On the other hand, the $N_{pc} = 100$ case can handle much larger power requests and can do much more to limit large battery current. However, these additional improvements come with a monetary cost, which is discussed more at the end of this section.

It is found across all three HESS sizing cases that the SDP-EC does the most to improve battery lifespan, offering a peak lifespan of 4.69 years at 10.15 MPGe, 5.16 years at 9.81 MPGe, and 5.72 years at 9.72 MPGe for the $N_{pc} = 10$, $N_{pc} = 40$, and $N_{pc} = 100$ cases, respectively.

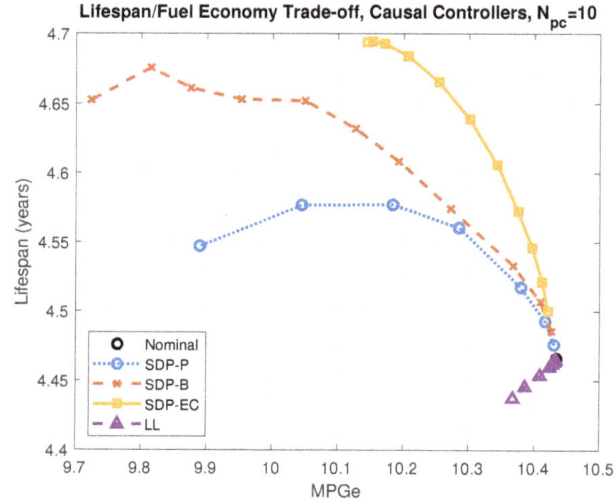

Figure 7. Comparison of energy consumption and battery aging for the four causal control methods, $N_{pc} = 10$.

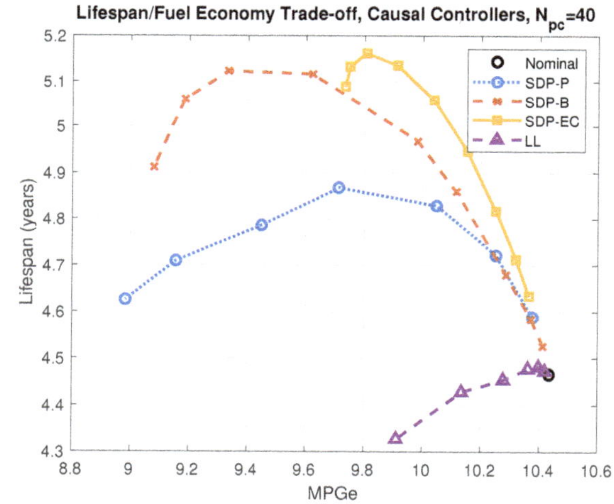

Figure 8. Comparison of energy consumption and battery aging for the four causal control methods, $N_{pc} = 40$.

The SDP-B offers substantial lifespan improvements as well, however, not to the degree of SDP-EC. SDP-B exceeds the performance of the the two non-SDP strategies but does not improve lifespan as well as the SDP strategies that include direct aging control. Although SDP-P does substantially increase battery lifespan, it does not "understand" the aging mechanics—such as the different effect of charging and discharging currents, or how damage from large currents is multiplied at high DoD—resulting in smaller lifespan increases than the strategies that control aging directly. The performances of SDP-EC and SDP-B relative to SDP-P clearly indicate the power of aging-aware energy management.

Not only does SDP-EC offer the best increase in battery lifespan, it offers the best improvements to the overall energy consumption/battery aging trade-off. That is, for all three HESS sizes and for any given rate of energy consumption, the SDP-EC strategy offers the largest improvements to battery lifespan; further increases to lifespan incur the smallest increases to energy consumption. Not only is the peak lifespan improvements for the SDP-EC method higher than SDP-B and SDP-P, it reaches that peak at a lower MPGe than the peaks of the SDP-B and SDP-P curves. The performance of SDP-EC compared to SDP-B indicates the value of controlling ultracapacitor aging and energy losses in addition to battery aging.

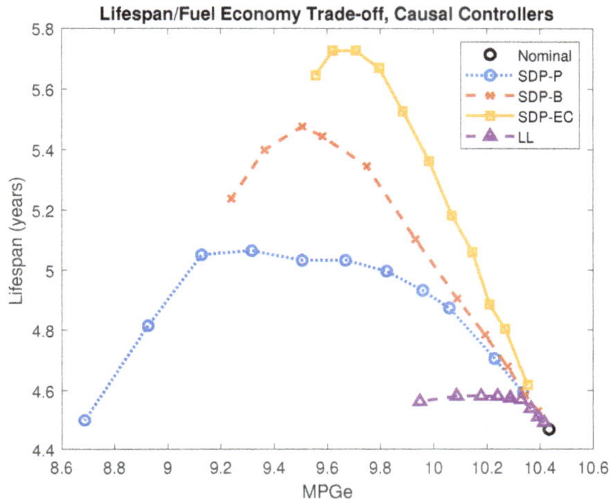

Figure 9. Comparison of energy consumption and battery aging for the four causal control methods, $N_{pc} = 100$.

The baseline case, Load Leveling, does not perform well: in the $N_{pc} = 10$ and $N_{pc} = 40$ cases, it offers virtually no lifespan improvement at all. With the large HESS, although it is able to match the SDP performance at low levels of UC usage, it quickly reaches its peak before dropping off. In this case, Load Leveling offers a peak lifespan of only 4.58 years.

The average change in ultracapacitor state of aging, measured at the end of battery life, is plotted in Figure 10 for $N_{pc} = 40$ as a representative case. Nominal aging—the aging of an ultracapacitor that is stored at the target SOC and at the temperature given in Section 2.3, and is otherwise unused—is found to be ΔSoA per year = 4.064%. At low degrees of UC usage, all three SDP methods are shown to have UC aging near the nominal. However, as UC usage increases, the SDP-B and SDP-P methods are seen to have the UC aging rate grow—SDP-B, in fact, reaches a peak UC aging rate of 4.415% at 9.19 MPGe. On the other hand, SDP-EC is shown to have measurably less UC aging at high levels of UC usage. This is expected, as SDP-EC seeks to limit UC aging while SDP-B does not.

The $N_{pc} = 10$ and $N_{pc} = 100$ cases are not presented here; however, similar trends in UC aging per controller type are observed.

Taken together, these results indicate a clear benefit to using strategies with predictive power, and that predictive power combined with energy storage aging models incorporated into the control strategy offers the best way to increase battery lifespan. The results of the four causal strategies are summarized in Table 5 for the $N_{pc} = 10$ case, Table 6 for the $N_{pc} = 40$ case, and Table 7 for the $N_{pc} = 100$ case.

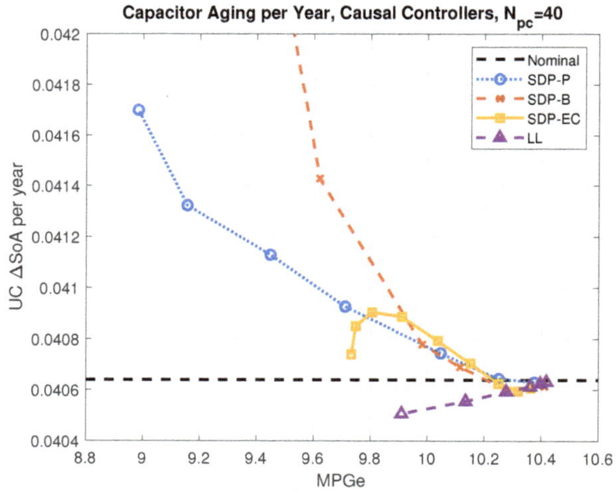

Figure 10. Comparison of energy consumption and UC aging for the four causal control methods, $N_{pc} = 40$.

Table 5. Comparison of Causal Controllers, small HESS ($N_{pc} = 10$).

Type	Peak Lifespan (Years)	MPGe at Peak (MPGe)	Lifespan vs. Nominal (%)	MPGe vs. Nominal (%)	UC ΔSoA Per Year at Peak
Nominal	4.47	10.43	–	–	4.064
SDP-B	4.68	9.81	4.7	−5.9	6.633
SDP-EC	4.69	10.14	5.1	−2.8	4.587
SDP-P	4.58	10.18	2.5	−2.4	4.105
LL	4.47	10.43	–	–	4.064

Table 6. Comparison of Causal Controllers, mid-sized HESS ($N_{pc} = 40$).

Type	Peak Lifespan (Years)	MPGe at Peak (MPGe)	Lifespan vs. Nominal (%)	MPGe vs. Nominal (%)	UC ΔSoA Per Year at Peak
Nominal	4.47	10.43	–	–	4.064
SDP-B	5.12	9.62	14.7	−7.8	4.138
SDP-EC	5.16	9.81	15.6	−6.0	4.091
SDP-P	4.87	9.71	9.0	−6.9	4.093
LL	4.48	10.40	0.4	−0.3	4.063

Table 7. Comparison of Causal Controllers, large HESS ($N_{pc} = 100$).

Type	Peak Lifespan (Years)	MPGe at Peak (MPGe)	Lifespan vs. Nominal (%)	MPGe vs. Nominal (%)	UC ΔSoA Per Year at Peak
Nominal	4.47	10.43	–	–	4.064
SDP-B	5.48	9.51	22.6	−8.9	4.100
SDP-EC	5.73	9.71	28.2	−7.0	4.060
SDP-P	5.07	9.32	13.4	−10.7	4.057
LL	4.58	10.09	2.6	−3.3	4.046

5.3. Ultracapacitor Overuse

In Section 2.4, it was established that there is necessarily a trade-off between battery aging and energy consumption when using a HESS to limit battery aging. A consequence of this is seen in every simulated controller, shown in Figures 6–9, at the tail end of each curve: as the ultracapacitor is used more and more extensively, energy consumption increases as more energy is lost from the ultracapacitor internal resistance. These losses must be made up for by discharging the battery more deeply—depth of discharge being a key aging factor, per the model presented in Section 2.2. At some point, increases to the DoD aging factor outweigh the impact of decreases in the other aging factors, and battery lifespan eventually begins to decrease rather than increase. Thus, in cases where the ultracapacitor is used very heavily, attempts to control battery aging can have the opposite of the intended effect.

The $N_{pc} = 100$ SDP-P results for energy losses are shown in Figure 11 as a representative example. Energy losses increase with increasing ultracapacitor usage, eventually leading to a decreased lifespan. This behavior emphasized the importance of tuning the energy management strategy properly. With poor tuning, it is possible for the HESS to do more harm than good. Similar trends are seen with the other control methods and other HESS sizes, but are not shown here.

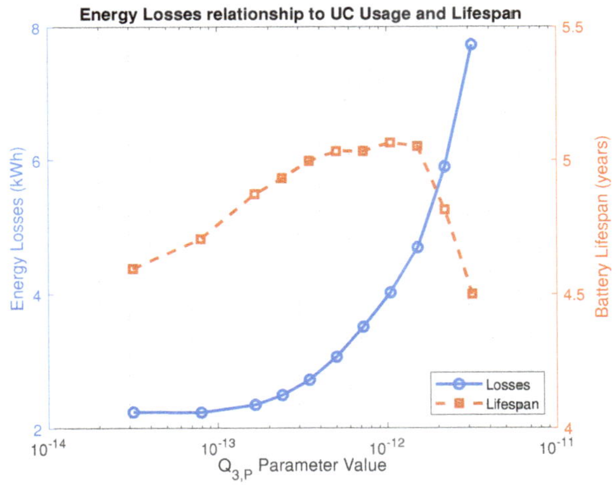

Figure 11. Comparison of energy losses (left axis) to battery lifespan (right axis) versus the $Q_{3,P}$ weighting parameter, where increasing $Q_{3,P}$ increases ultracapacitor usage.

5.4. Cost-Benefit Analysis

The cost–benefit of each simulated point is computed per Equations (58)–(61) and plotted versus MPGe in Figure 12 for the $N_{pc} = 10$ case, in Figure 13 for the $N_{pc} = 40$ case, and in Figure 14 for the $N_{pc} = 100$ case. Positive values indicate that value is added to the system, while negative values indicate a cost.

In general, it is observed that the SDP-EC method offers clear value over the other methods: for any HESS sizing and for any given MPGe, the SDP-EC method offers the highest benefit per mile. In the $N_{pc} = 10$ case, it was shown to be the only method that offered a positive return on investment. In the $N_{pc} = 40$ case, the SDP-P and SDP-B methods did offer a positive return: the SDP-EC's maximum benefit per mile was over 50% greater than SDP-B and over 120% greater than SDP-P. The $N_{pc} = 100$ case is similar to the $N_{pc} = 10$ case in terms of relative performance: only the SDP-EC offers notable benefit. Although the SDP-B does offer a small positive return for some tunings, the maximum benefit of the SDP-B method is less than a quarter of the maximum benefit of the SDP-EC method.

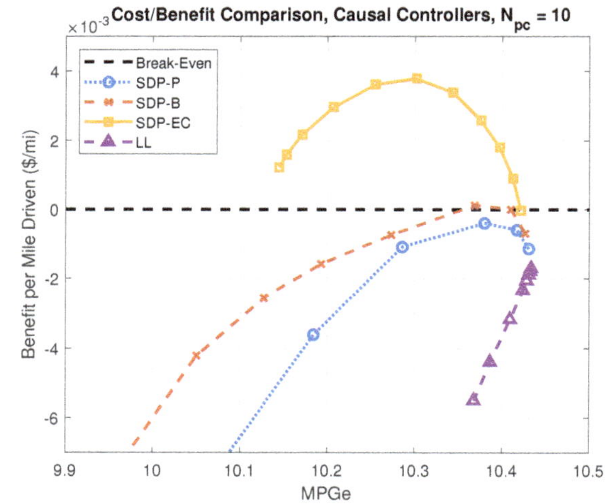

Figure 12. Cost–benefit analysis for the four causal control methods, $N_{pc} = 10$.

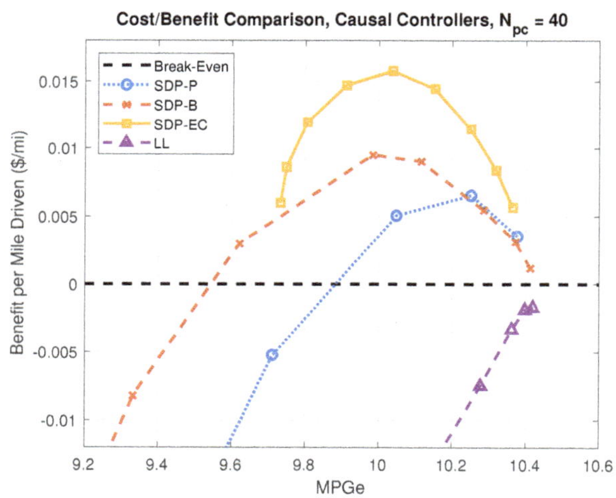

Figure 13. Cost–benefit analysis for the four causal control methods, $N_{pc} = 40$.

Another takeaway from Figure 12 is that the peak economic benefit occurs at a higher MPGe (lower ultracapacitor utilization) than the peak lifespan increase. This makes intuitive sense: lifespan improvements level off near the peak while energy consumption continues to grow. Therefore, near the lifespan peak, the marginal improvement battery-cost-per-mile is less than the marginal decrease in fuel economy. For the $N_{pc} = 100$ SDP-EC method, there is not much difference between the peak lifespan increase (occurring at 9.71 MPGe) and the peak benefit (occurring at 9.80 MPGe); however, for a smaller HESS or for weaker strategies, the difference can be substantial: in the $N_{pc} = 100$ case, the SDP-B peak lifespan increase is at 9.51 MPGe, which is effectively break-even in terms of value, while the peak benefit occurs at 9.75 MPGe. SDP-P has its peak benefit at 10.06 MPGe and peak lifespan at 9.32 MPGe; looking at the the SDP-EC method in the $N_{pc} = 10$ case, the benefit at the peak lifespan increase is approximately half of the maximum possible benefit. Clearly, the economic factors should be considered when deciding on the controller tuning.

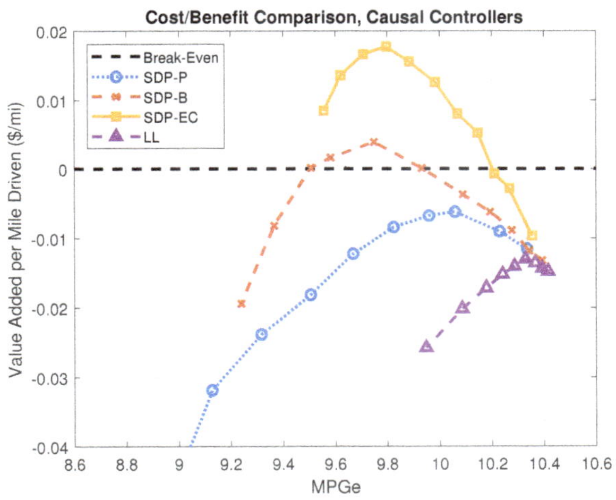

Figure 14. Cost-benefit analysis for the four causal control methods, $N_{pc} = 100$.

Finally, the estimated payback time is computed for the maximum benefit of the SDP-EC method for all three cases using Equation (62). It is found that, for the given UC, battery, and energy costs, the small HESS ($N_{pc} = 10$) has a payback time of 11.3 years, the mid-sized HESS ($N_{pc} = 10$) has a payback time of 15.6 years, and the large HESS ($N_{pc} = 100$) has a payback time of 21.6 years.

In order to observe the full trend of the payback period for different ultracapacitor sizes, additional simulations are run for N_{pc} equal to 2, 5, and all increments of 10 between 10 and 100. The optimal EMS and optimal $Q_{2,SOC}$ are recomputed and the vehicle is simulated again for the new EMS and new UC size. The payback time and battery lifespan at the most cost-effective tuning for each N_{pc} are then plotted in Figure 15. This shows that, although increasing the HESS does does improve the battery lifespan, the cost of the extra UCs exceeds the savings of that extra lifespan.

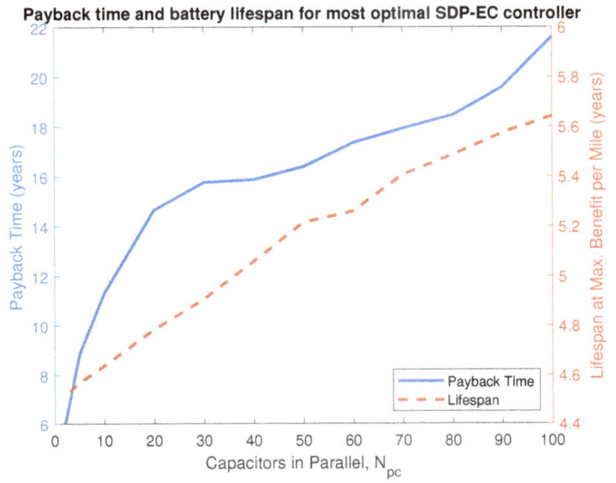

Figure 15. Estimated payback time for optimal SDP-EC controller with varying HESS size.

Finally, the authors note the sensitivity of the benefit per mile and the payback time period to assumptions about component pricing, energy pricing, and aging mechanisms.

For instance: this research assumes that energy is priced at the US average across all sectors, $V_{nrg} = \$0.1067/\text{kWh}$ [47]. If, instead, energy was priced at the transportation sector average for the state of Illinois (such as for a Chicago Transit Authority bus), the energy price of $V_{nrg} = \$0.0632/\text{kWh}$, also from [47], would reduce the payback time by approximately 30%. On the other hand, the California price $V_{nrg} = \$0.1280/\text{kWh}$ would increase the payback time by 30%.

Alternatively, we can consider that battery and UC components use the pricing of reference [50] rather than [46] while maintaining $V_{nrg} = \$0.1067/\text{kWh}$; the increased battery and ultracapacitor prices from [50] result in a payback time of 6.9 years for the $N_{pc} = 10$ UC and 13.7 years for the $N_{pc} = 100$ UC. On the other hand, the component prices of [51] would indicate that the HESS is not beneficial under any circumstance.

A different battery aging model in the literature [52], used in an array of battery and HESS control literature such as [12,15,19,23,53,54], models lithium ion phosphate batteries as aging at up to 3× the rate of the model used in this research. If the battery ages even 1.5× the modeled rate, then we would see a payback period of 4.8 and 8.8 years for $N_{pc} = 10$ and $N_{pc} = 100$, respectively.

Finally, although battery end-of-life can be considered a hard limit for battery use based on range constraints, the ultracapacitor, on the other hand, can continue to be used beyond 80% capacitance fade. This would not be unreasonable, considering how Figure 15 shows that the effectiveness of the proposed control method is maintained as the number of cells (and over UC pack capacitance) is decreased. Therefore, if, for instance, the UCs were used until 70% capacitance fade, then we would see a payback period of 7 and 12 years for $N_{pc} = 10$ and $N_{pc} = 100$, respectively. One could, alternatively, assume that the UC does not need to be replaced at all (setting $UCCPM$ to 0), as UC life exceeds the 12 year lifespan of an individual transit bus [55]. However, there is still value to considering these replacement costs from the perspective of an entire vehicle fleet.

All this is to say: an engineer must take caution that a HESS is economically appropriate for a given application; there may be circumstances where a HESS is highly beneficial, and others where it may be impractical. With that said, this research has demonstrated that, for any HESS sizing and for any given MPGe, the SDP-EC method offers a larger increase to battery lifespan and a higher benefit per mile than the other considered methods. The important takeaway of this analysis is how proper control of the HESS is critical for maximizing both battery lifespan and HESS value, and that joint control of battery aging, UC aging, and energy losses is the most effective method to manage the HESS.

6. Conclusions

This paper develops controllable battery and ultracapacitor aging models for a HESS. Various energy management strategies are developed for the purpose of minimizing battery aging. As a case study, these models and control strategies are applied to a simulated electric bus to determine the battery lifespan and energy consumption of each strategy. An array of different HESS sizes and controller tunings are simulated in order to determine the trade-off between battery aging and energy consumption for each strategy. Additionally, the cost–benefit of the HESS is analyzed to determine the relative economic benefit of the proposed control strategies.

Simulation results showed that the SDP-EC method, which controls a weighted combination of battery aging, ultracapacitor aging, and energy losses, offers the biggest improvement to the aging–energy consumption trade-off across all considered HESS sizes. At its peak, this strategy offered a 28.2% increase in battery lifespan and required only a 7.0% decrease in MPGe. The SDP-B method, which controls battery aging but neither ultracapacitor aging nor energy losses, was the next most effective controller, indicating the importance of including an aging model directly in the control.

Simulation results also demonstrated that excessive use of the ultracapacitor can, in fact, be detrimental to the lifespan of the battery. Ultracapacitor use incurs additional energy losses and, if the ultracapacitor is heavily used, then these losses can result in additional battery aging. Furthermore, the cost–benefit analysis showed that only the strategies that included direct aging control would reliably add value to the system; the SDP-EC method was the most proven manner of adding economic value to the HESS. These points, taken together, indicate the importance of control strategy selection and design.

Future work for this research includes the optimization of component sizing, given the proposed new methods of energy management. Additionally, work is ongoing to investigate the robustness of the control strategies for uncertainty in the battery and ultracapacitor models. Finally, other energy management strategies should be considered and compared to the methods here, such as DDP formed into a rule base or the Equivalent Consumption Minimization Strategy applied to aging control.

Author Contributions: K.M. and F.A. conceived and designed the experiments; K.M. developed the bus model performed the numerical experiments; K.M. and F.A. analyzed the data; K.M. wrote the paper. All authors have read and agreed to the published version of the manuscript.

Funding: This research received no external funding.

Institutional Review Board Statement: Not applicable.

Informed Consent Statement: Not applicable.

Acknowledgments: This work was supported by the University of California, Davis, Department of Mechanical and Aerospace Engineering.

Conflicts of Interest: The authors declare no conflict of interest.

Abbreviations

The following abbreviations are used in this manuscript:

APV	Alternative Powertrain Vehicle
EV	Electric Vehicle
HEV	Hybrid Electric Vehicle
UC	Ultracapacitor
HESS	Hybrid Energy Storage System
EMS	Energy Management Strategy
DP	Dynamic Programming
DDP	Deterministic Dynamic Programming
SDP	Stochastic Dynamic Programming
SOC	State of Charge
DOD	Depth of Discharge
SOA	State of Aging
PM	Palmgren-Miner
MPGe	Miles per Gallon Equivalent
BCPM	Battery Cost per Mile
UCCPM	Ultracapacitor Cost per Mile
ECPM	Energy Cost per Mile
GGE	Gasoline Gallon Equivalent
BPM	Benefit per Mile

References

1. Lee, T.K.; Filipi, Z. Impact of Model-Based Lithium-Ion Battery Control Strategy on Battery Sizing and Fuel Economy in Heavy-Duty HEVs. *SAE Int. J. Commer. Veh.* **2011**, *4*, 198–209. [CrossRef]
2. Moura, S.J.; Stein, J.L.; Fathy, H.K. Battery-Health Conscious Power Management in Plug-In Hybrid Electric Vehicles via Electrochemical Modeling and Stochastic Control. *IEEE Trans. Control Syst. Technol.* **2013**, *21*, 679–694. [CrossRef]
3. Suri, G.; Onori, S. A control-oriented cycle-life model for hybrid electric vehicle lithium-ion batteries. *Energy* **2016**, *96*, 644–653. [CrossRef]
4. Guzzella, L.; Sciarretta, A. *Vehicle Propulsion Systems*, 3rd ed.; Springer: Berlin/Heidelberg, Germany, 2013.
5. Williamson, S.S. *Energy Management Strategies for Electric and Plug-in Hybrid Electric Vehicles*; Springer: New York, NY, USA, 2013.

6. Kovaltchouk, T.; Multon, B.; Ahmed, H.B.; Aubry, J.; Venet, P. Enhanced Aging Model for Supercapacitors Taking Into Account Power Cycling: Application to the Sizing of an Energy Storage System in a Direct Wave Energy Converter. *IEEE Trans. Ind. Appl.* **2015**, *51*, 2405–2414. [CrossRef]
7. Kovaltchouk, T.; Ahmed, H.B.; Multon, B.; Aubry, J.; Venet, P. An aging-aware life cycle cost comparison between supercapacitors and Li-ion batteries to smooth Direct Wave Energy Converter production. In Proceedings of the 2015 IEEE Eindhoven PowerTech, Eindhoven, The Netherlands, 29 June–2 July 2015; pp. 1–6. [CrossRef]
8. Hammar, A.; Venet, P.; Lallemand, R.; Coquery, G.; Rojat, G. Study of Accelerated Aging of Supercapacitors for Transport Applications. *IEEE Trans. Ind. Electron.* **2010**, *57*, 3972–3979. [CrossRef]
9. Zhang, L.; Hu, X.; Wang, Z.; Sun, F.; Deng, J.; Dorrell, D.G. Multiobjective Optimal Sizing of Hybrid Energy Storage System for Electric Vehicles. *IEEE Trans. Veh. Technol.* **2018**, *67*, 1027–1035. [CrossRef]
10. Song, Z.; Zhang, X.; Li, J.; Hofmann, H.; Ouyang, M.; Du, J. Component sizing optimization of plug-in hybrid electric vehicles with the hybrid energy storage system. *Energy* **2018**, *144*, 393–403. [CrossRef]
11. Eldeeb, H.H.; Elsayed, A.T.; Lashway, C.R.; Mohammed, O. Hybrid Energy Storage Sizing and Power Splitting Optimization for Plug-In Electric Vehicles. *IEEE Trans. Ind. Appl.* **2019**, *55*, 2252–2262. [CrossRef]
12. Song, Z.; Li, J.; Hou, J.; Hofmann, H.; Ouyang, M.; Du, J. The battery-supercapacitor hybrid energy storage system in electric vehicle applications: A case study. *Energy* **2018**, *154*, 433–441. [CrossRef]
13. Wieczorek, M.; Lewandowski, M. A mathematical representation of an energy management strategy for hybrid energy storage system in electric vehicle and real time optimization using a genetic algorithm. *Appl. Energy* **2017**, *192*, 222–233. [CrossRef]
14. Wang, Y.; Wang, L.; Li, M.; Chen, Z. A review of key issues for control and management in battery and ultra-capacitor hybrid energy storage systems. *eTransportation* **2020**, *4*, 100064. [CrossRef]
15. Song, Z.; Li, J.; Han, X.; Xu, L.; Lu, L.; Ouyang, M.; Hofmann, H. Multi-objective optimization of a semi-active battery/supercapacitor energy storage system for electric vehicles. *Appl. Energy* **2014**, *135*, 212–224. [CrossRef]
16. Shen, J.; Dusmez, S.; Khaligh, A. Optimization of Sizing and Battery Cycle Life in Battery/Ultracapacitor Hybrid Energy Storage Systems for Electric Vehicle Applications. *IEEE Trans. Ind. Inform.* **2014**, *10*, 2112–2121. [CrossRef]
17. Akar, F.; Tavlasoglu, Y.; Vural, B. An Energy Management Strategy for a Concept Battery/Ultracapacitor Electric Vehicle With Improved Battery Life. *IEEE Trans. Transp. Electrif.* **2017**, *3*, 191–200. [CrossRef]
18. Carter, R.; Cruden, A.; Hall, P.J. Optimizing for Efficiency or Battery Life in a Battery/Supercapacitor Electric Vehicle. *IEEE Trans. Veh. Technol.* **2012**, *61*, 1526–1533. [CrossRef]
19. Zhao, C.; Yin, H.; Ma, C. Quantitative Evaluation of LiFePO Battery Cycle Life Improvement Using Ultracapacitors. *IEEE Trans. Power Electron.* **2016**, *31*, 3989–3993. [CrossRef]
20. Du, J.; Zhang, X.; Wang, T.; Song, Z.; Yang, X.; Wang, H.; Ouyang, M.; Wu, X. Battery degradation minimization oriented energy management strategy for plug-in hybrid electric bus with multi-energy storage system. *Energy* **2018**, *165*, 153–163. [CrossRef]
21. Hou, J.; Song, Z. A hierarchical energy management strategy for hybrid energy storage via vehicle-to-cloud connectivity. *Appl. Energy* **2020**, *257*, 113900. [CrossRef]
22. Zhang, S.; Hu, X.; Xie, S.; Song, Z.; Hu, L.; Hou, C. Adaptively coordinated optimization of battery aging and energy management in plug-in hybrid electric buses. *Appl. Energy* **2019**, *256*, 113891. [CrossRef]
23. Cordoba-Arenas, A.; Onori, S.; Guezennec, Y.; Rizzoni, G. Capacity and power fade cycle-life model for plug-in hybrid electric vehicle lithium-ion battery cells containing blended spinel and layered-oxide positive electrodes. *J. Power Sources* **2015**, *278*, 473–483. [CrossRef]
24. Tang, L.; Rizzoni, G. Energy management strategy including battery life optimization for a HEV with a CVT. In Proceedings of the 2016 IEEE Transportation Electrification Conference and Expo, Asia-Pacific (ITEC Asia-Pacific), Busan, Korea, 1–4 June 2016; pp. 549–554.
25. Mohan, G.; Assadian, F.; Longo, S. Comparative analysis of forward-facing models vs backwardfacing models in powertrain component sizing. In Proceedings of the IET Hybrid and Electric Vehicles Conference 2013 (HEVC 2013), London, UK, 6–7 November 2013; pp. 1–6. [CrossRef]
26. Zeng, X.; Yang, N.; Wang, J.; Song, D.; Zhang, N.; Shang, M.; Liu, J. Predictive-model-based dynamic coordination control strategy for power-split hybrid electric bus. *Mech. Syst. Signal Process.* **2015**, *60–61*, 785–798. [CrossRef]
27. Sangtarash, F.; Esfahanian, V.; Nehzati, H.; Haddadi, S.; Bavanpour, M.A.; Haghpanah, B. Effect of Different Regenerative Braking Strategies on Braking Performance and Fuel Economy in a Hybrid Electric Bus Employing CRUISE Vehicle Simulation. *SAE Int. J. Fuels Lubr.* **2008**, *1*, 828–837. [CrossRef]
28. Wang, B.H.; Luo, Y.G.; Zhang, J.W. Simulation of city bus performance based on actual urban driving cycle in China. *Int. J. Automot. Technol.* **2008**, *9*, 501–507. [CrossRef]
29. Markel, T.; Brooker, A.; Hendricks, T.; Johnson, V.; Kelly, K.; Kramer, B.; O'Keefe, M.; Sprik, S.; Wipke, K. ADVISOR: A systems analysis tool for advanced vehicle modeling. *J. Power Sources* **2002**, *110*, 255–266. [CrossRef]
30. Erdinc, O.; Vural, B.; Uzunoglu, M. A dynamic lithium-ion battery model considering the effects of temperature and capacity fading. In Proceedings of the 2009 International Conference on Clean Electrical Power, Capri, Italy, 9–11 June 2009; pp. 383–386.
31. Nelson, R.F. Power requirements for batteries in hybrid electric vehicles. *J. Power Sources* **2000**, *91*, 2–26. [CrossRef]
32. Fauvel, C..; Vikesh, N.; Aymeric, R. *Medium and Heavy Duty Hybrid Electric Vehicle Sizing to Maximize Fuel Consumption Displacement on Real World Drive Cycles*; Power (W): Los Angeles , CA, USA, 2012.

33. Dougal, R.; Gao, L.; Liu, S. Ultracapacitor model with automatic order selection and capacity scaling for dynamic system simulation. *J. Power Sources* **2004**, *126*, 250–257. [CrossRef]
34. Omar, N.; Monem, M.A.; Firouz, Y.; Salminen, J.; Smekens, J.; Hegazy, O.; Gaulous, H.; Mulder, G.; Van den Bossche, P.; Coosemans, T.; et al. Lithium iron phosphate based battery—Assessment of the aging parameters and development of cycle life model. *Appl. Energy* **2014**, *113*, 1575–1585. [CrossRef]
35. Safari, M.; Morcrette, M.; Teyssot, A.; Delacourt, C. Life-Prediction Methods for Lithium-Ion Batteries Derived from a Fatigue Approach I. Introduction: Capacity-Loss Prediction Based on Damage Accumulation. *J. Electrochem. Soc.* **2010**, *157*, A713–A720. [CrossRef]
36. Zhou, C.; Qian, K.; Allan, M.; Zhou, W. Modeling of the Cost of EV Battery Wear Due to V2G Application in Power Systems. *IEEE Trans. Energy Convers.* **2011**, *26*, 1041–1050. [CrossRef]
37. Mallon, K.R.; Assadian, F.; Fu, B. Analysis of On-Board Photovoltaics for a Battery Electric Bus and Their Impact on Battery Lifespan. *Energies* **2017**, *10*, 943. [CrossRef]
38. Xu, B.; Oudalov, A.; Ulbig, A.; Andersson, G.; Kirschen, D.S. Modeling of lithium-ion battery degradation for cell life assessment. *IEEE Trans. Smart Grid* **2016**, *9*, 1131–1140. [CrossRef]
39. Lee, D.H.; Kim, U.S.; Shin, C.B.; Lee, B.H.; Kim, B.W.; Kim, Y.H. Modelling of the thermal behaviour of an ultracapacitor for a 42-V automotive electrical system. *J. Power Sources* **2008**, *175*, 664–668. [CrossRef]
40. Mallon, K.; Assadian, F. Robustification and its Implication for Hybrid Electric Vehicle Energy Management Strategies. *J. Dyn. Syst. Control.* **2020**, *143*, 091001. [CrossRef]
41. Tate, E.D.; Grizzle, J.W.; Peng, H. Shortest path stochastic control for hybrid electric vehicles. *Int. J. Robust Nonlinear Control* **2008**, *18*, 1409–1429. [CrossRef]
42. Lin, C.C.; Peng, H.; Grizzle, J.W. A stochastic control strategy for hybrid electric vehicles. In Proceedings of the 2004 American Control Conference, Boston, MA, USA, 30 June–2 July 2004; Volume 5, pp. 4710–4715.
43. Johannesson, L.; Asbogard, M.; Egardt, B. Assessing the Potential of Predictive Control for Hybrid Vehicle Powertrains Using Stochastic Dynamic Programming. *IEEE Trans. Intell. Transp. Syst.* **2007**, *8*, 71–83. [CrossRef]
44. Vagg, C.; Akehurst, S.; Brace, C.J.; Ash, L. Stochastic dynamic programming in the real-world control of hybrid electric vehicles. *IEEE Trans. Control Syst. Technol.* **2015**, *24*, 853–866. [CrossRef]
45. Opila, D.F.; Wang, X.; McGee, R.; Grizzle, J. Real-time implementation and hardware testing of a hybrid vehicle energy management controller based on stochastic dynamic programming. *J. Dyn. Syst. Meas. Control* **2013**, *135*, 021002. [CrossRef]
46. Zhu, T.; Wills, R.G.A.; Lot, R.; Kong, X.; Yan, X. Optimal sizing and sensitivity analysis of a battery-supercapacitor energy storage system for electric vehicles. *Energy* **2021**, *221*, 119851. [CrossRef]
47. US Energy Information Administration. *July 2021 Monthly Energy Review*; Technical Report; U.S. Energy Information Administration: Washington, DC, USA, 2021.
48. Barlow, T.J.; Latham, S.; Mccrae, I.S.; Boulter, P.G. *A Reference Book of Driving Cycles for Use in the Measurement of Road Vehicle Emissions*; TRL Published Project Report; Transport Research Laboratory: Crothorne, UK, 2009.
49. Hughes-Cromwick, M. *2019 Public Transportation Fact Book*; World Transit Research; American Public Transportation Association: Chicago, IL, USA, 2019.
50. Min, H.; Lai, C.; Yu, Y.; Zhu, T.; Zhang, C. Comparison Study of Two Semi-Active Hybrid Energy Storage Systems for Hybrid Electric Vehicle Applications and Their Experimental Validation. *Energies* **2017**, *10*, 279. [CrossRef]
51. Mongird, K.; Viswanathan, V.V.; Balducci, P.J.; Alam, M.J.E.; Fotedar, V.; Koritarov, V.S.; Hadjerioua, B. *Energy Storage Technology and Cost Characterization Report*; Technical Report; Pacific Northwest National Lab. (PNNL): Richland, WA, USA, 2019.
52. Wang, J.; Liu, P.; Hicks-Garner, J.; Sherman, E.; Soukiazian, S.; Verbrugge, M.; Tataria, H.; Musser, J.; Finamore, P. Cycle-life model for graphite-$LiFePO_4$ cells. *J. Power Sources* **2011**, *196*, 3942–3948. [CrossRef]
53. Serrao, L.; Onori, S.; Sciarretta, A.; Guezennec, Y.; Rizzoni, G. Optimal energy management of hybrid electric vehicles including battery aging. In Proceedings of the 2011 American Control Conference, San Francisco, CA, USA, 29 June–1 July 2011; pp. 2125–2130. [CrossRef]
54. Lin, X.; Hao, X.; Liu, Z.; Jia, W. Health conscious fast charging of Li-ion batteries via a single particle model with aging mechanisms. *J. Power Sources* **2018**, *400*, 305–316. [CrossRef]
55. Laver, R.; Schneck, D.; Skorupski, D.; Brady, S.; Cham, L. *Useful Life of Transit Buses and Vans*; Number: FTA-VA-26-7229-07.1; National Academy of Sciences: Washington, DC, USA, 2007.

energies

MDPI

Article

A Comparative Analysis of Brake-by-Wire Smart Actuators Using Optimization Strategies

Ehsan Arasteh * and Francis Assadian *

Department of Mechanical and Aerospace Engineering, University of California, Davis, CA 95616, USA
* Correspondence: earasteh@ucdavis.edu (E.A.); fassadian@ucdavis.edu (F.A.)

Abstract: Since the automotive industry is shifting towards electrification, brake-by-wire technologies are becoming more prevalent. However, there has been little research comparing and optimizing brake-by-wire actuators in terms of their energy expenditure and response time. This paper investigates the comparison of three different smart brake-by-wire actuators, Electro-Hydraulic Brakes (EHB), Electro-Mechanical Brakes (EMB), and Electronic Wedge Brakes (EWB), first by defining an objective metric and then using both linear and nonlinear optimization techniques. Modeling of the actuators is performed using the bond graph method. Then, the controllers are designed using a robust control strategy, Youla parameterization. After designing the controllers, two types of optimization are performed on the actuators. Optimizations are performed in two ways: 1. by linearizing the plants and optimizing using their transfer functions and 2. by nonlinear optimization of the plants in the closed-loop following a specific clamp force target. The objective metrics or the cost functions for these optimizations are chosen to be the energy usage of the plants during the closed-loop operation, maximum power requirement, and their dynamic responsiveness. Using this optimization framework, we can show a significant improvement in the energy usage of the actuators and slight improvements in their responsiveness. In the end, the actuators are compared in terms of their energy usage for sets of initial and optimized physical parameters.

Keywords: EHB; EMB; EWB; system modeling; bond graph; optimization; control design; Youla parameterization; robust control; nonlinear optimization; brake-by-wire; actuator; electro-mechanical brake; electronic wedge brake; electro-hydraulic brake

Citation: Arasteh, E.; Assadian, F. A Comparative Analysis of Brake-by-Wire Smart Actuators Using Optimization Strategies. *Energies* **2022**, *15*, 634. https://doi.org/10.3390/en15020634

Academic Editor: Silvio Simani

Received: 18 November 2021
Accepted: 7 January 2022
Published: 17 January 2022

1. Introduction

By-wire technologies are becoming more in demand because of their contribution to the vehicle's fuel efficiency and electrification. They are well suited for autonomous vehicles, Electric Vehicles (EVs), and Hybrid Electric Vehicles (HEVs) due to their electronic interface and architectures. Between all these by-wire technologies, brake-by-wires are essential since they play a critical role in the vehicle's safety [1–3].

Brake-by-wire systems can reduce the overall component weight due to the reduction in the number of parts and integrated packaging. They also improve energy consumption because they would only use energy when required. The reduction of energy usage by different components is an essential aspect of reducing the CO_2 emissions of vehicles. This type of reduction is significant in reducing fleet CO_2 emissions. When the brake pedal is released, the brake pad may not release entirely, causing caliper drag. Brake-by-wire is even more energy-efficient by eliminating caliper drag through sensors and control methods. To enhance vehicle safety, individual wheel braking and faster activation times of BBW actuators can be utilized by the vehicle's Electronic Stability Control (ESC) system. The biggest barriers to brake-by-wire systems gaining popularity in the automotive industry are the reliability of new actuators, as well as the risk and cost of deploying new braking technology [4].

1.1. Brake-by-Wire Actuators

Wet and dry brakes are the two types of braking systems. Dry brakes are simply mechanical systems that rely on electric motors, whereas wet brakes employ fluids to produce pressure on the piston and the caliper. A type of wet brake is electro-hydraulic brakes, which include solenoid valves that can be operated with an electronic signal to vary the braking pressure. These types of brakes require an accumulator and a pump, and usually a vacuum booster to maintain their high pressure throughout the braking operation. There are also a few different types of dry brakes. In a pure electro-mechanical brake, a small electric motor, planetary gear set, and roller screw energizes the brake pad [5–8]. For this type of brake, however, a 42 volt motor is needed, which is energy intensive [4]. On the other hand, a wedge mechanism is used in electronic wedge brakes to create a brake caliper, which is drawn into the brake and uses less energy because it utilizes the wheel's rotation. They usually need a 14 V supply as opposed to the 42 V of the electro-mechanical brake. However, this comes at the cost of more complicated mechanics and control [5,8–16]. Vienna Engineering has created a brake system based on a crank-shaft mechanism that reduces the complexity of dealing with the reduction gears and roller screws [17].

Purely dry brakes are expensive compared to conventional electro-hydraulic brakes since they are new technologies and need more testing and research before they can be reliable enough to go into production. The reliability of purely dry brakes is a challenge since they have more electronic components, and these components need to work reliably in an environment where vibration, shock, and temperature can significantly affect their nominal performance [4].

On the other hand, EHBs have been used by manufacturers for a while. For example, an integrated electro-hydraulic brake system utilizes an electro-mechanical actuator (similar to an electro-mechanical brake) as a modulator of a master cylinder. This electro-mechanical actuator uses a motor to rotate a gear mechanism and a ball-screw that pushes the piston. This axial force pressurizes the brake fluid inside the master cylinder. The pressurized brake fluid is then transferred to the wheel chamber using a high-pressure pipeline, where this pressure displaces the caliper (similar to an EHB). This integrated electro-hydraulic brake uses most of the EHB parts that the automotive manufacturers are already familiar with and removes the need for a pump, vacuum booster, and accumulator [18,19]. It also seems that this type of brake actuator is gaining popularity among car manufacturers. Bosch GmbH developed the iBooster and ESP hev (electronic stability program for hybrid and electric vehicles), which are integrated electro-hydraulic and Hydraulic Control Units (HCU) in 2013. In 2017, Continental AG introduced MK C1, which is an integrated EHB with fast actuation and without any vacuum booster or accumulator along with emergency brake functionalities [20]. ZF TRW (with the IBC) and Hyundai Mobis (with iMEB) are among the more recent suppliers who developed and manufactured integrated electro-hydraulic brake actuators.

1.2. Objective Metrics for Brake-by-Wire Systems

Objective metrics are required for performing a comparative analysis of the systems under consideration. These metrics are utilized to measure each system's performance, robustness, and safety correctly. Similar metrics have already been used in other automotive applications to optimize or compare different topologies (different configurations). For example, Shankar et al. use several criteria for optimization and component sizing of plug-in hybrid electric vehicles. The objective functions in their optimization include all-electric range (AER), the CO_2 emission from the drive-cycle, and the cost of components [21].

Gombert et al. provide some basic metrics for brake-by-wire actuators and their vehicle configurations [4]. They provide some background for the objective metrics that need to be considered for Brake-By-Wire (BBW) actuators. Yao et al. consider a multi-objective optimization with a few constraints for their combined electromagnetic and electronic wedge brake-by-wire actuator. The objective comprises a time to braking at an acceptable slew rate, maximum initial braking torque, and electric power of the DC motor. Their constraints include the maximum power of the DC motor, brake slew rate, and maximum braking torque (maximum ground friction coefficient) [22,23]. Kwon et al. use a multi-objective formulation to optimize a caliper for the wedge brake. Their objective function includes the minimization of weight and the maximization of caliper stiffness. They then use the response surface model to optimize and find the best possible set of caliper parameters [24].

Metrics and metric-based optimization have also been used in the control architecture of brake-by-wire systems. Fengjiao et al. use multi-objective optimization for their control strategy of an electro-hydraulic brake system in an EV. Their objectives include 1. braking stability, which can be expressed as a quadratic function of friction adhesion on the rear and front wheels and the brake input, and 2. regenerative energy recovery. The constraints include battery charging power, motor peak torque, and the relationship between vehicle stability while braking and road surface friction [23]. Hielinger et al. used parameter optimization for an autonomous emergency braking system. Their cost function includes safety performance and customer acceptance. Safety performance is measured as the reduction of the impact speed (the speed at which the vehicle might collide to the nearest obstacle; if there is no collision, the cost becomes zero). Customer acceptance includes a sub-cost function for the brake profile (the deceleration of the vehicle summed over time) and braking the distance (minimum distance between the vehicle and the obstacle) [25]. Kelling et al. studied a distributed electronic and control architecture design for brake-by-wire systems and compared a conventional centralized architecture with a proposed fault-tolerant and distributed system in terms of safety and cost advantages [26].

1.3. Control Strategies for Brake-by-Wire Systems

Many researchers have used the sliding mode method to control the wheel slip for Anti-Lock Braking (ABS). Sliding Mode Controller (SMC) is a nonlinear control technique and an inherently non-continuous control law, which requires additional filtering to suitably smooth out this discontinuous control law, to force the system to operate on a sliding surface which defines the system's closed-loop dynamic. Compared to bang-bang control, SMC has the benefits of smaller actuation and added robustness. Anwar utilized a sliding mode controller to control slip in a hybrid BBW system that resulted in a good slip regulation in low friction surfaces and a smooth operation of the ABS, and reduced noise, vibration, and harshness (NVH) in EHB systems [27]. Tanelli et al. use pseudo-sliding mode control combining slip-deceleration (MSD), which continuously controls slip and deceleration while avoiding chattering and is robust against measurement noise and low sampling frequency [28]. However, SMC is not widely used in the automotive industry due to its design complexity, calibration difficulties, proper consideration of actuator delays, and difficulties with addressing robustness. Actuators have delays that can make the sliding mode lead to chatter, energy loss, and the excitation of unmodeled dynamics. However, this is not as much of a problem in the continuous control design [29]. Soltani et al. use a linearized model of EHB and synthesize closed-loop shaping Youla parameterization for the wheel slip control. The stability and performance of the controller were tested on an HiL (hardware in the loop) setup [30].

1.4. Contribution and Paper Structure

This paper discusses a novel approach to optimize three different brake-by-wire actuators. The novelty of this paper is as follows:

1. To the best of the authors' knowledge, this is the first paper on the optimization and comparison of brake-by-wire actuators' energy usage and responsiveness;
2. Use of transfer functions as a way to optimize a nonlinear plant;
3. The optimization of the brake-by-wire actuators operating in closed-loop (and following a target) has not been investigated before;

4. The use of a robust control method (Youla parameterization) to control an EHB brake with build and dump valves (the use of Youla parameterization for EMB and EWB has already been investigated in another paper by the authors [31]);.

The structure of this paper is as follows: In the materials and methods section, the procedures used to achieve the results are discussed. The actuator modeling subsection discusses how each actuator has been mathematically modeled. The model-based control synthesis subsection discusses the robust control design. The optimization section discusses the transfer function and nonlinear plant optimization. In the section results and discussion, the optimization results are presented and discussed. Moreover, in the final section, the conclusions, the final conclusion is drawn, and the benefits and the pitfalls of the optimization framework are discussed.

2. Materials and Methods

2.1. Actuator Modeling

The schematics of EHB, EMB, and EWB brakes are shown in Figure 1a–c.

EHB model consists of a high-pressure source (master cylinder), hydraulic lines, build and dump valves, a brake cylinder chamber, and brake pads. The master cylinder provides pressure into the high-pressure line, controlled by the build and dump valves. Build and dump valves are considered to have varying states between their fully open and fully closed states. This is as opposed to the valves that can either be fully open or fully closed at a given time. In practice, these are solenoid valves that can be controlled with pulse width modulation. For the sake of initial comparison between these smart brake actuators, a vehicle model containing only one-wheel is utilized. When the pressure increases in the brake cylinder chamber, this pressure will move the brake pad forward. This forward movement of the braking pad stops the brake disk as a result. Upon stopping, the dump valve opens, decreases the pressure, and releases the brake pads, bringing them back to their original position.

The EMB comprises a small electric motor, planetary gear set, ball-screw mechanism, brake pad, and caliper. A planetary gear set and a ball screw mechanism move the brake pad when the motor rotates. This movement will result in a clamp force that is denoted by F_{cl} as illustrated in Figure 1b.

The EWB actuator converts the motor's rotation to a linear force on the wedge by using a planetary gear set (not depicted in the schematic) and a roller screw. The motor shaft's axial stiffness and resistance are also considered in modeling this actuator. K_{cal} represents the combined caliper stiffness and the stiffness between the wedge and the disk. This is similar to the EMB configuration, except that the caliper is shaped like a wedge, which, by inserting it inside the brake casing, creates a self-reinforcing mechanism.

Bond graph is a graphical modeling approach for dynamical systems based on the flow/exchange of power, and therefore, energy. Among the many benefits of bond graphs, they are suitable for the systems with multiple energy domains such as mechatronic systems that usually include various electronic, electrical, mechanical, and hydraulic components [32]. Bond graphs are multi-energy domain and open architecture, which means one can easily add and expand the models with minimum effort compared to other modeling techniques. Furthermore, the monitoring and processing power and energy consumption of various components and parts are conducted with ease when using bond graphs. Given the mentioned benefits of this modeling technique, this method is adopted here to study and model BBW systems.

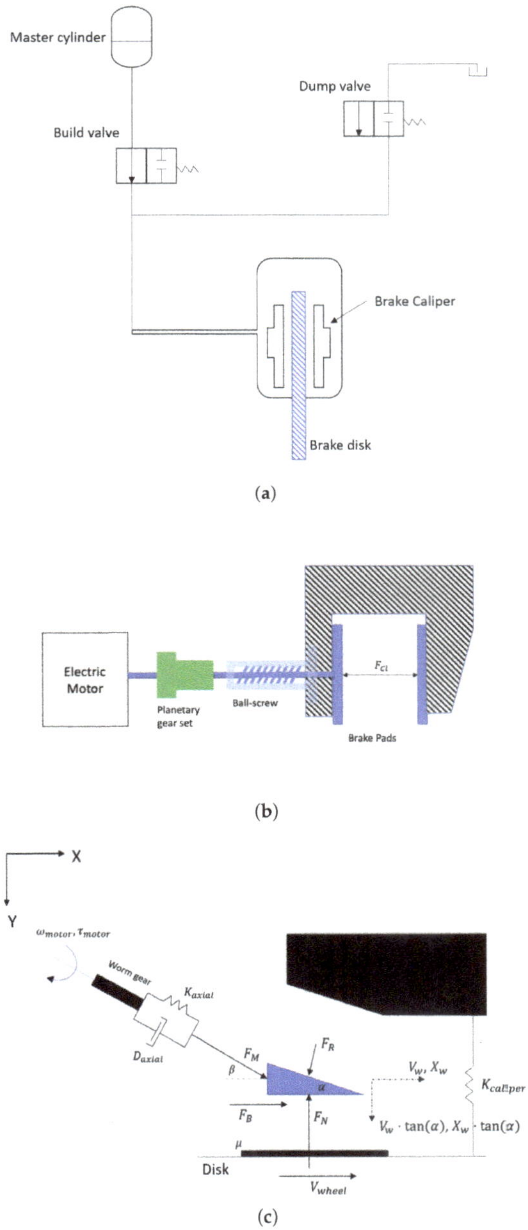

Figure 1. Schematics of brake-by-wire actuators. (**a**) Electro-Hydraulic Brake [33]; (**b**) Electro-Mechanical Brake; (**c**) Electronic Wedge Brake [9].

Figure 2a–c show the bond graph of EHB, EMB, and EWB, respectively. A one-wheel vehicle model is included in all the actuator bond graph models. The wheel has rotational inertia and is connected to a point mass. For the preliminary studies of brake actuators and their algorithms (for example, the Anti-Lock Braking System, ABS, and Traction Control System, TCS), this simple one-wheel model can be used and is easy to implement later on a hardware-in-the-loop test. Models such as this can be used for studying longitudinal dynamics in the vehicle. Since it focuses only on the longitudinal dynamics of the vehicle, it is perfectly suited for studying brake-by-wire actuators and ABS technologies [34].

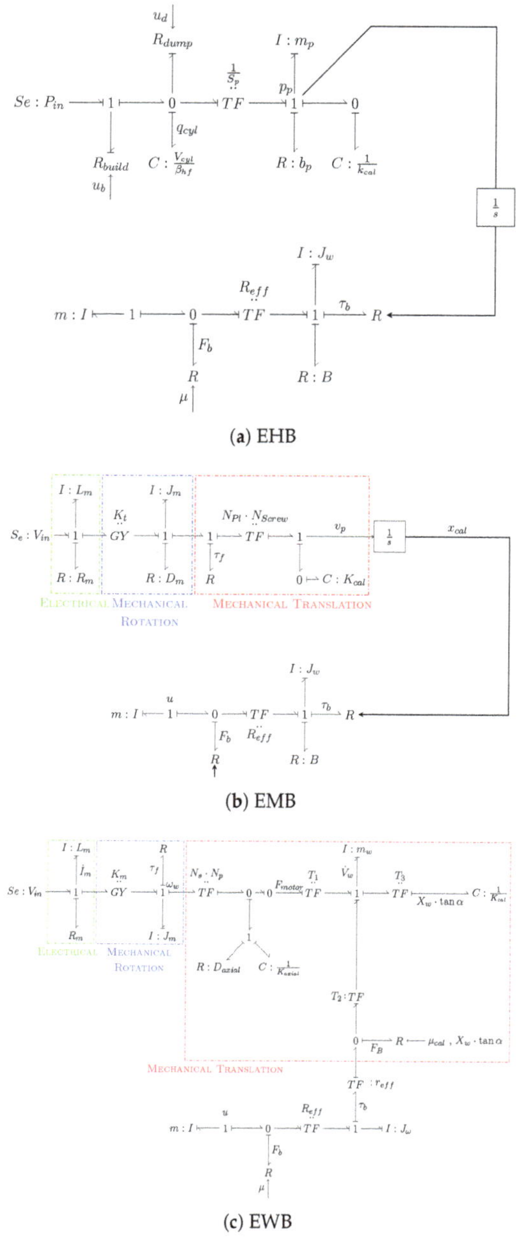

(a) EHB; (b) EMB; (c) EWB

Figure 2. Bond graphs of brake-by-wire actuators.

Based on the bond graphs in Figure 2a–c, the equations of motion for EHB, EMB, and EWB can be written. Equations (1)–(4) represent the equations of motion for the EHB. q_{cyl}, p_p, x_{cal}, P_{in}, u_b, and u_d are the volumetric displacement of the cylinder fluid, momentum of the caliper, caliper displacement, pressure of the master-cylinder (high pressure input), duty ratio of build valve, and duty ratio of dump valve (between 0 and 1), respectively.

C_d, S_b, S_d, ρ, β_{hf}, V_{cyl}, S_p, b_p ,m_p, x_0, and k_{cal} are the maximum flow coefficient of the valve, cross-sectional area of the build valve when fully open, cross-sectional area of the dump valve when fully open, density of the brake fluid, bulk modulus of the brake fluid, cylinder's volume, cylinder's cross-section surface, damping coefficient, brake pad's mass, brake clearance, and caliper stiffness, respectively [33]. Since these equations are highly nonlinear because of the valves, a linearized version, for the purpose of control development, is given in Equations (5)–(7). In this linearization, it is assumed that $u_d = 1 - u_b$, and this means that when one valve is open, the other is closed.

EHB equations of motion are as follows:

$$\dot{q}_{cyl} = C_d S_b u_b \sqrt{\frac{2}{\rho}(P_{\text{in}} - \frac{\beta_{hf}}{V_{cyl}} q_{cyl})} - C_d S_d u_d \sqrt{\frac{2}{\rho}(\frac{\beta_{hf}}{V_{cyl}} q_{cyl})} - \frac{S_p}{m_p} p_p \tag{1}$$

$$\dot{p}_p = S_p \frac{\beta_{hf}}{V_{cyl}} q_{cyl} - b_p \frac{p_p}{m_p} - k_{\text{cal}} \max(x_{cal} - x_0, 0) \tag{2}$$

$$\dot{x}_{cal} = \frac{1}{m_p} p_p \tag{3}$$

$$P_{cyl} = \frac{\beta_{hf}}{V_{cyl}} q_{cyl} \tag{4}$$

Linearized EHB equations are as follows:

$$A = \begin{bmatrix} C_d S_b u_0 \sqrt{\frac{1}{2\rho}} \frac{-\frac{\beta_{hf}}{V_{cyl}}}{\sqrt{(P_{in} - \frac{\beta_{hf}}{V_{cyl}} q_{c0})}} - C_d S_d (1 - u0) \sqrt{\frac{1}{2\rho} \frac{\beta_{hf}}{V_{cyl}} \frac{1}{\sqrt{q_{c0}}}} & \frac{-S_p}{m_p} & 0 \\ S_p \frac{\beta_{hf}}{V_{cyl}} & -\frac{b_p}{m_p} & -k_{cal} \\ 0 & \frac{1}{m_p} & 0 \end{bmatrix} \tag{5}$$

$$B = \begin{bmatrix} C_d S_b \sqrt{\frac{2}{\rho}(P_{\text{in}} - \frac{\beta_{hf}}{V_{cyl}} q_{c0})} + C_d S_d \sqrt{\frac{2}{\rho}(\frac{\beta_{hf}}{V_{cyl}} q_{c0})} & 0 & 0 \end{bmatrix} \tag{6}$$

$$\begin{bmatrix} \dot{q}_{cyl} \\ \dot{p}_p \\ \dot{x}_{cal} \end{bmatrix} = A \begin{bmatrix} q_{cyl} \\ p_p \\ x_{cal} \end{bmatrix} + B\, u_b \tag{7}$$

Similarly, equations of motion for the EMB can be written using Equations (8a)–(8d). Note that the same nonlinear friction model has been used for the EMB and EWB models. I_m, V_{in}, and ω_m are current, voltage input, and angular velocity of the shaft, respectively. L_m, R_m, K_t, J_m, D_m, N_s, N_p, and K_{cal} are the inductance of the electric motor, electrical resistance, electromotive force constant, total moment of inertia of the rotational parts (including the shaft and gears), axial viscous friction, planetary gear reduction ratio, ball-screw gear reduction ratio, and caliper stiffness, respectively.

EMB equations of motion are as follows:

$$\dot{I}_m = \frac{1}{L_m} \times (V_{in} - R_m \times I_m - K_t \times \omega_m) \tag{8a}$$

$$\dot{\omega}_m = \frac{1}{J_m} \times (K_t \times I_m - D_m \times \omega_m - \tau_f - N_p \times N_s \times K_{cal} \times \max(X_{cal} - x_0, 0)) \tag{8b}$$

$$\dot{X}_{cal} = N_s \times N_p \times \omega_m \tag{8c}$$

$$F_{Cal} = \begin{cases} K_{cal}(X_{cal} - x_0), & \text{if } \quad X_{cal} \geq x_0 \\ 0, & \text{otherwise} \end{cases} \tag{8d}$$

Equations (9a)–(9h) show the equations of motion for the EWB, where N, q_{ax}, K_{ax}, D_{ax}, X_w, V_w, F_m, α, and μ_{cal} are combined gear reduction, shaft axial displacement, shaft axial stiffness, shaft axial viscous resistance, wedge displacement, wedge velocity, motor force exerted to the wedge, wedge angle, and friction coefficient between the pad and the wheel, respectively.

EWB equations of motion are as follows:

$$\dot{I}_m = \frac{1}{L_m} \times (V_{in} - i_m \times R_m - K_m \times \omega_m) \tag{9a}$$

$$\dot{q}_{ax} = LN\omega - \frac{V_w}{\cos(\alpha)} \tag{9b}$$

$$F_m = K_{ax}q_{ax} + D_{ax}\dot{q}_{ax} \tag{9c}$$

$$\dot{\omega} = \frac{1}{J_m}\{K_m I_m - D_m \times \omega - \tau_f - L \times N \times F_m\} \tag{9d}$$

$$\dot{V}_w = \frac{1}{m_w(1+\tan^2(\alpha))} \times \{\frac{F_m}{\cos(\alpha)} + (K_{cal} \times X_w \times \tan(\alpha) \times (\mu_{cal} - \tan(\alpha))\} \tag{9e}$$

$$\dot{X}_w = V_w \tag{9f}$$

$$F_B = \mu_{cal}K_{cal}X_w\tan(\alpha) \tag{9g}$$

$$F_{Cal} = \begin{cases} K_{cal}(X_{cal} - x_0), & \text{if} \quad X_w \geq x_0 \\ 0, & \text{otherwise} \end{cases} \tag{9h}$$

τ_f is the lumped nonlinear frictions present in the shaft, planetary gears, and worm gear. The Lugre friction model has been used to model this nonlinear friction. The Lugre model is used for modeling the frictions in actuators since it offers a dynamical model which captures the dynamics very well while needing a lower number of parameters. Other types of friction models can be used as well to represent the frictions. Equation (10a) represents the Lugre friction model [35] where σ_0, σ_1, σ_2, ω_s, j, τ_c, and τ_s are the contact (bristle) stiffness, damping coefficient of the bristle, viscous friction coefficient, Stribeck velocity, shape factor, Coulomb friction, and static friction, respectively. Equation (10d) shows that there is a linear relationship between the Coulomb friction and the clamping force, which is usually derived through experiment. As clamping force increases, the normal forces inside the gears increase as well, which results in increasing the friction torque [6].

Lugre dynamic friction model for EMB and EWB is as follows:

$$g(v) = \tau_c + (\tau_s - \tau_c) \times e^{-|\frac{\omega}{\omega_s}|^j} \tag{10a}$$

$$\dot{z} = \omega - \sigma_0 \times \omega \times \frac{Z}{g(v)} \tag{10b}$$

$$\tau_f = \sigma_0 \times z + \sigma_1 \times \dot{z} + \sigma_2 \times \omega \tag{10c}$$

$$\tau_c = C + G \times F_{Cal} \tag{10d}$$

2.2. Model-Based Control Synthesis

Youla parameterization is a robust control method that leverages closed-loop frequency shaping to attain the desired closed-loop behavior. These closed-loop transfer functions consist of (T_y), known as a complementary sensitivity transfer function, sensitivity transfer function S_y, and Youla Y transfer function (explained below). This method shapes closed-loop transfer functions while ensuring internal stability along with disturbance rejection at low frequencies and sensor noise and unmodeled disturbance rejections at high frequencies. This method was selected because of its ease of control design using the model-based approach for designing appropriate low-level controllers based on the developed bond graph models [36].

The central notion in this method is to form a closed-loop transfer function (T_y) with a transfer function named Youla ($Y(s)$). Multiply Youla by the plant transfer function (G_p) to create the desired closed-loop transfer function (Equation (11)). For good tracking performance in steady-state, the magnitude of $T_y(s)$ should be set to one at low frequencies. To ensure high-frequency noise rejection, $T_y(s)$ should be small at high frequencies:

$$T_y(s) = Y(s) \times G_p(s) \tag{11}$$

As a result, we can shape the closed-loop transfer function using Equation (11) (given we meet all the interpolation conditions for ensuring internal stability mentioned below). It should be noted that the Youla transfer function maps the desired reference signal to the actuator effort. With good target following, such as $|T_y(s)| = 1$ at low frequency, the Youla transfer function is approximately equal to the inverse of the plant transfer function at low frequency and equal to the controller $G_c(s)$ transfer function at high frequency. Thus, keeping Youla's magnitude small at high frequencies would reduce actuator effort and minimize the impact of sensor noise on the actuator.

The closed-loop transfer function ($T_y(s)$) and the sensitivity transfer function ($S_y(s)$) are complementary to each other, as shown by (Equation (12)). Due to this algebraic constraint, the sensitivity transfer function should be small at low frequencies (to reject low-frequency disturbances) and equal to one in magnitude at high frequencies:

$$S_y(s) = 1 - T_y(s) \tag{12}$$

If G_p is stable, the feedback loop would be internally stable if and only if $Y(s)$ is selected to be a stable transfer function. In this regard, $Y_y(s)$, $S_y(s)$, $T_y(s)$, and $G_p \times S_y$ should all be stable to make the feedback loop internally stable. Consequently, to meet these conditions in case of an unstable pole (α_p) which is repeated n-times in the plant (G_p), Equations (13) and (14) define rational interpolation conditions, which must be met to enforce internal stability. If it is a single unstable pole (not repeated), Equation (13) is the only interpolation condition that needs to be satisfied:

$$T_y(\alpha_p) = 1, S_y(\alpha_p) = 0 \tag{13}$$

$$\frac{d^k T_y}{ds^k}(\alpha_p) = 0, \frac{d^k S_y}{ds^k}(\alpha_p) = 0, \quad \forall k \in [\![1, n]\!] \tag{14}$$

If there is a repeated non-minimum phase zero (α_z), zeros in the RHP (Right Half Plane), the interpolation conditions are met by Equations (15) and (16). If the unstable zero is only repeated once, Equation (15) is the only interpolation condition that must be satisfied [36]:

$$S(\alpha_z) = 1, T(\alpha_z) = 0 \tag{15}$$

$$\frac{d^k S_y}{ds^k}(\alpha_z) = 0, \frac{d^k T_y}{ds^k}(\alpha_z) = 0, \quad \forall k \in [\![1, n]\!] \tag{16}$$

Once we ensure that the conditions in the Equations (13)–(16) are met, we can acquire the controller using Equation (17):

$$G_c(s) = Y(s) \times S_y(s)^{-1} \tag{17}$$

Cascaded Control

Since the brake-by-wire smart actuators are Single Input Multiple Output (SIMO) problems, we consider the cascaded control scheme. Cascaded control enables systems with relatively more nonlinearities to perform better and be more robust. Therefore, the controllers were designed using cascaded control to mitigate different nonlinearities in the brake actuators (e.g., mechanical friction, pressure nonlinearities). Each inner closed-loop is an open-loop for their outer loop controller design. The controller design of each plant is conducted through the Youla parameterization approach, as discussed in the previous section. Figure 3 shows this cascaded control design for EMB and EWB actuators. In the EMB and EWB, for the first loop, motor voltage is input, and motor current is the output. In the second loop, the motor's desired current is input, and the motor's angular velocity is output. Finally, for the outermost loop, the input is the desired motor angular velocity, and the clamping force is the output (the normal force of the brake pad on the wheel). The shaft's current and angular velocity can be measured directly and is readily available, but the clamping force must be estimated or measured with a force sensor.

One important part of the EMB/EWB plant, which is not shown in Figure 3, is the current and voltage saturation. Current is saturated at ±25 A, and voltage is set to saturate at ±42 Volts. Because of these, the controllers might saturate and, therefore, make the plant unstable. A simple gain anti-windup was used to address the current saturation and to mitigate this issue (Figure 4). There could also be another anti-windup for the voltage saturation; however, normally, the voltage does not reach saturation levels if the current saturation has been addressed. Furthermore, adding an extra anti-windup may result in limiting the bandwidth of the closed-loop system. In addition, other anti-windup strategies such as the one in [37] or [38], which is specifically for cascaded controllers, could have been utilized.

In the EHB control design, a SISO controller was designed based on the linearized equations mentioned in Equations (5)–(7). The operating points taken for this linearization are $u_0 = 0.3$ and $q_{c0} = 0.3 \times q_0$. q_0 is the steady-state value of q_{cyl}. In addition, as mentioned before, it is assumed that $u_d = 1 - u_b$. This continuous control law works well when building and dumping the pressure in the cylinder chamber. However, in the case of keeping constant pressure during the steady-state, we might run into the issue of having both build and dump valves open partially at the same time and therefore losing some of the master cylinder's pressure which wastes energy. For example, $u_d = 0.7$ and $u_b = 0.3$ would hold the constant cylinder pressure, but this is not energy efficient as the pumps keep running. For this reason, a switching logic was added to the continuous Youla controller. This switching statement changes the values of u_b and u_d to zero once the clamping force error is within the desired threshold. Otherwise, it passes the same values from the controller to the plant as shown in Figure 5.

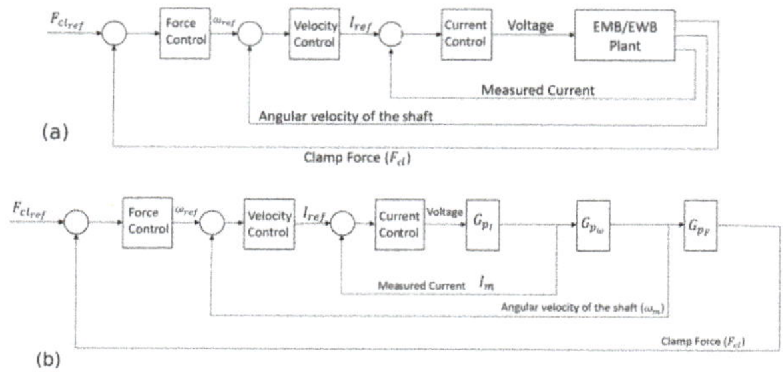

Figure 3. Cascaded control scheme for the EMB/EWB (**a**); The bottom Figure shows the decomposed system for control design (**b**).

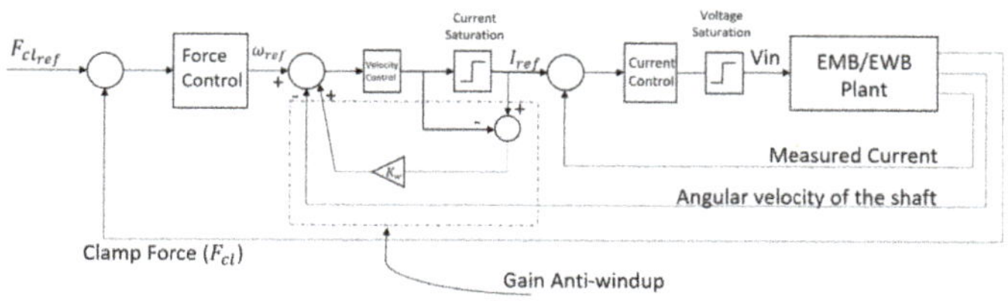

Figure 4. Anti-windup gain used to compensate for the current saturation.

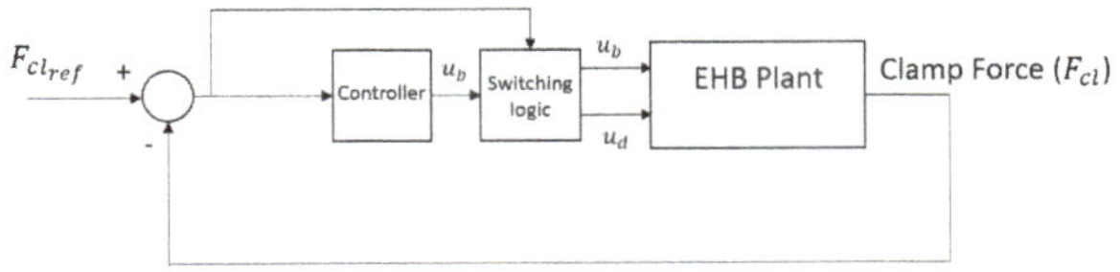

Figure 5. Control scheme for the EHB.

2.3. Optimization

2.3.1. Linear Optimization: Using Transfer Functions

In this section, a multi-objective optimization scheme for brake-by-wire actuators based on their transfer functions is considered. This optimization considers dynamic responsiveness and the actuator's effort as objective metrics.

After linearizing the plants, we can obtain their transfer functions. We then synthesize the controllers using the Youla parameterization technique discussed in Section 2.2. There are a few assumptions made when designing the controllers during the optimization process. Controllers are designed to create closed-loop transfer functions to be in a certain form. Equations (18), (19), (20), (21a), (21b), (22), (23a) and (23b) describe the form of Plant transfer function (Gp), Youla transfer function (Y), and closed-loop transfer function (T) for the first, second, and the third loop of EMB (EWB follows a similar control design pattern, and EHB follows the same pattern, but it only has one control loop, and therefore, the design choice is similar to the one in Equation (19)). The goal here is to design a closed-loop transfer function with the frequency shape of a second-order Butterworth filter and add extra first-order filters whenever necessary. For example, in Equation (21a), the second open-loop transfer function has an integrator that should not be canceled by the Youla transfer function. Hence, a high pass filter was added to the Youla transfer function ($\frac{1}{s+W_1}$). W_1 is the pole for the filter that is added to the Youla transfer function. This can be chosen in such a way that it does not affect the bandwidth of the closed-loop system . In this equation, G'_{p2} represents the plant transfer function of the second loop without the s in the numerator. Moreover, in the case of Equation (23a), a repeated first-order transfer function was added to the Youla transfer function to make this transfer function proper (G_{p_3} is a fourth-order transfer function; for simplicity, we choose to use first-order poles. N-th order Butterworth filter could also be used in this case). More details on the design of these controllers are provided in [31]. It should be noted that ω_{n_1}, ω_{n_2}, and ω_{n_3} are chosen for each loop to have a specific bandwidth ($\omega_{n_1}, \omega_{n_2}$, and ω_{n_3} are the Butterworth filter's cut-off frequencies for the different added Butterworth filter to Youla transfer functions). In the case of EMB, this is 200 Hz, 10 Hz, and 2 Hz for the first, second, and last loop, respectively. For the EWB, they are chosen to be 500 Hz, 400 Hz, and 2 Hz. Finally, for EHB, it is chosen to be 2 Hz. Therefore, all of the brake-by-wire actuators have the same final closed-loop of 2 Hz for the clamp force loop. This is a deliberate choice to make sure all the brake-by-wire actuators have the same bandwidth (for the clamping force) for the final comparison in terms of energy and responsiveness metrics. The chosen control parameters mentioned here (such as ω_n and ξ) will remain the same over the course of all optimizations. This is performed to have fixed control design for the optimization procedure, and the only change will be the physical parameters of the system. Figure 6 shows the Bode magnitude plots of T, S, and Y for this type of control design.

$$G_{p_I} = \frac{J_m s^2 + D_m s + (N_s N_p)^2 K_{cal}}{(L_m J_m) \times s^3 + (R_m J_m + L_m D_m) \times s^2 + (R_m D_m + L_m (N_s N_p)^2 K_{cal} + K_t^2) \times s + R_m (N_s N_p)^2 K_{cal}} \tag{18}$$

$$Y_1 = \frac{1}{G_{p_I}} \times \frac{\omega_{n_1}^2}{s^2 + 2 \times \xi \times \omega_{n_1} \times s + \omega_{n_1}^2}, T_1 = \frac{\omega_{n_1}^2}{s^2 + 2 \times \xi \times \omega_n \times s + \omega_{n_1}^2} \tag{19}$$

$$G_{p_2} = T_1 \times G_{p_\omega} = \frac{\omega_{n_1}^2}{s^2 + 2 \times \xi \times \omega_n \times s + \omega_{n_1}^2} \times \frac{K_t \times s}{J_m s^2 + D_m s + (N_s N_p)^2 K_{cal}} \quad (20)$$

$$Y_2 = \frac{1}{G'_{p_2}} \times \frac{\omega_{n_2}^2}{s^2 + 2 \times \xi \times \omega_{n_2} \times s + \omega_{n_2}^2} \times \frac{1}{s + W_1} \times (\frac{W_2}{s + W_2})^2 \quad (21a)$$

$$T_2 = \frac{\omega_{n_2}^2}{s^2 + 2 \times \xi \times \omega_{n_2} \times s + \omega_{n_2}^2} \times \frac{s}{s + W_1} \times (\frac{W_2}{s + W_2})^2 \quad (21b)$$

$$G_{p_3} = T_2 \times G_{p_F} = \frac{\omega_{n_2}^2}{s^2 + 2 \times \xi \times \omega_{n_2} \times s + \omega_{n_2}^2} \times \frac{s}{s + W_1} \times (\frac{W_2}{s + W_2})^2 \times \frac{K_{cal} N_s N_p}{s} \quad (22)$$

$$Y_3 = \frac{1}{G_{p_3}} \times \frac{\omega_{n_3}^2}{s^2 + 2 \times \xi \times \omega_{n_3} \times s + \omega_{n_3}^2} \times (\frac{W_3}{s + W_3})^4 \quad (23a)$$

$$T_3 = \frac{\omega_{n_3}^2}{s^2 + 2 \times \xi \times \omega_{n_3} \times s + \omega_{n_3}^2} \times (\frac{W_3}{s + W_3})^4 \quad (23b)$$

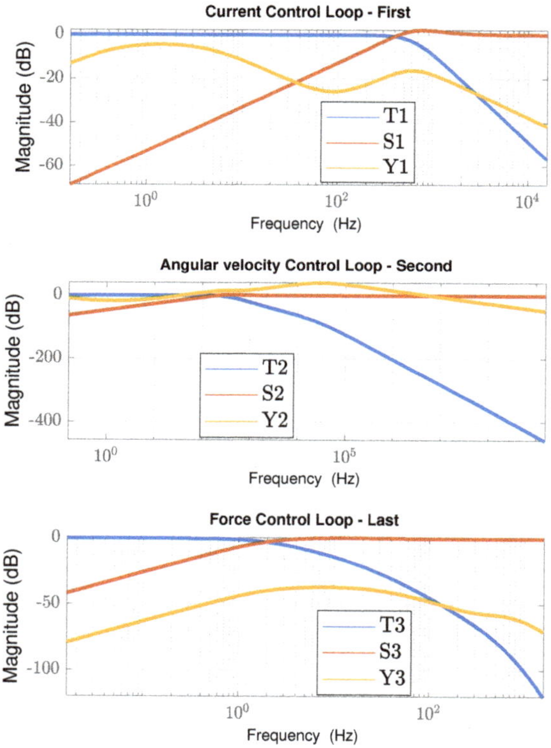

Figure 6. An example design of cascaded controllers for the EMB/EWB.

After designing the controllers, we then utilized the aforementioned transfer functions to optimize actuator response and actuator usage. The process of optimization starts with designing the controllers based on the physical parameters of the system and deriving these transfer functions as a function of the physical system parameters (since the parameters change, the G_ps change, and therefore, these transfer functions will be different for each set of physical parameters).

The bandwidth of a plant is related to its dynamic response, and therefore, increasing the bandwidth would result in faster system response. The bandwidth of a system/plant is defined as the frequency range where the magnitude of the system gain does not drop below −3 dB. For EMB/EWB, the plant transfer function, which is chosen (denoted as G_p here) for calculating the system bandwidth, is from voltage input to the clamping force output. For EHB, the plant transfer function maps u_b to the clamping force. The bandwidth of the plant is denoted as $Bandwidth\{G_p\}$.

Another factor to consider is the actuator's power usage which is related to the actuator's effort, and hence, it is related to the Youla transfer function, denoted as $Y(s)$. The Youla transfer function for the overall system in case of cascaded control design would become $Y_{sys} = Y_1 \times Y_2 \times Y_3$. In here, Y_{sys} stands for the Youla transfer function from the clamping force reference to the voltage input (or to the u_b for the EHB), which is the Youla transfer function of the overall control system. It can be shown if the magnitude of $T_y(s)$ or the gain of closed-loop transfer function is one at low frequencies, the Youla transfer function at low frequencies is inversely related to the plant transfer function $G_p(s)$ (From Equation (11), if $T_y = 1$, then $Y = \frac{1}{Gp}$). By increasing the plant gain at low frequency (approximately its DC gain, denoted as $DC\{G_p\}$ in this section), $Y(s)$ will decrease at low frequencies.

Furthermore, we need to lower the overall values of the Youla transfer function magnitude at other important frequencies, especially around the plant bandwidth. This will ensure the reduction of the actuator effort in all possible frequencies. To this end, we can use an H2 norm of this transfer function. H2 norm is related to the output signal energy when the system input is an impulse [36]. Since we are interested in a specific frequency region of the Youla transfer function, a band-pass filter, see Figure 7, is used to emphasize the frequency region of interest. ω_L and ω_H are chosen to be $0.1 \times Bandwidth\{G_p\}$ and $1e4 \times Bandwidth\{G_p\}$. This will ensure that the Youla transfer function magnitudes at low and mid-range frequencies stay low. The optimization problem is then formulated by combining all these costs as given in Equation (24):

$$\begin{aligned} \underset{x}{\text{minimize}} \quad & f(x) = \alpha_1 \times ||Y_{sys} \times W_Y||_2 + \alpha_2 \times \frac{1}{DC\{G_p\}} + \alpha_3 \times \frac{1}{Bandwidth\{G_p\}} \\ \text{subject to} \quad & x \in [x_{min}, x_{max}], \end{aligned} \tag{24}$$

where α_1, α_2, and α_3 are tuning parameters, and x is the vector of physical parameters of the system that can be changed during the design process (e.g., gear ratios, moments of inertia, and motor's inductance). x_{min} and x_{max} denote the minimum and maximum of the parameter set, respectively. W_Y is the frequency weighting function for H_2 norm optimization. It should be noted that each cost in Equation (24) is normalized by its nominal value to ensure the minimization of the three costs is done without bias. A choice of physical parameters for each actuator, their initial and optimized value are given in Table 1.

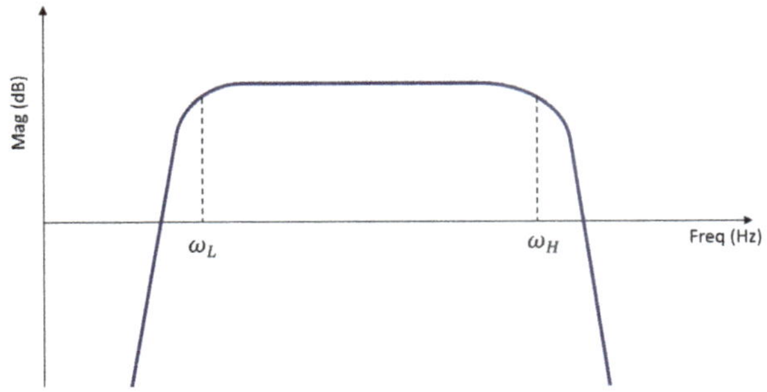

Figure 7. Band-pass filter, W_Y, is used to emphasize specific frequency region of Youla transfer function in the H_2 norm optimization.

Table 1. Initial and optimized physical parameter values of EMB, EWB, and EHB and the range of the parameters.

	Parameter	Units	Lower Bound	Upper Bound	Initial	TF-Based Opt.	Nonlinear Opt.
EMB	L_m	H	4.48×10^{-5}	5×10^{-3}	5.6×10^{-5}	6.36×10^{-5}	2.8×10^{-3}
	R_m	Ω	2.50×10^{-2}	1×10^{-1}	5×10^{-2}	2.5×10^{-2}	3.76×10^{-2}
	J_m	$\mathrm{kg\,m^2\,s^{-2}}$	6.0×10^{-5}	5.8×10^{-4}	2.9×10^{-4}	7.19×10^{-5}	1.03×10^{-4}
	D_m	N m s	2.0×10^{-4}	1.5×10^{-2}	9×10^{-3}	2.02×10^{-4}	9.0×10^{-4}
	N_s	-	7.96×10^{-5}	1.3×10^{-3}	6.37×10^{-4}	1.3×10^{-3}	1.2×10^{-3}
	N_p	-	6/266	18/266	4.14×10^{-2}	6.74×10^{-2}	6.26×10^{-2}
	K_{cal}	$\mathrm{N\,m^{-1}}$	2.3×10^{7}	4.3×10^{7}	3.35×10^{7}	4.3×10^{7}	4.19×10^{7}
	K_t	$\mathrm{N\,m\,A^{-1}}$	5.0×10^{-2}	5.2×10^{-1}	6.97×10^{-2}	1.59×10^{-1}	4.3×10^{-1}
EWB	L_m	H	4.48×10^{-5}	5×10^{-3}	5.6×10^{-5}	4.7×10^{-3}	4.48×10^{-5}
	R_m	Ω	2.50×10^{-2}	1×10^{-1}	5×10^{-2}	2.5×10^{-2}	2.6×10^{-2}
	J_m	$\mathrm{kg\,m^2\,s^{-2}}$	6.0×10^{-5}	5.8×10^{-4}	2.9×10^{-4}	5.8×10^{-4}	9.26×10^{-5}
	D_m	N m s	2.0×10^{-4}	1.5×10^{-2}	9×10^{-3}	2.0×10^{-4}	2.1×10^{-4}
	N_s	-	7.96×10^{-5}	7.96×10^{-4}	4.77×10^{-4}	7.96×10^{-4}	7.89×10^{-4}
	N_p	-	6/266	18/266	4.17×10^{-2}	6.77×10^{-2}	6.76×10^{-2}
	K_{cal}	$\mathrm{N\,m^{-1}}$	2.3×10^{7}	4.3×10^{7}	3.35×10^{7}	4.3×10^{7}	4.29×10^{7}
	K_t	$\mathrm{N\,m\,A^{-1}}$	5.0×10^{-2}	5.2×10^{-1}	6.97×10^{-2}	5.0×10^{-2}	5.88×10^{-2}
	α	degrees	10	24.5	10	24.5	24
	m_w	kg	0.1	0.5	0.3	2.9×10^{-1}	3.15×10^{-1}
EHB	V_{cyl}	$\mathrm{m^3}$	1.6×10^{-5}	1.28×10^{-4}	1.6×10^{-5}	1.6×10^{-5}	7.93×10^{-5}
	S_b	$\mathrm{m^2}$	1×10^{-7}	4×10^{-7}	4.0×10^{-7}	4.0×10^{-7}	2.16×10^{-7}
	S_p	$\mathrm{m^2}$	6.38×10^{-4}	1.02×10^{-2}	1.6×10^{-3}	1.7×10^{-3}	3.7×10^{-3}
	m_p	kg	5×10^{-1}	2	1.973	1.967	1.25
	K_{cal}	$\mathrm{N\,m^{-1}}$	2.3×10^{7}	4.3×10^{7}	4.3×10^{7}	4.3×10^{7}	3.69×10^{7}

2.3.2. Nonlinear Optimization

Using linear transfer functions to optimize the plants provide us with an optimized initial parameter set. Although this set might be good enough for primarily linear plants; most brake actuators are nonlinear due to different factors such as friction, plant saturation, and dead-zone. Therefore, to further optimize the plants, we should perform optimization with the nonlinear plants to consider all the nonlinear effects. Since all the plants are going to run in the feedback control environment in practice, the nonlinear optimization is done on the closed-loop systems. A 10 kN step clamp force reference target is chosen for the brake's closed-loop system. The controllers are designed in the same way explained in the Sections 2.2 and 2.3.1. The control parameters are fixed in the same way as the transfer function optimization while the physical parameters of the systems change. Since the physical parameters of the system are changed, we need to recalculate the controllers at each step of function evaluation in the optimization (G_p changes, and so does Y and G_c). Since the optimization is performed on nonlinear plants, a different objective function should be used. The objective function for the nonlinear optimization consists of four parts: energy usage, maximum power, settling time, and overshoot percentage. Energy usage (E_{usage}) is the total amount of energy used by the actuator to follow the target in two seconds (enough for the actuators to reach and hold the target). Maximum power ($\max \mathbb{P}_{usage}$) is the maximum power used by the actuator during the 2 s that the actuator follows the 10 kN step reference. Settling time (T_s) is the time that it takes for the caliper force to build up to near ±2% of the steady-state value. Overshoot percentage ($OS\%$) is the percentage that the maximum value of the caliper force deviates from the 10 kN reference target. Power usage for the EMB and EWB is defined as current multiplied by the voltage.

For the EHB, we are adding up the amount of power loss (denoted as $\mathbb{P}_{build}$ and $\mathbb{P}_{dump}$) to be equal to the power usage (Figure 2a):

$$\mathbb{P}_{build} = e_{build} \times f_{build} = (P_{in} - \frac{\beta_{hf}}{V_{Cyl}} \times q_{cyl}) \times \{C_d S_b u_b \sqrt{\frac{2}{\rho}(P_{\text{in}} - \frac{\beta_{hf}}{V_{cyl}} q_{cyl})}\} \tag{25a}$$

$$\mathbb{P}_{dump} = e_{dump} \times f_{dump} = (\frac{\beta_{hf}}{V_{Cyl}} \times q_{cyl}) \times \{C_d S_d u_d \sqrt{\frac{2}{\rho}(\frac{\beta_{hf}}{V_{cyl}} q_{cyl})}\} \tag{25b}$$

Taking all of these into account, the cost function for nonlinear optimization is given in Equation (Section 2.3.2). Note that each cost is normalized by its nominal value:

$$\begin{aligned} \underset{x}{\text{minimize}} \quad & f(x) = \alpha_1 \times E_{usage} + \alpha_2 \times \max \mathbb{P}_{usage} + \alpha_3 \times T_s + \alpha_4 \times OS\% \\ \text{subject to} \quad & x \in [x_{min}, x_{max}] \end{aligned} \tag{26}$$

3. Results and Discussion

Figures 8–10 show the results of a 10 kN clamp force step response and the linear and non-linear optimization of brake-by-wire actuators. These simulations are performed using the nonlinear plant of the actuators. The nonlinearities that exist in the EMB and EWB include motor current and voltage saturation, brake caliper saturation, and the Lugre friction model. The nonlinearities in the EHB include the valve nonlinearities and the dead-zone. The "Initial" represents the initial set of parameters of the plant before the optimization. The initial setting for each actuator is compared with a similar setting in the literature to make sure the results are sound and follow other researchers' results. However, for the optimized plants' results, since this is the first study that discusses optimization on these physical parameters, there are no other papers to compare the results with. The "TF-based Opt." represents the linear transfer function optimization, and "Nonlinear Opt." represents the results for the set of plant parameters after the optimization is performed using the nonlinear plants as discussed in Section 2.3.

For the EHB, Figure 8 shows that the transfer function-based and the nonlinear optimization both have reduced the cylinder pressure. It must be noted that the results for the initial set of parameters are consistent with [33]. The cylinder's pressure in the Zhao et al. reaches steady-state around 0.3 s, similar to the results for this study. This shows that robust control along with the linearization is working for this EHB actuator. The readers have to note that the difference between the EHB model studied here is that the valves are considered to change continuously, and therefore, a continuous Youla control scheme is used to control the valves; however, in reality, this needs to be taken care of using a digital controller and pulse width modulation technique. The clamping force response time for the optimized simulation has also decreased from 0.5 s to around 0.3 s and 0.2 s for TF-based and nonlinear optimization, respectively. The power usage plot shows that, in all the cases, the power consumption stops once the actuators reach the steady-state target. This is because of the switching logic that closes both valves once they reach the steady-state value of the clamping force. Since the optimized plants reach the steady-state faster, and they use less actuation to do so, their energy and power usage is reduced (see Table 2).

Table 2. Comparison of the amount of energy used in the 10 kN step response for 2 seconds (the amounts are in Joules).

	EMB	EWB	EHB
Initial set	15.5	60.13	109.73
TF-Based optimized set	2.73	1.91	44.36
Nonlinear optimized set	1.69	2.14	29.70

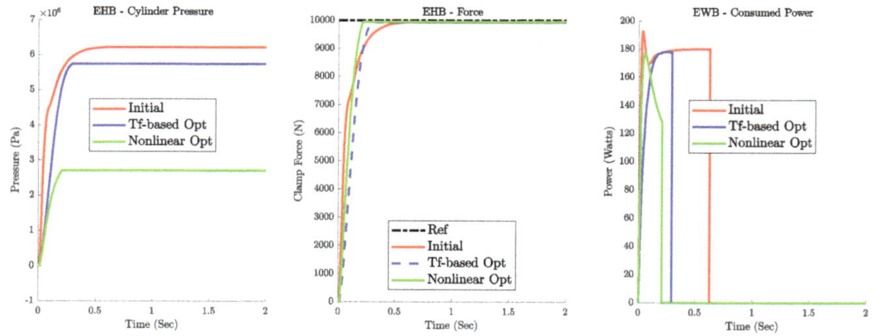

Figure 8. Comparison of the initial parameter sets vs. optimized in an EHB for a 10 kN step input. The clamp force plot for the initial parameter setting was consistent with Zhao et al. [33]. The optimized results are the novelty of this paper.

For the EMB, Figure 9 illustrates that the clamping force step response is about the same for all plants. This is because the controllers are set to have the same bandwidth; they all have the same response. In this case, the nonlinearities are mitigated by robust controllers, and the current/voltage saturation is taken care of by the anti-windup compensator. As shown, the current reaches its saturation level for the initial EMB plant, and the gain anti-windup is shown to be working. However, the difference between the plants manifests itself in the power consumption plot. The initial plant uses a lot more power and energy to perform the same task as the optimized plants. Tf-based and nonlinear optimized plants both have a significantly smaller power usage, with the nonlinear optimized plant consuming a slightly lower amount of power. The overall energy consumption for these plants is summarized in Table 2. Comparing to Line et al., for the initial parameter setting, the clamping force also reaches the steady-state around 0.2 s [6]. The current is higher than the results shown in Line et al.; however, the voltage is not plotted for their results. One explanation is that a higher amount of current would result in lower voltage and vice versa.

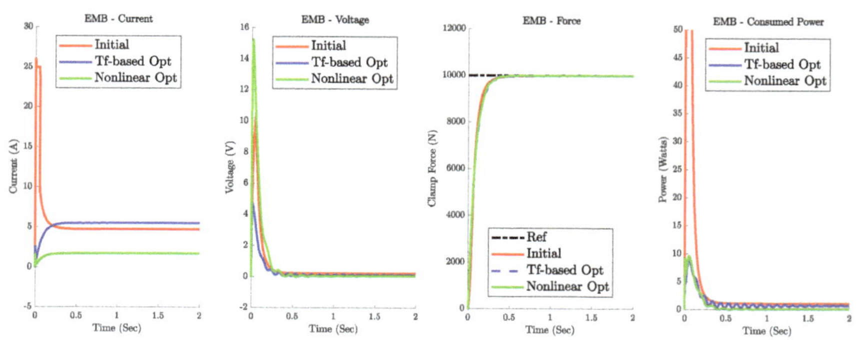

Figure 9. Comparison of the initial parameter sets vs. optimized in an EMB for a 10 kN step input. The clamp force plot for the initial parameter setting was consistent with Line et al. [6]. The optimized results are the novelty of this paper.

For the EWB, Figure 10 shows the step response for all the plants. For the initial parameter setting, the clamping force reaches to steady-state in around 0.5 s. Compared to Che Hasan et al., which uses a PID controller, this is around 0.2 s faster [12]. Additionally, using the Youla parameterization along with cascaded control, the overshoot is also smaller. The voltage peak for both are around the same, although the voltage for this paper is slightly higher. The nonlinear optimized plant is showing a faster response than the transfer function optimized and the initial plant. The overshoot in the nonlinear optimization of the plant has also slightly decreased when compared with the Tf-based optimization plant. As shown, the current once again saturates for the initial plant. For this plant, the current was saturated for around 0.3 s, and the anti-windup compensation has taken care of this; however, this has negatively impacted the closed-loop response and made it slower. Looking at the power consumption, it is clear that the Tf-based opt. has used a slightly lower amount of power, and the nonlinear opt. has used a significantly lesser amount of actuation power. It should be noted that the voltage and current have undershot in all the plants, which comes from the fact that overshoots in the clamping force, as shown in Figure 10, are being compensated by these undershoots.

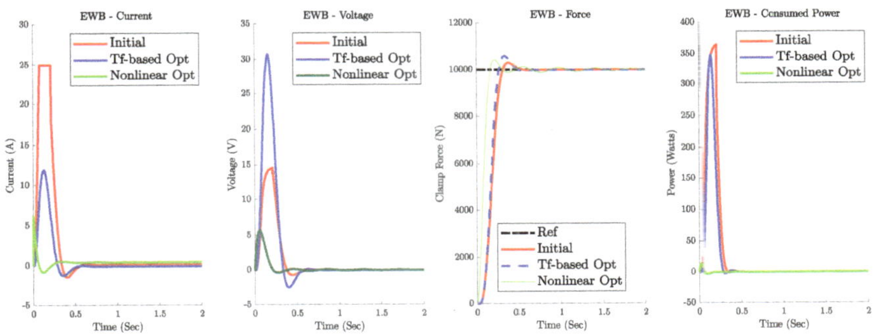

Figure 10. Comparison of the initial parameter sets vs. optimized in an EWB for a 10 kN step input. The clamp force plot for the initial parameter setting was consistent with Che Hassan et al. [12]. The optimized results are the novelty of this paper.

Similar to the step response, we have performed a ramp response of 10 kN/s with the saturation of 10 kN for the given plants. Figures 11–13 show similar results to the ones of the step response as discussed previously. It should be noted that the ramp response is only discussed in this paper, and the cited papers above did not mention performing this test on the actuators.

Table 3 shows the amount of energy usage by each plant with a different set of parameters. Comparing the energy usage of brake-by-wire actuators in Tables 2 and 3, we can conclude that EMB and EWB use significantly lower amounts of energy. However, looking once again at the Figures 8–10, we can see that EHB has at least a 0.1–0.2 s faster response than the dry brake-by-wire actuators such as EMB and EWB.

Table 3. Comparison of the amount of energy used in the ramp response for 2 seconds (the amounts are in Joules).

	EMB	EWB	EHB
Initial set	5.14	18.06	174.42
TF-Based optimized set	2.17	0.83	128.67
Nonlinear optimized set	1.41	0.82	89.72

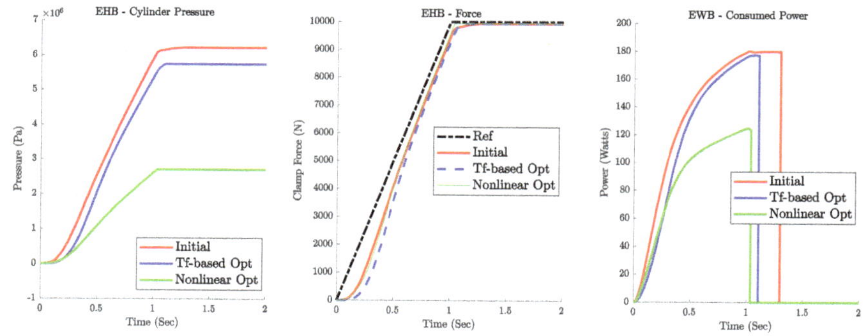

Figure 11. Comparison of the initial parameter sets vs. optimized in an EHB for a ramp input.

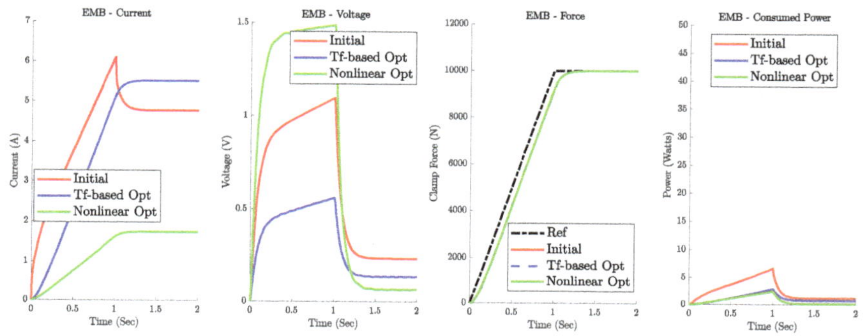

Figure 12. Comparison of the initial parameter sets vs. optimized in an EMB for a ramp input.

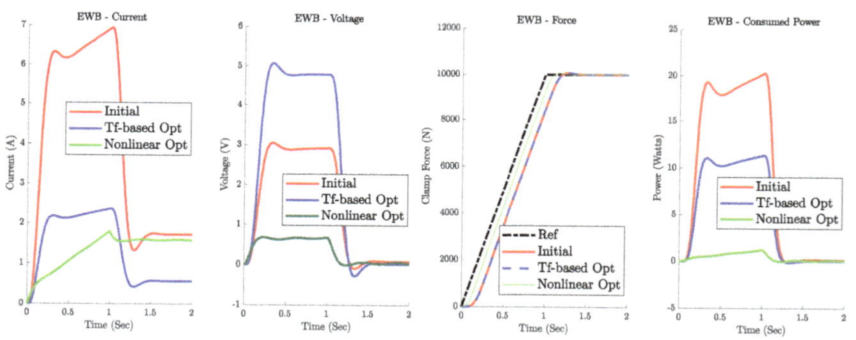

Figure 13. Comparison of the initial parameter sets vs. optimized in an EWB for a ramp input.

4. Conclusions

In this paper, we presented the modeling and a new control strategy for three different brake-by-wire actuators. The physical optimization of these plants using linear transfer functions, and nonlinear plants were discussed, and the results were presented. The optimized results show a promising energy reduction when compared with the nominal parameters. EHB's, EMB's, and EWB's energy consumption were reduced to around 10%, 3%, and 20% of their original sets of parameters, respectively. This method can be effectively utilized for other brake-by-wire actuators to reduce their energy consumption while increasing their dynamic response. It should be noted that in practice, other criteria such as structural, electrical, and heat transfer measures should be added to the optimization method. This can result in more constraints for the optimization problem, which in turn will alter the final results. However, the optimization framework and the objectives will stay the same. Usually, having more constraints will result in a lesser deviation from the initial results. However, as shown in the results, the gains in energy consumption and dynamic responsiveness are high enough to be considered even with added constraints to the optimization problem. This calls for more future studies.

This paper aims to create a framework for optimizing brake-by-wire actuators by considering the problem from the perspective of energy consumption and actuator dynamic response. This framework can be further expanded to add other measures such as cost, weight, and reliability.

Author Contributions: Conceptualization, E.A. and F.A.; methodology, E.A. and F.A.; software, E.A.; validation, E.A. and F.A.; formal analysis, E.A. and F.A.; supervision: F.A.; writing—original draft preparation, review and editing: E.A. and F.A.; funding acquisition: F.A. All authors have read and agreed to the published version of the manuscript.

Funding: This research received no external funding.

Institutional Review Board Statement: Not applicable.

Informed Consent Statement: Not applicable.

Data Availability Statement: Not applicable.

Conflicts of Interest: The authors declare no conflict of interest.

Abbreviations

The following abbreviations are used in this manuscript:

Notation	Definition
F_{cl}	Clamping force
q_{cyl}	Volumetric displacement of the cylinder fluid
p_p	Momentum of the caliper
x_{cal}	Caliper displacement
P_{in}	Pressure input
u_b	Duty ratio of the build valve
u_d	Duty ratio of the dump valve
C_d	Maximum flow coefficient of valve
S_b	Cross-sectional area of the build valve when fully open
S_d	Cross-sectional area of the dump valve when fully open
ρ	Density of the brake fluid
β_{hf}	Bulk modulus of the brake fluid
V_{cyl}	Cylinder's volume
S_p	Cylinder's cross-section surface
b_p	Damping coefficient
m_p	Brake pad's mass
x_0	Brake clearance
k_{cal}	Caliper stiffness
I_m	Electric current
V_{in}	Voltage input
ω_m	Angular velocity of the shaft
L_m	Inductance of the electric motor
R_m	Electrical resistance in the electric motor
K_t	Electromotive force constant
J_m	Total moment of inertia of the rotational parts
D_m	Axial viscous friction
N_s	Planetary gear reduction ratio
N_p	Ball-screw gear reduction ratio
N	Combined gear reduction ($N_s \times N_p$)
q_{ax}	Shaft axial displacement
K_{ax}	Shaft axial stiffness

D_{ax}	Shaft axial viscous resistance
X_w	Wedge displacement
V_w	Wedge velocity
F_m	Motor force exerted to the wedge
α	Wedge angle
μ_{cal}	Friction coefficient between the pad and the wheel
τ_f	Lumped nonlinear frictions present in the actuator
σ_0	Contact (bristle) stiffness (Lugre friction model)
σ_1	Damping coefficient of the bristle (Lugre friction model)
σ_2	Viscous friction coefficient (Lugre friction model)
ω_s	Stribeck velocity (Lugre friction model)
j	Shape factor (Lugre friction model)
τ_c	Coulomb friction (Lugre friction model)
τ_s	Static friction (Lugre friction model)
$T_y(s)$	Complementary sensitivity transfer function
$G_p(s)$	Plant transfer function
$S_y(s)$	Sensitivity transfer function
$Y(s)$	Youla transfer function
α_p	Unstable pole
α_z	Non-minimum phase zero
G_c	Controller transfer function
q_{c0}	The operating point of q_{cyl} (for the purpose of linearization)
u_0	The operating point of u (for the purpose of linearization)
G_{p_I}	Plant transfer function for the current loop ($\frac{I_m}{V_{in}}$)
G_{p_ω}	Plant transfer function for the omega loop ($\frac{\omega}{I_m}$)
G_{p_F}	Plant transfer function for the Force loop ($\frac{F_{cl}}{\omega}$)
G'_{p_2}	Plant transfer function of the second loop without the s in the numerator
Y_i	Youla transfer function for the i-th loop
T_i	Closed-Loop transfer function of the i-th loop
W_i	Constants of the first order transfer functions added to Youla
ω_{n_i}	Butterworth filters' cut-off frequency
ξ	Damping ratio of Butterworth filter
Y_{sys}	Youla transfer function of the system ($Y_1 \times Y_2 \times Y_3$)
W_Y	Filter used to emphasize specific frequency region of Y_{sys} in the H2 norm
$DC(G_p)$	DC gain of plant tranfer function (G_p)
$Bandwidth\{G_p\}$	Bandwidth of plant transfer function (G_p)
x_{min}	Minimum of the parameter set
x_{max}	Maximum of the parameter set
$\mathbb{P}_{build}$	Amount of power loss in the build valve
$\mathbb{P}_{dump}$	Amount of power loss in the dump valve
e_{build}	Effort in the build valve (refers to Figure 2a)
f_{build}	Flow in the build valve (refers to Figure 2a)
e_{dump}	Effort in the dump valve (refers to Figure 2a)
f_{dump}	Effort in the dump valve (refers to Figure 2a)
$\mathbb{P}_{usage}$	Power usage of the actuator
$OS\%$	Overshoot percentage
T_s	Settling time

References

1. Lamnabhi-Lagarrigue, F.; Annaswamy, A.; Engell, S.; Isaksson, A.; Khargonekar, P.; Murray, R.M.; Nijmeijer, H.; Samad, T.; Tilbury, D.; Van den Hof, P. Systems & Control for the future of humanity, research agenda: Current and future roles, impact and grand challenges. *Annu. Rev. Control* **2017**, *43*, 1–64. [CrossRef]
2. Yuan, Y.; Zhang, J.; Li, Y.; Lv, C. Regenerative Brake-by-Wire System Development and Hardware-In-Loop Test for Autonomous Electrified Vehicle. 2017. Available online: https://www.researchgate.net/publication/315859566_Regenerative_Brake-by-Wire_System_Development_and_Hardware-In-Loop_Test_for_Autonomous_Electrified_Vehicle (accessed on 7 January 2022). [CrossRef]
3. Sinha, P. Architectural design and reliability analysis of a fail-operational brake-by-wire system from ISO 26262 perspectives. *Reliab. Eng. Syst. Saf.* **2011**, *96*, 1349–1359. [CrossRef]

4. Gombert, B.D.; Schautt, M.; Roberts, R.P. The Development of Alternative Brake Systems. In *Encyclopedia of Automotive Engineering*; Crolla, D., Foster, D.E., Kobayashi, T., Vaughan, N., Eds.; John Wiley & Sons, Ltd.: Chichester, UK, 2014; pp. 1–11. [CrossRef]
5. Cheon, J.S. Brake by Wire System Configuration and Functions using Front EWB (Electric Wedge Brake) and Rear EMB (Electro-Mechanical Brake). *Actuators* **2010**. [CrossRef]
6. Line, C.; Manzie, C.; Good, M. Electromechanical Brake Modeling and Control: From PI to MPC. *IEEE Trans. Control Syst. Technol.* **2008**, *16*, 446–457. [CrossRef]
7. Line, C.; Manzie, C.; Good, M. Robust Control of an automotive electromechanical brake. *IFAC Proc. Vol.* **2007**, *40*, 579–586. [CrossRef]
8. Sakamoto, T.; Hirukawa, K.; Ohmae, T. Cooperative control of full electric braking system with independently driven four wheels. In Proceedings of the 9th IEEE International Workshop on Advanced Motion Control, Istanbul, Turkey, 27–29 March 2006; pp. 602–606. [CrossRef]
9. Hartmann, H.; Schautt, M.; Pascucci, A.; Gombert, B. eBrake®—The Mechatronic Wedge Brake. *SAE Trans.* **2002**, 2146–2151. [CrossRef]
10. Fox, J.; Roberts, R.; Baier-Welt, C.; Ho, L.M.; Lacraru, L.; Gombert, B. Modeling and Control of a Single Motor Electronic Wedge Brake. *SAE Trans.* **2007**, 321–331. [CrossRef]
11. Han, K.; Kim, M.; Huh, K. Modeling and control of an electronic wedge brake. *Proc. Inst. Mech. Eng. Part C J. Mech. Eng. Sci.* **2012**, *226*, 2440–2455. [CrossRef]
12. Che Hasan, M.H.; Khair Hassan, M.; Ahmad, F.; Marhaban, M.H. Modelling and Design of Optimized Electronic Wedge Brake. In Proceedings of the 2019 IEEE International Conference on Automatic Control and Intelligent Systems (I2CACIS), Selangor, Malaysia, 29 June 2019; pp. 189–193. [CrossRef]
13. Cheon, J.S.; Kim, J.; Jeon, J. New Brake By Wire Concept with Mechanical Backup. *SAE Int. J. Passeng. Cars Mech. Syst.* **2012**, *5*, 1194–1198. [CrossRef]
14. Emam, M.A.A.; Emam, A.S.; El-Demerdash, S.M.; Shaban, S.M.; Mahmoud, M.A. Performance of Automotive Self Reinforcement Brake System. *J. Mech. Eng.* **2012**, *1*, 7.
15. Ahmad, F.; Hudha, K.; Mazlan, S.; Jamaluddin, H.; Aparow, V.; Yunos, M.M. Simulation and experimental investigation of vehicle braking system employing a fixed caliper based electronic wedge brake. *Simulation* **2018**, *94*, 327–340. [CrossRef]
16. Ho, L.M.; Roberts, R.; Hartmann, H.; Gombert, B. *The Electronic Wedge Brake—EWB*; SAE International: Warrendale, PA, USA, 2006. [CrossRef]
17. Putz, M.H. *VE Mechatronic Brake: Development and Investigations of a Simple Electro Mechanical Brake*; SAE International: Warrendale, PA, USA, 2010. [CrossRef]
18. de Castro, R.; Todeschini, F.; Araújo, R.E.; Savaresi, S.M.; Corno, M.; Freitas, D. Adaptive-robust friction compensation in a hybrid brake-by-wire actuator. *Proc. Inst. Mech. Eng. Part I: J. Syst. Control Eng.* **2014**, *228*, 769–786. [CrossRef]
19. Gong, X.; Ge, W.; Yan, J.; Zhang, Y.; Gongye, X. Review on the Development, Control Method and Application Prospect of Brake-by-Wire Actuator. *Actuators* **2020**, *9*, 15. [CrossRef]
20. Büchler, R. Future Brake Systems and Technologies-MK C1®-Continental's brake system for future vehicle concepts. In *7th International Munich Chassis Symposium 2016*; Pfeffer, P.D.P.E., Ed.; Springer Fachmedien Wiesbaden: Wiesbaden, Germany, 2017; pp. 717–723. [CrossRef]
21. Shankar, R.; Marco, J.; Assadian, F. The Novel Application of Optimization and Charge Blended Energy Management Control for Component Downsizing within a Plug-in Hybrid Electric Vehicle. *Energies* **2012**, *5*, 4892–4923. [CrossRef]
22. Yao, M.; Miao, J.; Cao, S.; Chen, S.; Chai, H. The Structure Design and Optimization of Electromagnetic-Mechanical Wedge Brake System. *IEEE Access* **2020**, *8*, 3996–4004. [CrossRef]
23. Fengjiao, Z.; Minxiang, W. Multi-objective optimization of the control strategy of electric vehicle electro-hydraulic composite braking system with genetic algorithm. *Adv. Mech. Eng.* **2015**, *7*, 168781401456849. [CrossRef]
24. Kwon, Y.; Kim, J.; Cheon, J.S.; Moon, H.; Chae, H.J. Multi-Objective Optimization and Robust Design of Brake By Wire System Components. *SAE Int. J. Passeng. Cars Mech. Syst.* **2013**, *6*, 1465–1475. [CrossRef]
25. Hierlinger, T.; Dirndorfer, T.; Neuhauser, T. A Method for the Simulation-Based Parameter Optimization of Autonomous Emergency Braking Systems. In Proceedings of the 25th International Technical Conference on the Enhanced Safety of Vehicles (ESV), Detroit, MI, USA, 5–8 June 2017. Available online: https://trid.trb.org/view/1483876 (accessed on 7 January 2022).
26. Kelling, N.A.; Heck, W. *The BRAKE Project—Centralized Versus Distributed Redundancy for Brake-by-WireSystems*; SAE International: Warrendale, PA, USA, 2002. [CrossRef]
27. Anwar, S. Anti-Lock Braking Control of a Hybrid Brake-by-Wire System. *Proc. Inst. Mech. Eng. Part D: J. Automob. Eng.* **2006**, *220*, 1101–1117. [CrossRef]
28. Savaresi, S.M.; Tanelli, M. *Active Braking Control Systems Design for Vehicles*; Advances in Industrial Control; Springer: London, UK, 2010. [CrossRef]
29. Khalil, H.K. *Nonlinear control*; Prentice Hall, 2015.
30. Soltani, A.; Assadian, F. New Slip Control System Considering Actuator Dynamics. *SAE Int. J. Passeng. Cars Mech. Syst.* **2015**, *8*, 512–520. [CrossRef]
31. Arasteh, E.; Assadian, F. Advanced Applications of Hydrogen and Engineering Systems in the Automotive Industry. In *New Robust Control Design of Brake-by-Wire Actuators*; IntechOpen: London, UK, 2020. [CrossRef]

32. Karnopp, D.; Margolis, D.L.; Rosenberg, R.C. *System Dynamics: Modeling and Simulation of Mechatronic Systems*, 5th ed.; Wiley: Hoboken, NJ, USA, 2012.
33. Zhao, J.; Song, D.; Zhu, B.; Chen, Z.; Sun, Y. Nonlinear Backstepping Control of Electro-Hydraulic Brake System Based on Bond Graph Model. *IEEE Access* **2020**, *8*, 19100–19112. [CrossRef]
34. Arasteh, E.; Assadian, F.; Filipozzi, L. Algorithm to Generate Target for Anti-Lock Braking System using Wheel Power. *J. Mech. Aerosp. Eng.* **2021**, *2*, 2–7.
35. Wang, X.; Lin, S.; Wang, S. Dynamic Friction Parameter Identification Method with LuGre Model for Direct-Drive Rotary Torque Motor. *Math. Probl. Eng.* **2016**, *2016*, 1–8. [CrossRef]
36. Assadian, F.; Mallon, K.R. *Robust Control: Youla Parameterization Approach.*; John Wiley: Hoboken, NJ, USA, 2021.
37. Zaccarian, L.; Teel, A.R. A common framework for anti-windup, bumpless transfer and reliable designs. *Automatica* **2002**, *38*, 1735–1744. [CrossRef]
38. Todeschini, F.; Formentin, S.; Panzani, G.; Corno, M.; Savaresi, S.M.; Zaccarian, L. Nonlinear Pressure Control for BBW Systems via Dead-Zone and Antiwindup Compensation. *IEEE Trans. Control Syst. Technol.* **2016**, *24*, 1419–1431. [CrossRef]

MDPI
St. Alban-Anlage 66
4052 Basel
Switzerland
Tel. +41 61 683 77 34
Fax +41 61 302 89 18
www.mdpi.com

Energies Editorial Office
E-mail: energies@mdpi.com
www.mdpi.com/journal/energies

www.ingramcontent.com/pod-product-compliance
Lightning Source LLC
Chambersburg PA
CBHW040835250926
51850CB00007BA/1318